HUMAN IMPACT RESPONSE
MEASUREMENT AND SIMULATION

PUBLISHED SYMPOSIA

Held at the
General Motors Research Laboratories
Warren, Michigan

HUMAN IMPACT RESPONSE

MEASUREMENT AND SIMULATION

Proceedings of the
Symposium on Human Impact Response
held at the
General Motors Research Laboratories
Warren, Michigan
October 2-3, 1972

Edited by
WILLIAM F. KING and HAROLD J. MERTZ
General Motors Research Laboratories

PLENUM PRESS • **NEW YORK–LONDON** • **1973**

Library of Congress Catalog Card Number 73-80138
ISBN 0-306-30745-6

© 1973 Plenum Press, New York
A Division of Plenum Publishing Corporation
227 West 17th Street, New York, N.Y. 10011

United Kingdom edition published by Plenum Press, London
A Division of Plenum Publishing Company, Ltd.
Davis House (4th Floor), 8 Scrubs Lane, Harlesden, London, NW10 6SE, England

PREFACE

This book contains the papers and discussions presented at the *Symposium on Human Impact Response*, which was held at the General Motors Research Laboratories on October 2 and 3, 1972. This symposium was the sixteenth in an annual series presented by the Research Laboratories. Each symposium has covered a different technical discipline, selected as timely and of vital interest to General Motors as well as to the technical community at large.

The subject of the 1972 symposium was considered to be an appropriate one in view of current widespread concern about vehicle safety and the protection of the occupant during impact situations. In recent years much research has been accomplished on the mechanisms of impact injury and on specifying human injury threshold. This knowledge of human injury threshold has been used as the basis for Federal Motor Vehicle Safety Specifications, in which maximum permissible loads or accelerations are specified during barrier testing with human simulations. The early assumption that a test dummy could validly represent a human in an impact situation has been discredited. It is now realized that the human simulation must have equivalent mechanical properties of resilience and damping to duplicate the impact, as well as equivalent articulation to reproduce the kinematics of the accident.

In this situation it was particularly appropriate to expose simultaneously the key personnel of the research community, government, the automotive industry, and the dummy manufacturers to current knowledge of biomechanics and human simulation. The meetings covered the current state of the art, impact response and appraisal criteria for the head, neck, and chest, plus kinematic and mobility aspects. Specific simulations to obtain biomechanically correct performance were presented and discussed. As a result of this interchange of information, a better understanding of the mutual problem was obtained, and hopefully some progress made toward improved fidelity in human simulation.

The small group of engineers and scientists invited to the symposium were chosen because of their expertise and current activity in the field of biomechanics and vehicle safety. The number was limited both by the desire to keep the group intimate and by the limitations of the physical facilities. For each session papers were presented by currently active authorities in that particular technical area. To assure competent direction during the technical sessions, four men of recognized reputation in the vehicle safety field acted as the session chairmen. Ample time was allotted during the meeting for the presentation of the papers and for subsequent informal discussion. The oral discussions were transcribed and edited lightly, then submitted to the discussors for corrective action and approval. Both the formal papers and oral discussions are reproduced in this volume.

Although this book records the original contributions of the authors and much of the discussion they elicited, the most important contribution of this symposium may not have been recorded; that is, the many individual discussions and interactions that occurred at the many social functions included in the total symposium format.

This symposium could not have been held or these proceedings published without the valuable assistance of many people: Mr. R. L. Scott, who efficiently handled the myriad details involving the physical arrangement of the symposium: Mr. D. N. Havelock and Mr. R. L. Mattson, for assistance in editing the symposium volume; Mrs. G. Grant, Mrs. K. A. Kirksey, Mrs. B. M. Lavender, and Mrs. D. J. Splan, who participated variously as "ladies Friday" in the correspondence, preparation of manuscripts, transcription of tapes, and reception of guests at the technical and social functions.

Finally, sincere thanks are again extended to all the contributors and participants for creating a stimulating and worthwhile symposium.

H. J. Mertz
W. F. King
March 1973

CONTENTS

SESSION I

THE STATE OF THE ART

Session Chairman
R. C. HAEUSLER

*Chrysler Corporation
Highland Park, Michigan*

TRAUMA EVALUATION NEEDS

J. D. STATES, M.D.

University of Rochester, Rochester, New York

ABSTRACT

The personal bias of the author should be clearly understood so that the opinions given in this presentation may be kept in proper context. The author is a practicing orthopaedic surgeon, medical school faculty member and principal investigator of a U.S. DOT Multidisciplinary Accident Investigation team. Reduction or prevention of long-term disability which frequently follows severe musculoskeletal injuries is the overriding goal of the author.

The nature of accidents causing life threatening and/or disabling injuries appears to have shifted since enactment of the 1966 National Highway and Traffic Safety Laws because of improved vehicle design. In the past, head-on impacts were the most common source of serious injury but in recent years, side impacts, pedestrian and motorcycle accidents appear to be the principal causes of serious injury accidents. Additionally, whiplash remains a problem in spite of its minor nature and the introduction of head restraints. Belt type restraint systems appear to be causing some injury which may be reduced by improved design of the restraint system or vehicle seat. Particular attention is given to restraint systems because the motoring public is unjustly critical of even minor failures in safety equipment function and does not understand safety equipment injury trade-offs.

The head and chest remain the most common causes of fatal injury. Trauma indicating simulations for these areas have been under development for many years and will not be considered in this presentation. The neck is possibly the most commonly injured structure because of the high frequency of rear-end impact accidents resulting in hyperextension (whiplash) injury. Recent studies by the Insurance Institute for Highway Safety and by the author's MDAI team indicate that

References pp. 12–13

head restraints reduce neck injury approximately 15%. This disappointing figure is apparently caused by two factors; failure of users to properly adjust head restraints and inadequate design of head restraints. Fixed head restraints appear to be a desirable ultimate goal so that active participation of the occupant is not necessary.

Differential rebound appears to occur because of differences in the spring rates of the head rest and seat back construction. Improved knowledge of human neck and dorsal spine kinematics and attention to seat design may resolve this problem. Present anthropomorphic dummy designs are primitive. A body of human volunteer experiments exists which, with further study, may reveal some of the details of neck kinematics which need to be incorporated in the dummy neck design.

Belt restraint systems particularly in the abdominal area appear to be failing occasionally because of submarining although the frequency of serious injury remains remarkably low. More accurate simulations of the human abdomen, pelvis and hips in conjunction with more extensive dummy testing in vehicles with suspension systems and production seats may elucidate the causes of failure of the belt system to better protect the occupants. The lap belt appears to slip off the bony pelvis and sweep upwards across the abdomen rupturing vital organs or forcing them into the chest cavity.

Serious lower extremity injuries continue to occur with alarming frequency and cause serious permanent disability which occasionally prevents a patient from returning to gainful occupation. Motorcycle and pedestrian accidents now appear to be the most common cause of injuries; because most automobiles are now equipped with occupant protecting dashboards, windshields, steering wheels and steering columns. Direct impacts against the lower extremities are the typical injury mechanism in pedestrian and motorcycle accidents. In-vehicle injury mechanisms include direct impact on the knee in the axis of the femoral shaft causing injury to either the patella, knee, femur or hip. Direct impact injuries of the lower leg may also occur inside the vehicle. Forced dorsiflexion of the foot and ankle against the knee fixed in the dashboard early in the accident sequence in head-on impact accidents may result in serious foot and ankle injuries if the engine elevates the toepan under the foot.

Disability in these situations occurs primarily from joint injury; bones virtually always heal although healing time may take several years. Traumatic arthritis is the result of joint injury and is a poorly defined little recognized entity. Initially, joint cartilage may not appear damaged and injury becomes clinically evident only a year or more after injury. Disability resulting from traumatic arthritis is caused by painful, stiff joints requiring the permanent use of crutches or canes. More sophisticated simulations of the lower extremities are necessary to measure energy input into joints. Range of motion and pressure measurement must be built into dummies to permit accurate injury detection.

INTRODUCTION

The personal bias of the author should be clearly understood so that the opinion given in this presentation may be kept in proper context. The author is a practicing orthopaedic surgeon, medical school faculty member, chief of the Orthopaedic service of an affiliated medical school hospital and Principal Investigator of a U.S. DOT Multidisciplinary Accident Investigation team. Reduction or prevention of long-term disability which frequently follows severe musculoskeletal injuries has been an overriding goal of the author's since accident investigation studies were initiated in 1959.

The head and chest remain the most common causes of fatal injury. Trauma-indicating simulations for these areas have been under development for many years and will not be considered in this presentation. To be considered are trauma similuations for those injuries of the abdomen and musculoskeletal system which are life threatening or which cause prolonged or permanent disability.

The nature of accidents causing life threatening and/or disabling injuries appears to have shifted since enactment of the 1966 National Highway Traffic Safety Laws because of improved vehicle design. In the past, head-on impacts were the most common cause of serious injury but in recent years side impacts, pedestrian, and motorcycle accidents appear to have become the principal causes of serious injury accidents. Additionally, whiplash remains a problem in spite of its minor nature and in spite of the introduction of head restraints. Belt-type restraint systems appear to be causing some injury which may be reduced by improved design of the restraint system and of the vehicle seat.

STATISTICAL DEFINITION OF TRAUMA EVALUATION NEEDS

Analysis of accident case studies which include vehicular damage and human injury has in some instances made possible the identification of commonly occurring injuries, or injuries which because of their severity deserve the attention of vehicle designers for the possibility of injury reduction or prevention. Whiplash neck injury caused by rear-end impact accidents results in injury in to between 28 and 40 percent of the occupants of rear-end impacted vehicles as reported by recent studies of States (1) and of O'Neill and Haddon, et al. (2). Less easily defined statistically are injuries caused by belt-type restraint systems and injuries of the musculoskeletal system. Williams reported a series of over 100 belt restraint system injuries in 1970 at the Stapp Car Crash Conference but this series was collected only through an exhaustive search of the literature (3).

Statistical definition of permanent disability producing musculoskeletal injuries is more difficult. Large accident case study series have reveaied the incidence of injuries to various parts of the musculoskeletal system. A recent study by Nelson (4) of 24,000 accidents in Australia for a one-year period ending 5/31/72 revealed that lower extremity injuries occurred in 37.9%, upper extremity injuries in 27.5% and spine injuries in .5% of the cases. However, this study revealed only the gross anatomical location of the injury and did not distinguish between joint and intermediate area injury. Disability appears to be related primarily to joint injuries and not to shaft injuries of long bones.

The U.S. Department of Transportation now has more than 1,200 cases of detailed accident investigations performed by Multidisciplinary Accident Investigation teams and by Cornell, the University of Michigan and UCLA accidents investigators in the University of Michigan Highway Safety Research Institute Data Bank. To some extent these case studies contain sufficient detail to permit better identification of injuries of various anatomical parts of the human body, particularly the extremities. A serious deficiency of the HSRI Data Bank is the lack of random case selection. The selection criteria for MDAI cases are that one vehicle in the accident be less than three years old and that either the vehicle be sufficiently damaged so that it can not be driven or that an occupant or pedestrian have sustained injury evident at the accident scene. Similar, but somewhat different criteria were used for cases investigated by other contributors. The net result is that bias exists in case selection but the precise nature of the bias has not defined and can not be weighed so that statistical significance can be given to the cases.

Rather than undertake the nearly impossible task of establishing the frequency and importance of various life threatening and disabling injuries, the author has reviewed his own experience based on an orthopaedic practice in a metropolitan area of a one-million population and experience in accident investigation and now numbering 488 cases, 153 of which were done under contract from the U.S. Department of Transportation. In most of these accidents, at least one victim sustained a musculoskeletal system injury.

NECK INJURY

Whiplash, as noted in the introduction, is probably the most common traffic accident injury of all occurring in 40% of rear-end impact accidents and comprises the largest single group of accidents in New York State (5,6). The injury mechanism occurring in uncomplicated rear-end collisions is almost certainly hyperextension of the cervical spine and backward rotation of the head as demonstrated by experimental studies of McNab (8), Wickstrom (9), Patrick (10) and Tuell (11). Tuell, an orthopaedic surgeon, allowed himself to be rear-ended by a 1948 DeSoto while he was seated in a 1951 Ford. High speed motion pictures demonstrated that his head and neck hyperextended 120° over the seat back. He stated afterwards that his neck was sore for a few days but he had no significant disability.

The pathology of whiplash is far more complex than a simple joint strain. Animal autopsy studies by Wickstrom (9) and human autopsy studies by Luongo (12) have demonstrated a multitude of injuries of the soft tissues of the brain stem and cervical spine. Additionally, central nervous system or brain injury also plays a major role in the injuries sustained by some occupants in rear-end impacts (13). Complex kinematics of the cervical spine involving delayed muscle contraction in the response to stretch reflexes and other reflexes may well aggravate rather than protect the neck as is commonly believed to be the function of muscle tone in energy absorption (14).

The symptoms of whiplash or hyperextension neck injury are often delayed several hours or even several days as demonstrated by States (1). Police reports indicated an injury incidence of only one-half that which was determined by follow-on telephone interviews of accident vehicle occupants.

Head restraints required on passenger vehicles sold in the U.S. since January 1969 are not as effective in preventing whiplash injury as would be expected or hoped because of their obvious protection against extreme hyperextension. Studies by O'Neill and Haddon (2) and States, etal (1) revealed a whiplash injury frequency reduction of not more than 15%. This disappointing figure may be attributed to two factors; failure of users to properly adjust head restraints and inadequate design of head restraints. Hyperextension can obviously occur when a tall occupant does not elevate an adjustable head restraint allowing the head to hyperextend over the top of the head rest.

Case Study Number 1 — A 22-year-old-man, 6'1" tall sustained a moderately severe whiplash injury while driving a 1969 Cougar which was struck from behind. His head rest was in the downmost position. His wife, 5'6" tall, did not sustain neck injury.

States (15) has hypothesized a second injury mechanism in which differential rebound may occur even with properly adjusted head restraints because of the difference in spring rates of the head rest and shoulder back padding. This difference allows the shoulders to rebound while the head remains in the rearward most position.

Case Study Number 2 — A 39-year-old professional flyer and airline pilot sustained a typical whiplash neck injury when his 1969 Volkswagen Variant station wagon was struck from behind by a walk-in type delivery truck. He was wearing a lap and shoulder belt at the time of the accident. The vehicle was equipped with a fixed head restraint. The padding material of the head restraint had a very low spring rate in contrast to the seat back which had a high spring rate typical of seat padding materials and construction.

The injury mechanisms of more serious neck injuries involving the bony structures and/or the spinal cord appear to be equally complex and perplexing. Roll-over accidents characteristically cause such injuries (16). Experimental studies by Schneider et al. (17) utilizing monkeys reveals the importance of head position at the moment of impact with respect to the injury mechanism.

Present designs of simulations for the human cervical spine are relatively crude. In addition to failing to take into consideration the energy absorbing characteristics of the soft tissue and the possible role played by muscle stretch reflexes and other as yet

References pp. 12–13

unidentified injury attenuating devices of the human body, the simulations do not satisfactorily provide the stretch characteristics as determined by Ewing. In a series of human volunteer studies (18) he demonstrated that the cervical spine stretches two centimeters. This is a significant energy absorbing mechanism and in part may account for the difficulties manufacturers have had in complying with the current Federal requirements for the Federal Motor Vehicle Safety Standard 208-Restraint Systems.

LAP BELT INJURIES, ABDOMINAL INJURIES AND SEAT DESIGN

Williams (3) collected over 100 cases of injuries caused by belt type restraint systems. Most of these are minor in nature although some are life threatening or fatal. Most recently a case was reported in which the lap belt swept upwards across the abdomen forcing the abdominal contents into the chest through a rent in the diaphragm (19).

Case Study Number 3 (RAI 144) — A 6'1" man weighing 190 pounds sustained a traumatic diaphragmatic hernia through which most of his adominal organs were forced into his chest cavity. He submarined beneath his lap belt while driving a 1971 Chevrolet which struck a gore. The guard rails in the gore at the apex were supported by vertical railroad rails deeply imbeded in the earth.

The injuries in Case Study Number 3 were caused by submarining under a lap belt. In an earlier study presented in 1969 the author reviewed 243 racing accidents and noted that submarining occurred in only four (20). An analysis of the factors preventing submarining in racing accidents revealed that seat design was probably one of the most important factors for preventing submarining.

Racing seats represent the other end of the spectrum in passenger car seat design in that they are form-fitting, lightly padded and do not have springs. In spite of this, competition drivers consider them comfortable and less fatiguing although they are used in cars with far stiffer suspensions and with tires operated at high pressures. Personal experience with the use of such a seat in a production car confirmed the fact that the additional support reduced fatigue because the lateral support reduces the constant adjustment of the flank and back musculature necessary to keep the torso upright. Additionally, better vehicle "feel" is possible with a bucket seat. Loss of traction and drifting of the vehicle are more easily detected by a driver in snug-fitting seats. Lastly, a bucket seat potentiates the effectiveness of belt-type restraint systems. The thin padding in the seat cushion portion does not allow downward excursion of the pelvis and reduces the risk of submarining. The lateral support, particularly of the pelvis, maintains the driver in position in front of the steering wheel and protects against lateral intrusions and prevents the occupants from being thrown against the side structure of the car.

In recent years much progress has been made in dummy design particularly with respect to location of the hip joint. Early dummies articulated the hip joint at the

intersection of the midline of the torso and lower leg which placed it one and a half
to two inches behind the anatomical location of the hip joint axis and created an
artificial situation with respect to submarining (see Fig. 1).

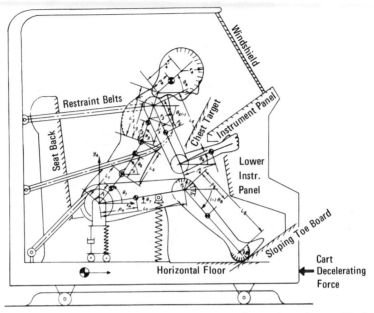

Fig. 1. Diagram of a mathematical model using dimensions of an early anthromophic dummy
(Swedish Model DMV) reveals that the hip axis dislocated at the junction of the midlines of the
torso and the thigh — from Technical Report Number YM-2250-V-1 by Raymond R. McHenry and
Kenneth M. Nab, Cornell Aeronautical Laboratory, Inc. December, 1966.

   The additional two inches resulted in leverage below the lap belt which allowed the
pelvis to slip from beneath the lap belt with relative ease and did not truly represent
the human body (see Fig. 2).

   Additional factors should be considered in human simulations with respect to seat
design and belt restraint system performance. The musculature of the back, hips and
lower extremities almost certainly plays an important part in the kinematics of a
restrained occupant. Additionally, the abdominal contents may belly forward over
the belt locking the lap belt in position low over the pelvis. Aldman in his early work
in 1959 and 1960 recognized this and incorporated a bag in the abdominal area of his
dummy (21).

   Other simulations for the study of submarining do not appear satisfactory. The
anatomy of the lower torso and hips and the relative weights of body segments of
animals otherwise suitable and available for experimental work are so different from
human beings as to make them unsatisfactory simulations for most experimental
work. Cadavers are more useful but postmortem soft tissue changes affect the
dynamic performance of the abdominal contents profoundly limiting their usefulness.

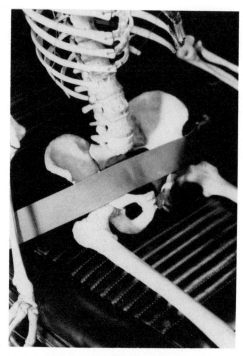

Fig. 2. The human hipjoint is located remarkably close to the anterior iliac spine and to the normal loading points of a lap belt restraint system. Anteversion of the femoral neck permits location of the hipjoint anterior to the junction of the midlines of the torso and the thigh.

Improved simulation of the lower extremities is necessary to develop protective vehicle designs for the reduction of lower extremity injuries. Such injuries continue to occur with alarming frequency and frequently cause serious permanent disability which may prevent a patient from returning to gailful occupation. States (22) demonstrated an injury mechanism in which axial compression of the lower leg occurred because the knee is fixed in the dash and forced dorsiflexion of the foot from toepan deformation causes serious injury to the midtarsal bones and profound disability of the foot. This occurs in head-on collisions in which the engine is forced backwards elevating or crumpling the toepan under the foot.

Case Study Number 4 — A 25 year old man driving a 1967 Ford was struck head-on by a 1967 Pontiac GTO. He sustained fractures of the right talus and cuboid because of forced dorsiflexion of his foot caused by fixation of his knee in the dashboard and upward deformation of the toepan. These fractures healed but the patient developed traumatic arthritis in his foot requiring fusion three years after injury. He was out of work nearly three years because of this automobile accident.

The frequency of this injury is not known but ten cases have occurred in the author's personal case studies and others are seen almost monthly by professional associates.

Traumatic Arthritis, that arthritis which follows impact injuries of joint surfaces, remains difficult to diagnose and define early in its onset. Typically, traumatic arthritis does not become clinically evident until two to five years after an injury. Initially the only evidence of injury may be soft tissue swelling and joint effusion. X-rays of the joint do not reveal disruption of the bony structures. Articular cartilage is radiolucent and is not revealed by x-ray except with the use of special contrast media which are not part of routine examinations. In instances where direct examination of joint surfaces soon after injury is made possible by surgical intervention for some other purpose, cracks, fissures and lossenings of the articular cartilage have been noted (23). Direct examination through surgical approaches, however, is rarely possible, most traumatic arthritis is initially undiagnosed.

Case Study Number 5 — A 35 year old woman contused her right knee while driving a 1960 Chevrolet which was struck head-on in the right front corner. Her knee became painful and swollen and she was seen in a local hospital emergency department where she was examined, x-rays taken but no bony injury found. She continued to have symptoms and sought orthopaedic care three years after the accident. At that time x-rays revealed irregularity of the articular cartilage of the patella. The patella was removed surgically revealing almost complete degeneration of the articular cartilage.

The long-term consequences are serious. Such arthritis results in profound disability because of stiffness, pain and weakness. Treatment is complex requiring the use of prosthetic devices for joint replacement and does not restore a joint to normal in spite of an optimal recovery. The weight-bearing joints; the hip, knee, foot and ankle, are most commonly involved.

The tolerance of human joints for injury is not defined and some individuals have very fragile joints, particularly the elderly, significantly below that for a fracture of the adjacent bony structures. For this reason, more sensitive and improved simulations of the joints of the lower extremities are necessary to permit design of vehicle structure which will reduce or prevent joint injury.

Animal joints have been successfully used as injury indicating joint simulations. Nagel used bovine knees for dynamic testing of knee structures (24) and, more recently, Radin has used a variety of animals for studying energy absorption in joints in the living animal (25).

LOWER EXTREMITY INJURIES OF MOTORCYCLISTS AND PEDESTRIANS

Motorcycle and pedestrain accidents appear to have replaced passenger cars as the most common source of serious lower extremity injuries (24). The design of injury reducing devices for motorcycles is in its infancy but appears hopeful.

Case Study Number 6 — A 19 year old recent high school graduate and son of a motorcycle police officer was injuried while riding his own motorcycle home at 11:00

p.m. He was struck by a car making a left-hand turn across his path. The driver of the car denied seeing the motorcyclist although the motorcyclist had his headlight on. The motorcyclist sustained compound fractures of the left lower leg and a simple left femoral shaft fracture. He is still under care 15 months after the accident and will have permanent disability because of persistent stiffness in his left knee and ankle.

Pedestrian protection also appears possible through modification of vehicle front-end design but will be a more difficult goal to attain. In both instances, more sophisticated simulations of the lower extremity will facilitate the necessary research.

CONCLUSIONS AND SUMMARY

Neck injuries remain one of the most common accident injuries and in some instances cause profound permanent disability. The nature of the injuries is far more complex than has been appreciated until very recently. Kinematics and injury mechanisms are currently under research scrutiny and hopefully will permit the design and construction of improved injury indicating simulations.

Restraint systems continue to cause injury which may be reduced or prevented by introducing some of the design features of bucket seats used in racing vehicles. Improved simulations of the human abdomen and hips are essential for the development of such seats for passenger car use.

Joint injury causes traumatic arthritis which is not easily diagnosed in its early stages but may cause profound permanent disability. The joints of the lower extremity are primarily affected; improved injury indicating simulations of joints are necessary for further research and vehicle design efforts to prevent joint injuries. Motorcyclists and pedestrians now appear to have become the most frequent victims of serious lower extremity injuries. Improved lower extremity simulations and the use of animal joints will facilitate the development of injury reducing designs for motorcycles and vehicle front ends.

REFERENCES

1. J. D. States et al., "Injury Frequency and Head Restraint Effectiveness in Rear-End Impact Accidents", Sixteenth Stapp Car Crash Conference, SAE, November 8-10, 1972, to be published.
2. B. O'Neill, W. Haddon, Jr., A. B. Kelley, and W. W. Sorenson, "Automobile Head Restraints: Frequency of Neck Injury Insurance Claims in Relation to the Presence of Head Restraints", Preliminary Report received, September, 1971.
3. J. S. Williams, "The Nature of Seat Belt Injuries", Fourteenth Stapp Car Crash Conference Proceedings, pp. 44-65, SAE, New York, N.Y., 1970.
4. P. Nelson, "The Pattern of Injuries Received in Motor Car Accidents", Proceedings of the Fourth International Congress on Accident Traffic Medicine, pp. 254-255, Paris, France, 1972.
5. HIT Lab Reports 2:9, 1972. HSRI, University of Michigan.
6. New York State Department of Motor Vehicles Annual Report, 1970.
7. Ford Motor Company, Automotive Research Office, Report No. S70-1, 1970.
8. I. McNab, "Whiplash Injuries of the Neck", Proceedings of the Annual Meeting of the AAAM, pp. 11-15, 1965.

9. J. D. Wickstrom, J. L. Martinez, D. Johnston, and N. C. Tappen, "Acceleration-Deceleration Injuries of the Cervical Spine in Animals", Seventh Stapp Car Crash Conference Proceedings, pp. 284-301, C. C. Thomas, Springfield, Illinois, 1963.
10. H. J. Mertz and L. M. Patrick, "Investigation of the Kinematics and Kinetics of Whiplash", SAE Paper No. 670919, pp. 175-206, presented at the Proceedings of the Eleventh Stapp Car Crash Conference, 1971.
11. J. I. Tuell, M.D., Member of the American Academy of Orthopaedic Surgeons, 1305 Seneca, Seattle, Washington 98101. Personal Communication, 1969
12. M. A. Luongo, "Pathologist's Sub-Lethal Determinations", Proceedings of the Collision Investigation Methodology Symposium, Warrenton, Virginia. Sponsored by the U.S. Department of Transportation, National Highway Safety Bureau and the Automobile Manufacturer's Association, August 1969.
13. A. K. Ommaya, F. Faas, and P. Yarnell, "Whiplash and Brain Damage", Journal of the American Medical Association 204: pp. 285-287, 1968.
14. U. Pontius, Division of Orthopaedics, Tulane University, New Orleans, Louisiana, "Computer Analysis of the Mechanics of the Cervical Spine". Presentation-Computers in Orthopaedic Research and Education, March 23, 1972, Atlanta, Georgia. Unpublished.
15. J. D. States et al., "The Enigma of Whiplash Injury", Proceedings of the Thirteenth Annual Conference of the AAAM, pp. 83-108, 1969.
16. J. D. States and R. Sweetland, "The Injury Risk of Roll-over Accidents in Racing and Highway Driving", Proceedings of the Third Trienial Congress on Medical and Related Aspects of Motor Vehicle Accidents, pp. 203-212, University of Michigan, Ann Arbor, 1971.
17. H. H. Gosch, E. Gooding, and R. C. Schneider, "An Experimental Study of Cervical Spine and Cord Injuries", The Journal of Trauma, 12: pp. 570-576, 1972.
18. C. L. Ewing, et al. "Dynamic Response in the Head and Neck of the Living Human to $-G_x$ Impact Acceleration", Twelfth Stapp Car Crash Conference Proceedings, pp. 424-439, SAE, New York, N.Y., 1968.
19. Case Number RAI 144, University of Rochester Accident Investigation Team, Available through U.S. Clearing House, Springfield, Illinois.
20. J. D. States, "Restraint System Effectiveness in Racing Crashes", Proceedings of the Eleventh Annual Meeting of the AAAM, pp. 173-191, C. C. Thomas, Springfield, Illinois, 1970.
21. B. Aldman, "Biodynamic Studies on Impact Protection", Vol. 56 Supplement 192, Acta Physiologica, Scandinavica, Stockholm, 1962.
22. J. D. States, et al., "Obscure Injury Mechanisms in Automobile Accidents", Proceedings of the Fifteenth Annual Conference of the AAAM, pp. 47-57, SAE, New York, N.Y., 1972.
23. J. D. States, "Traumatic Arthritis, A Medical and Legal Dilemma", Proceedings of the Fourteenth Annual Conference of the AAAM, pp. 21-28, University of Michigan, Ann Arbor, 1971.
24. F. W. Cook and D. A. Nagel, "Biomechanical Analysis of Knee Impact", Thirteenth Stapp Car Crash Conference Proceedings, pp. 117-133, SAE, New York, N.Y., 1969.
25. E. L. Radin and I. L. Paul, "Response of Joints to Impact Loading: In-Vitro Wear", Arthritis and Rheumatism 14: pp. 356-362, 1971.
26. Personal Communication, Citywide Orthopaedic Rounds, October, 1971, University of Rochester School of Medicine, Division of Orthopaedic Surgery.

DISCUSSION

R. E. Vargovick *(Ford Motor Company)*

I have two questions; 1.) how do you define submarining and 2.) in the accident you investigated on US 15, did you determine that the belt was positioned initially correctly or incorrectly?

J. D. States

We define submarining as having a belt slide off the boney pelvis. In other words, if it rises above the anterior iliac spine we consider it submarining. We don't know the position of the belt in this man's case. He said he had it low on his pelvis. He wasn't overweight, and it occurred when he had just a suitcoat jacket on; not an overcoat and heavy clothes. It still could have been up above his boney pelvis, however, and I can't answer that for you.

H. E. VonGierke *(Aerospace Medical Research Laboratory)*

Have you found, what was shown in some cadaver experiments, that in cases when the backrest is not high enough for the occupant that the backrest actually contributes to the whiplash injury making it more severe by providing sharper curvature for the bending of the neck? I'm thinking of the cadaver experiments by Dr. Lange at the Max Planck Institute in Dortmund, Germany.

J. D. States

We've been aware of this but have not been able to quantitate the severity of whiplash. I'm working on this right now, but I find that I have to see the patient. I have not been able to depend on my own partners' to quantitate the severity of a neck injury. There are no objective guidelines except the limitation of motion. I think we will duplicate these findings in clinical accident investigations but at this point I don't have enough statistics to say that it does make any difference. We've seen cases where the neck injury appears worse because the seat was a little higher than it would be ordinarily, but there are only 10 or 12 in our series and we are unable to draw any conclusions at this point.

Y. K. Liu *(Tulane University)*

I'd like to make some comments on what Dr. States said. One is that the effect of neuromusculature which he mentioned on the dynamics of whiplash is the topic of a recent dissertation of precisely the same title, "The Effects of Neuromusculature on the Dynamics of Whiplash." This research will be presented on Wednesday at the Annual Conference of Engineering in Biology and Medicine in Bal Harbor by myself and one of my students. The research report pertaining to this work is in the process of being prepared and should be available in December of this year. I will not try to pre-empt what he has to say at Bal Harbor at this particular point. There isn't time, nor is this the proper place for it.

Secondly, in trying to arrive at certain subclinical indicators of this class of injury, where you don't see anything in the process of normal clinical examination, is there a better way of indicating this class of injury? We recently have been awarded a grant by the National Institute of Neurological Diseases and Stroke to study the use of an optimally filtered EEG as an indicator of this subclass of injury.

It has been shown by the Japanese orthopedic surgeon Tsuchiya and his colleagues that, in fact, there is subcortical spikings associated with whiplash in rabbits. We feel that these spikes cannot be detected on the usual scalp EEG because of the normal on-going EEG activity. The only way you could, in fact, detect these spikes is to build an optimal filter which would filter out the normal EEG activity. We have completed the feasibility of scalp detection of subcortical spikes. The question now is to detect the spikes induced by whiplash and to see whether, in fact, just by a scalp recording alone you can infer the occurrence subcortical spiking.

J. D. States

Do you think these spikes are due to brain injury or are they due to the neck musculature and some of the reflex mechanisms that are operating on the neck? I was aware of these abnormal spikes and concluded it was a brain injury caused by whiplash that produced them. I'd like to define these reflex mechanisms that affect the neck muscles better, but never really conceived of an approach. I wonder what you attributed those spikes to.

Y. K. Liu

We don't have any idea what causes the spikes. The spikes are observed mainly in the brain stem area. The one thing we do know is that spiking is not a normal activity. Now, in those cases where there is clinical correlation, i.e., the electroencephalographer can make a clinical diagnosis, we are interested only as a check on our scheme for spike detection. What we are interested in, in fact, are those cases in which he is not sure in his diagnosis. As to why this is the case, we are far from being able to pin it down in terms of etiology. The system response to the trauma in the spiking, lets put it that way. That's all we know.

D. A. Nagel (Stanford University)

One question and one comment; I believe that it's Rolf from England who said that if the neck was rotated one way or the other, there was a greater chance of producing ligamentous and boney injury to the cervical spine with known quantitative force applications. I'm wondering if in your rear end study you have cases where, perhaps, the wife in the front seat was turned talking to her husband when they were rear-ended and if indeed some of your passengers whose necks were rotated got more injuries than the drivers.

The comment is on arthritis from knee injuries. In setting up an experimental situation where we were trying to find the force necessary to produce fractures in the knee we did use cattle legs and we found that there were several hundred pounds less force necessary to produce articular cartilage damage than was necessary to produce fracture. Yet, I'm not really sure this always leads to degenerative arthritis in the human situation because I've looked in on joints where there was significant articular cartilage damage and followed these patients for many years and they really didn't develop traumatic arthritis. I think that articular cartilage damage is certainly a factor but perhaps ligamentous instability also caused by the injury, and perhaps not picked up, would be another factor that would enter in here so that it gets even more complicated as we go into it further and further.

J. D. States

We have some patients who appeared to have more severe neck injuries because they were sideways in the car at the moment of impact. We also have two people who were wearing shoulder harnesses and it appeared as though they had a torsional effect because their shoulder went forward and their head remained straight ahead during the rebound from the seat into the shoulder harness. Statistically, how significant this is I don't know.

COMPARISON OF DYNAMIC RESPONSE
OF HUMANS AND TEST DEVICES (DUMMIES)

L. M. PATRICK

Wayne State University, Detroit, Michigan

INTRODUCTION

Comparison of the dynamic response of humans and dummies is an ambiguous task since the inference is that a simple comparison is possible with a quantitative value or finite number of quantitative values available for comparison, while in fact it consists of comparing a finite number of variables with an infinite number. The dummy has a fixed number of parts and can, supposedly, be adjusted to have a fixed measurable force and/or torque displacement characteristics between the parts. The individual human, on the other hand, has a far greater number of parts that are connected by tissues of infinite adjustability. There are a relatively few dummies available, while the variety of humans is endless with new variations being produced daily. Thus, for the comparison to be meaningful it is essential that the dynamics of a single dummy type be compared to the dynamics of some limited, representative human under predetermined muscle tonus. Further, the comparison must be made under a range of identical acceleration environments.

With such a large number of variables involved, it is obvious that many of them must be held invariant if a realistic, practical comparison is to be made. The most obvious step in that direction is the decision to represent the adult population by three sizes of dummies based on anthropometry encompassing approximately 95 percent of the adult population. They are the 5th percentile female, the 50th percentile male and the 95th percentile male with dimensions and mass distribution taken from measurements on a large population sample. Even with the dimensions and mass distributions established, there is considerable difference between the dummies manufactured to the available specifications by different manufacturers. Furthermore, the normal tolerances within a group of dummies manufactured by the same manufacturer to the same drawings will produce some measurable variation in the dynamic response. Joint adjustment is another important variable in keeping the dynamic response constant. While adjustments can be made on the major joints, most of the adjustments will vary with joint motion. That is, if the joint is set for the 1 g level by the time the dummy is placed in the vehicle and the joints are moved to obtain the exact position, the forces required to move the joints change. Thus, the dummies which are far more uniform than the human population have a rather large number of variables in their design. If repeatable results are to be obtained under the same impact conditions, it is essential that the dummies have uniformity, both in the construction and also in the operation of the joints.

While the need for standardization in the dummy is obvious, the human has so many variables affecting dynamic response that describing the complete range is impossible. Consequently, the response of the human must be established as an average for some part of the spectrum which is acceptable for the intended purpose. In most instances, it is not possible to use human volunteers in the acceleration environment of major interest which is that of minor injury or injury threshold. Extropolation of human volunteer data to the threshold injury level is not satisfactory since there is an unknown discontinuity at the injury threshold. Several attempts have been made to establish the injury threshold by comparing laboratory data with real data from accident investigations. In most instances, the real data are not adequate in sample size to permit accurate interpretation. Also, the conditions under which the accident occurs are unknown, and the laboratory reproduction of the collision is, therefore, of questionable accuracy. However, this appears to be the most productive method of establishing the human tolerance and, thereby, the response of the dummy in a collision environment in which the human has been involved.

Other factors which affect the dynamics of both the human and the dummy are the dynamics of the vehicle, the interior geometry, the load deflection characteristics of the interior components, and the initial position of the occupant. Each of these potential variables plays a role in the comparison of the dynamic response of humans and dummies.

BASIS FOR COMPARISON

Two major categories of comparison will be considered. The first is the gross motion of the entire body which is primarily measured by photographic means, and the second is the impact of the components in the vehicle with measurement made by transducers in the test devices. Photographic techniques are satisfactory for the gross motion in which the framing rate is adequate for measuring the displacements involved. During an impact, the displacements are small and of such short duration that normal high-speed photography (up to 5000 fps) is not adequate for measuring the rapid changes. Also, the resolution required for the minute changes is not adequate when it must also measure the gross motions prior to impact.

General trajectories of the human or the dummy occupant can be measured by locating suitable targets on the subjects and assuming rigid body motion between the targets. A rough approximation of the trajectory can be achieved by following a single target on each of the major body components. For more complete identification of the motion, it is necessary to have additional points and/or lines marked on the body components to permit the angular and translational motion of the component to be recorded in three planes.

The general trajectory of the occupant is important primarily to ensure duplication of the human impact by the dummy. If the trajectory is grossly different, the dummy might not hit the same interior component, or if it does hit the same component it might hit in a different attitude or on a different part of the vehicle or dummy component.

The qualitative comparison of humans, cadavers and dummies shows that the human has greater mobility between body components than the cadaver, which in turn has greater mobility than the dummy. While there is merit in trying to duplicate the human insofar as possible by the dummy, it may be necessary to sacrifice some of the degrees of freedom for repeatability. This digression from the human may appear to eliminate the reproduction of the motions and responses of the human. However, when the variation in the human is considered it shows a wide displacement-time envelope. If the simplified dummy is more repeatable and still remains within the envelope, the net result may be beneficial. Human occupants assume a wide variety of positions in the vehicle, causing a different trajectory for any given environment. The goal of the simulation with the dummy should be to stay somewhere within the envelope and, hopefully, in the middle of the envelope representing variable exposure in the impact population. A compromise is required between the degrees of freedom for realistic simulation and the repeatability due to the degrees of freedom.

The dummy must be suitable for all modes of impact, or more than one dummy will be required to cover the collisions to be studied such as unrestrained frontal force, restrained frontal force, unrestrained lateral force, restrained lateral force, rear, roll-over and oblique collisions.

Impact simulation requires transducer measurement to permit comparison between the human and the dummy to be made by transducers. Generally, this is accomplished by mounting accelerometers and/or force transducers in or on the dummy. The transducers are necessary since the impact is not amenable to photographic analysis due to insufficient framing rates with the high-speed cameras used for this type of research.

Accelerometers are generally preferred over force measuring devices, since minimum modification is required to the occupant or the vehicle. The force transducers require the insertion of the device between the head or other mass on which the force is being measured and the object struck. If it is placed on the vehicle component, it changes the characteristics of the component, and it is difficult to install at the exact point of impact. The accelerometers have the advantage of permitting a force to be calculated from the acceleration, and also to allow the well-known relationships between velocity, stopping distance and acceleration to be used in determining the conditions required for the impact to stay within known human tolerance.

BODY COMPONENT RESPONSE

When considering the impact of the human body to a rigid object, it must be determined whether rigid body dynamics prevail or whether the body deforms enough to require use of nonrigid body dynamics. In addition to considerations of rigidity of the component under consideration, the possibility of fracture of the human component is a necessary consideration.

Bone fracture in the human results in a discontinuity in the acceleration or force curve which is a finite indication of failure. The ductile metal structure of the dummy does not fail in the same manner to provide the characteristic discontinuity in the transducer record observed in the human. When fractures occur in the human or human cadavers during the impact, the force and acceleration generally decreases rapidly, while with the dummy the forces continue to increase as the impact severity increases.

Bones of the human are more highly damped than the metal structure of the dummy. Consequently, the dummy structures often resonate if the rate of application of the force reaches a critical level. The resonating continues for a longer time in the dummy than it does in the cadaver due to the higher damping in the latter case. Noise and ringing of the structure also occurs in the dummy when metal-to-metal contact is made during extreme excursions of adjacent parts. This is especially true in the dummies with metal cervical vertebra and other joints in which metal-to-metal contact occurs.

Resistance to rotation in human joints can vary from essentially zero to some more or less constant resistance until the point is reached where the tissues have reached the end of their extensions, or until interference starts when the resistance increases rapidly. In the dummy the joints are usually designed to have approximately constant resistance until the extreme is reached, at which time a metal stop causes the resistance to increase extremely fast. If the force is great enough, the metal stop fractures and the resistance then drops to zero. In the cadaver as the resistance increases to the failure point, tissues tear and the change in resistance is less and takes place over the greater angle than it does in the dummy. In the human the failure at the point is progressive, while in the dummy it is usually instantaneous.

In the human, impact to the organs can cause damage with no outward evidence, or the impact can cause external injury to the soft tissue with or without the internal organ injury. Little is known about the forces or accelerations required in an impact to cause injury to the internal organs. With the exception of brain injury as deduced from the acceleration to the head and some knowledge of the likelihood of injury to the thoracic organs from known chest impact, the forces or accelerations to produce injury to other internal organs are not well defined.

OCCUPANT ENVIRONMENT

An examination of the occupant environment provides insight into the conditions under which the comparison of humans and dummies must be made. Prior to the collision, the occupants are in a seated position and traveling at the same speed as the vehicle. When the collision occurs, the vehicle decelerates while the occupants move forward in the original direction and position until they hit the interior. In a minor collision, the forces which can be exerted by the muscles have an appreciable effect upon the motion of the occupant. In severe collisions the forces exerted by the muscles are not adequate to materially effect the occupant's motion. Assuming a severe collision (for example, a 30 mph barrier collision), the occupant continues forward in his original position in accordance with Newton's laws until he hits the vehicle interior.

For the passenger, the knees often strike the instrument panel before any other part of the body hits anything, resulting in a deceleration of the lower limbs and pelvis and the jackknifing action about the hips. The body continues forward with rotation about the hips until the head hits the windshield or the torso hits the instrument panel. Thus, the motion to be considered prior to impact is linear motion with essentially zero acceleration. Immediately after the impact, parts of the body are subjected to linear deceleration and rotation followed by a second, third, or more impacts as the body continues to move forward striking different interior vehicle components. Each secondary impact must be considered separately with its effect on the immediate body component, plus its effect on adjacent components by the interaction between the connective tissues taken into account.

The driver follows the same pattern with either the knees striking first or the abdomen striking the steering wheel first. In either case, the same series of events takes place with multiple impacts as different parts of the body contact the interior.

An occupant restrained by a conventional harness usually has a substantial force applied to his body through the webbing before any part of his body strikes the vehicle interior. Consequently, the impact velocity is lower and the impact severity is less. However, even with the harness fitted snugly, there is a minimum of eight to ten inches of forward motion of the occupant in a 30 mph barrier impact. This sometimes permits the knees to strike the instrument panel and the head may strike the windshield, instrument panel, or steering wheel. The lap and shoulder belt causes severe flexion of the neck, and the lap belt only results in flexion at the waist during the collision. The harness webbing causes concentrated forced to be applied to torso and pelvis. If the lap belt rides over the iliac crests, there are concentrated forces applied to the abdomen with potential injury.

With a three- or four-point harness, the time over which the occupant is decelerated might be as high as 200 ms. The harness loads are relatively smooth, approximately half sine wave curves with no abrupt peaks such as occur when striking a windshield or other more rigid object. The accelerations on the head of the dummy show high peaks as the neck flexes and reaches an extreme of motion in which the

References pp. 32–33

chin hits the chest or other bottoming-out takes place. The high magnitude, short duration spikes are not present on volunteers or cadavers under the same exposures.

During the impact of the occupant with the interior component the area, geometry, rigidity, location of the impact on the body, and the direction of the impact all play important roles on the potential injury to the body and, consequently, the importance of comparing the dynamic response of the human and the dummy. There are several basic rules which will minimize injury, such as distributing the force over a large part of the body, providing cushions for avoiding localized injuries, elimination of abrasions and lacerations through a soft outer surface of the component, and providing ample deformation to decelerate the body without exceeding the tolerance in terms of acceleration. These items must all be considered in comparing the human and dummy response in the vehicle.

In a head-to-windshield impact, the acceleration might go from zero to 150 g's and back to zero in three or four milliseconds as the glass fractures. The initial acceleration peak is followed by a 30 or 40 millisecond plow-in as the head bulges the windshield and deforms the plastic interlayer. Following the windshield impact, the head may be deflected downward onto the instrument panel for an additional impact of 20 to 40 millisecond duration. During the windshield impact, the direction of the applied force changes radically. Also, the neck is extended so the force applied through the body to the head offers potential neck injury. Fortunately, accident investigations show that the number of neck injuries is small, but the reaction on the neck should be known to ensure that the new designs will not cause injury.

INSTRUMENTATION

Accurate definition of the dynamic response of a vehicle occupant during a collision is hampered by lack of adequate measuring instrumentation. The complete instrumentation complement required for showing compliance with FMVSS 208 (1) consists of a triaxial accelerometer mounted at the CG of the head of the dummy, a triaxial accelerometer mounted at the CG of the torso of the dummy, and an axial force in each femur. Before considering the instrumentation required to provide the more complete evaluation of the dynamic response of the occupant, it should be pointed out that even with the limited number of transducers, interpretation is difficult and repeatability is not adequate. The goal for instrumentation of the dummy should be to obtain the necessary data to establish the desired results with a minimum of transducers.

The current requirement for instrumentation in the dummies to comply with FMVSS 208 is for eight channels of instrumentation on the dummy: Three linear accelerations in the head, three accelerations in the chest, and two linear force measurements on the femurs. Table 1 shows the instrumentation required for a fairly complete measure of the impact response on the human or the dummy. It is observed that the number of channels required would be well in excess of 50 if several of the limbs were included. It is not proposed that all of the transducers included in Table 1 be incorporated into the dummy. They are presented to illustrate the problem of measuring the dynamic response completely with transducers.

TABLE 1

Instrumentation Required for Measuring Impact Response on Human Volunteers, for Human Cadavers, and Anthropomorphic Dummies.

Body Component	Transducer Type	No.	Comments
Head	Accel	6	One triaxial accelerometer at C.G. generally used — 6 components required for complete motion.
Neck	Force or	3	No transducer regularly used.
	Torque	2	
Thorax	Accel	6	One triaxial at C.G. currently used.
	Linear Deformation	2	A-P and lateral.
Pelvis	Accel	6	One triaxial accelerometer generally used — 6 components required for complete motion.
Femur	Force	1	Force transducer near knee — does not consider bending loads or torsional loads.
	Bending	3	
Abdomen	Force	2	No transducer generally used.
	Pressure	1	
	Deformation	2	
Limbs	Force	1	Axial force, bending moment, and torque required for complete description of each segment.
	Bending	3	
Joints	Torque	2	No transducers available.
Soft tissue	Pressure	1	None commerically available.
	Laceration	1	Replaceable "skin".

Considering the head as a rigid body requires six accelerometers to completely describe the dynamic motion. This permits the linear acceleration at a point (triaxial) and three linear accelerometers positioned to measure the angular acceleration around the three axes in combination with the triaxial acceleration at a point. Cumulative linear and angular velocities and linear and angular displacements can be calculated from the accelerometer output by single and double integration. An initial position is required to serve as the base on which to start accumulating the velocity and displacement. The initial position can be obtained from the high-speed film. For this purpose and others, it is essential that a synchronization be established between the transducer record and the high-speed film timing.

Acceleration measurements on the head of the dummy are generally made in the hollow cavity. When using human volunteers or human cadavers, it is not feasible to

References pp. 32–33

mount accelerometers inside the head. Consequently, they are mounted externally. On volunteers, a bite plate is commonly used with a trixial accelerometer mounted on it, and additional accelerometers are mounted on a tight-fitting helmet molded to the head. Accelerometers have been mounted on the forehead in the temple area and on the occiput in this manner (2, 3, 4). A computer program has been established for calculating the linear acceleration at any point on the head and the angular acceleration of the head.

The dynamic motion of the neck is affected by the position of the head at any instant and the restraint on the torso. Mertz (2, 3) and Ewing (4) have made calculations on the motion of the neck during acceleration of human volunteers. Mertz (2, 3) calculates a torque, or moment, at the occipital condyles and shows it to be a measure of the potential injury to the neck. Attempts have been made to calculate the torque directly from a torque transducer mounted in the neck, but there is no commerically available transducer for this purpose.

The thorax is normally instrumented with a single triaxial accelerometer mounted at the CG which does not provide the angular rotation of the thorax. Additional information can be obtained by adding three more accelerometers in strategic positions to provide the angular acceleration during the impact. Since the torso is nonrigid, however, this is not as important as the same calculation for the head. Linear deformation of the rib cage in two directions would be beneficial in measuring the potential injury from the impact. Linear potentiometer are used for this purpose in the dummy. Chest deflection in a cadaver can be measured by inserting a rod through the chest and measuring the amount of protrusion during the impact (5).

Triaxial acceleration of the pelvis was considered in an early standard. However, it has been eliminated in the current FMVSS 208. Six linear accelerations will be required to measure the complete motion of the pelvis.

Axial femur force in each femur is currently required. It is measured by force transducers inserted near the distal end of the femur of the dummy. Most measurements in the human cadaver are difficult to make, and any modification to the femur to insert a load cell results in a gross change in the reaction of the femur. Some measurements have been made (5) of the femur loads during the impact of the knee to a padded target. For complete measurement of the forces in the femur, it would be desirable to include a measurement of the bending moment. A minimum of three transducers would be required for this purpose. Torsion of the femur might be of some interest, although it appears that joint damage would occur before femoral injury from torsion.

There is no provision for measuring forces on the abdomen, although it is an area of frequent injury in vehicle collisions. The driver often suffers abdominal injury when striking the lower part of the steering wheel rim. An occupant wearing a lap belt that is improperly adjusted and slides over the iliac crest is also subject to abdominal injuries. A force transducer in the webbing is one way to determine the force applied.

Another important measurement would be the pressure distribution over the abdomen. ITT (6) has developed a method for measuring the pressure distribution over an area of the body. It is in the early development stages and will require considerable refinement before it will be suitable for this purpose. A deformation measurement might be desirable in which the distance the wall of the abdomen is deflected would be the injury criterion.

The force on limbs can be measured in an axial direction as described under the femur transducer. Bending is also important in any of the limbs' segments, and if complete data are required, at least three force measurements will be required on each limb segment.

Joints should be measured for the maximum torque applied. In general, a joint is not injured until excessive motion is encountered. There are no transducers available commerically for this purpose, and no easy way to evaluate the potential injury to the joints.

Soft tissue injury has not been measured in the general dummy impact environment, except when glazing material is involved. Reiser (7), Patrick (8) and others have used a removable skin over the head of the dummy to measure the laceration potential. Generally, a multiple layer of moist chamois is used with the degree of laceration measured by the length of lacerations through one or both layers of chamois. Qualitative evaluation of abrasions can also be obtained with the chamois technique. A pressure measuring technique for soft tissue would be desirable to provide a quantitative measure of both load concentration and contusion injury potential.

Table 2 shows what the author considers to be a minimum of instrumentation for a reasonable evaluation of the dynamic response of the dummy. With six accelerometers strategically located in the head, complete motion of the head can be recorded by using the method of Mertz (2). The reaction at the occipital condyles, which is a measure of the injury potential of the neck, can also be calculated.

TABLE 2

Suggested Minimum Instrumentation for Measuring
Dummy and Human Dynamic Response to Impact.

Body Component	Transducer		Comments
	Type	No.	
Head	Accelerometer	6	One triaxial and 3 linear
Thorax	Accelerometer	3	
	Linear potentiometer	2	
Pelvis	Accelerometer	3	One triaxial
Femur	Force	1	One in each femur
Abdomen	Linear potentiometer	2	A-P and lateral

References 32–33

The thorax is a deformable body and, consequently, it is more important to know the acceleration of the CG and the deflection in two directions rather than the angular motion from additional accelerometers.

A single triaxial accelerometer mounted in the pelvis provides information on the inertia forces applied to the femurs. It also assists in determining whether the loads through the femurs are due only to the inertia forces or whether there is an additional force applied at the sacrum or back of the dummy with a crushing action on the femurs.

The abdomen is probably the most neglected part of the dummy insofar as evaluating injury potential is concerned. One method of improving the situation is to install a high hysteresis, deformable material in the abdomen. By measuring the deformation of the material with two linear potentiometers, the force can be deduced knowing the characteristics of the material. If the hysteresis is high enough, it will also provide at least a qualitative indication of the concentration of the force on the material if it is inspected immediately after the impact.

Table 2 represents instrumentation that requires eighteen channels of recording. The difficulty of evaluating all eighteen channels is great enough that it is essential to have an automatic data handling system. A tape recorder with a direct A-D converter and suitable programs provides a quick method of handling the data. In addition to the computer print out, a computer plotting procedure is also recommended.

DYNAMIC COMPARISON

The literature search revealed that there are very few studies in which human volunteers, human cadavers and anthorpomorphic dummies have been evaluated under identical impact conditions. There are many studies with dummies only, and several studies with volunteers and cadavers. However, the conditions for the separate studies are not identical and, consequently, a direct comparison cannot be made. A summary of several of those studies where a direct comparison is available follows.

Head injury has more serious consequences and occurs more often than injury to any other single part of the body. Most investigators find that 70 to 75 percent of the serious injuries in automobile collisions are to the head. Hodgson (9) has conducted a study in which he compared cadavers and dummies in a horizontal, pendulum impact. The subject under test was strapped to a frame hinged at the feet. The head was then lifted a predetermined distance and dropped onto a rigid surface with impact to the frontal bone. Accelerometers were mounted on the cadavers heads and in the dummy heads to measure the acceleration and allow the Severity Index (Gadd) to be calculated. Six cadavers were used with increasing drop heights until the threshold of linear fracture was defined. The six cadavers had an average drop height of 14 inches at fracture, with an average Severity Index of 1150. The Severity Index varied from 600 to 2000 for the cadavers. The dummies were dropped from a height of 15 inches with the Alderson VIP-50 recording a Severity Index of 2000, and the Sierra 1050 recording a Severity Index of 2700. It was found that, typically for rigid surface impact, the dummy response is shorter in duration and higher in amplitude at the drop height at which cadaver fracture occurred.

A direct comparison of the dynamic response of a human volunteer, cadavers and an anthropomorphic dummy (10) provides a direct comparison of the three under simulated collision conditions with harness and no direct impact with the interior of the vehicle. Fig. 1 is a composite diagram including frames from the high-speed

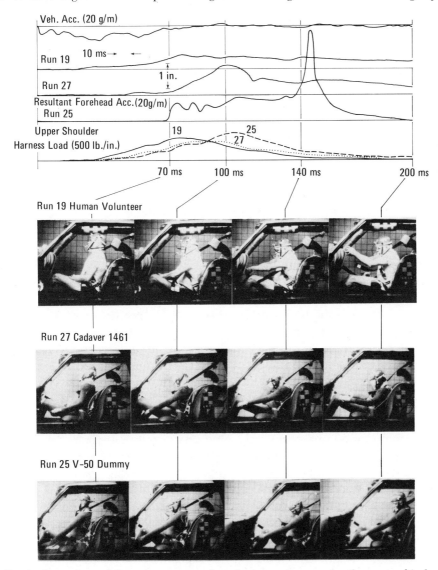

Fig. 1. Comparison of dynamic response of a volunteer, cadaver, and anthropomorphic dummy wearing a four-point harness in an 11 mph simulated barrier collision.

References pp. 32–33

movies for each of the occupants, and the forehead accelerations for each. The four frames from each of the high-speed movies were taken at 70 ms, 100 ms, 140 ms and 200 ms after the initiation of vehicle deceleration. Comparison of the three subjects at identical times during the impact shows similar movements. At 70 ms all three occupants have moved forward in an upright position. At 100 ms the cadaver and dummy appear to be slightly farther forward, and the knees are somewhat higher. At 140 ms the dummy and cadaver have about the same flexion which is somewhat greater than that of the volunteer. The volunteer was in a relaxed condition. Even so, the head did not flex as far as that of the dummy or the cadaver. At 200 ms the occupants are back in the seat, with the arms of the volunteer slightly above horizontal while the arms of the dummy and cadaver are lower. The cadaver appears to be slouched in the seat more than either the dummy or the volunteer, and the dummy is farther back than the volunteer. On the whole, it appears that all three occupants have followed a similar motion from a qualitative standpoint.

At the top of Fig. 1, a graph shows the acceleration of a vehicle, the forehead acceleration of the three subjects, and the upper shoulder harness load of the three subjects. The volunteer has an approximately constant acceleration from 70 ms to 200 ms. The cadaver had a peak acceleration at 100 ms, about twice that of the plateau for the volunteer. The dummy had a peak acceleration of short duration at about 150 ms of over six times that of the maximum value for the volunteer.

Comparison of the upper shoulder harness loads, also graphed in Fig. 1, shows the volunteer load to be between the cadaver and dummy in magnitude, with the peak for the volunteer occurring at about 75 ms while the peaks for the dummy and cadaver occurred at 95 to 110 ms. It should be pointed out that for this run with the volunteer relaxed, there was no slack in the shoulder harness, while for the dummy and cadaver there was the traditional one fist of slack. Run 19 was used since it was thought that the relaxed volunteer more nearly simulated the dummy and cadaver condition. If the shoulder harness load for the tense condition with slack is substituted for the relaxed condition of Run 17, the upper shoulder harness loads were 230, 350, and 520 pounds for the volunteer, cadaver and dummy — in that order. Also, the peak loads occurred at about the same time.

Mertz (3) compared human volunteer, dummy and cadaver results under simulated whiplash conditions. The torque at the occipital condyles was much higher for the dummies than the volunteers. In one case, the cadaver had approximately the same torque as the dummy, while Dummy No. 2 had considerably higher torques than either the volunteer or the cadaver. At higher impact velocities, the dummies showed much higher occipital condyle torque than the two cadavers. These values are shown in Figs. 2 and 3. Fig. 4 shows the extension of the neck of the two dummies and the two cadavers in the severe whiplash simulation. Dummy No. 2 had a considerably higher neck extension at 105 degrees than the cadavers and the other dummy at 86 degrees. A qualitative evaluation of the comparative motion of the volunteer, cadavers and dummies showed the volunteer to be more mobile than the

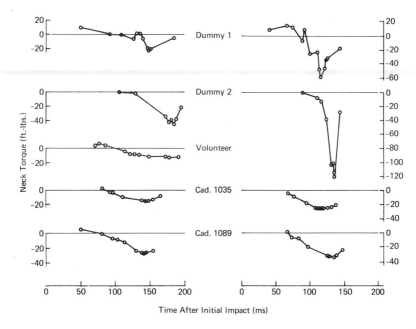

Fig. 2. Computed neck torques acting on the head at the occipital condyles as a function of time for (from left to right) 10 and 23 mph simulations, no headrest, rigid seat back.

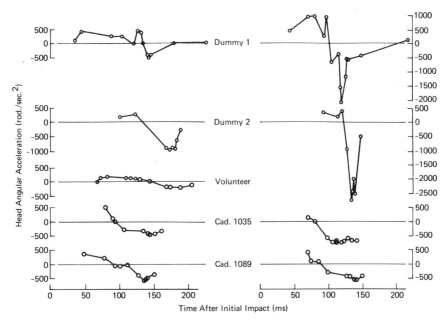

Fig. 3. Computed head angular accelerations as function of time (from left to right) 10 and 23 mph simulations, no headrest, rigid seat back.

References pp. 32–33

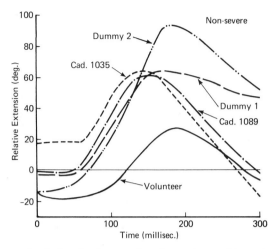

Fig. 4a. Comparison of relative head extension for various subjects for nonsevere 10 mph simulation, no headrest, rigid seat back.

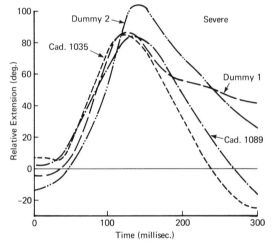

Fig. 4b. Comparison of relative head extension for various subjects for severe 23 mph simulation, no headrest, rigid seat back.

dummies or cadavers, with greater motion between the limbs and the torso; this, in spite of the fact that the volunteer was tense.

It is significant to note than when a head support was used in the simulated rear-end collision to prevent the extension of the neck, that the volunteer, dummy and cadaver load on the head support was approximately the same as shown in Fig. 5. In this case, the head was acting as a rigid body with little or no reaction through the neck to modify the rigid body reaction to an applied deceleration. The load deflection characteristics of the neck are obviously the important characteristics to consider in dummy neck design.

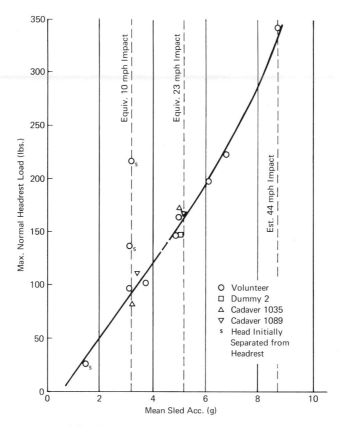

Fig. 5. Maximum normal headrest loads as function of mean sled acceleration for various subjects, flat headrest, belted, rigid seat back.

CONCLUSIONS

Based upon the references cited herein plus additional published and unpublished data, the following conclusions are drawn:

1. There is a paucity of data available for accurately comparing the dynamic response of humans and dummies.

2. The rapid changes taking place in dummy design make comparisons of dynamic response of the dummy and human of limited value.

3. Current dummies often show short duration, high amplitude acceleration or force peaks on the transducer records during impact that are not present in comparable human exposures.

4. More complete instrumentation is required to compare the human and dummy response.

5. Automatic data handling will be required to interpret the output of the more sophisticated instrumentation.

References pp. 32–33

6. Emphasis should be placed on repeatability of results.

7. Attempts to duplicate the skeletal structure have not been successful, and efforts should be directed toward developing a test device based on performance. It does not necessarily have to resemble a human.

8. It may be necessary to reduce the number of degrees of freedom in the test device to achieve reproducible results even at the expense of deviating from the dynamic response of the human.

9. Provision should be made for measuring potential neck injury.

10. Provision should be made for measuring potential abdominal injury.

11. Soft tissue injury (contusion, abrasion and laceration) should be measured in vulnerable body areas such as the neck and face.

12. Gross Comparison of dynamic response of current dummies with humans as obtained from high-speed film analysis shows better agreement than obtained by comparing transducer results during short duration impacts.

RECOMMENDATIONS

Improvement of the dummy to make the response more human-like can be accomplished if research programs are undertaken to:

1. Establish human and human cadaver responses under a set of known, reproducible conditions to serve as a standard for evaluating new dummies.

2. Optimize dummy design by compromising on the number of components and joints in the design.

3. Concentrate on performance rather than attempting to duplicate the structure of the human body.

4. Establish injury criteria for the abdomen, pelvis and limbs with methods for accurately comparing the human and test device results under dynamic impact conditions.

5. Use a replaceable skin with multiple layers or other means to measure the degree of laceration and/or soft tissue damage.

6. Develop a method of indicating bone fracture or other tissue failure levels in the test device.

7. Conduct a study to determine the type and minimum number of transducers required to establish correlation between the human and the dummy.

REFERENCES

1. *FMVSS 208 – 571.208 Standard No. 208; Occupant Crash Protection (effective January 1, 1972), Federal Register, Vol. 36, No. 232, December 2, 1971.*

2. H. J. Mertz, Jr. – "The Kinematics and Kinetics of Whiplash", Ph.D. Disertation, Wayne State University, 1967.
3. H. J. Mertz and L. M. Patrick – "Investigation of the Kinematics and Kinetics of Whiplash", Eleventh Stapp Car Crash Conference Proceedings, pp. 267-317, SAE Paper No. 670919, October 10-11, 1967, Anaheim California.
4. C. L. Ewing and D. J. Thomas – "Human Head and Neck Response to Impact Acceleration", Report to Naval Aerospace Medical Laboratory, USAARL 73-1. August 10, 1972, Pensacola Florida.
5. L. M. Patrick, C. K. Kroell and H. J. Mertz – "Forces on the Human Body in Simulated Crashes", Ninth Stapp Car Crash Conference Proceedings, Nolte Center for Continuing Education, University of Minnesota, 1966.
6. IIT – "Design, Development and Fabrication of a Full Scale Anatomical Display", Final Report to U.S. Army Natick Laboratories, Natick Mass., Contract No. DAAG17-70-C-0161.
7. R. G. Rieser, J. Chabal and C. W. Lewis – "Low Velocity Impacts and Temperature Sensitivity of Automotive Windshields", Fifteenth Stapp Car Crash Conference Proceedings, pp. 613-144, SAE Paper No. 710869, November 17-19, 1971, Coronado California.
8. L. M. Patrick and R. P. Daniel – "Comparison of Standard and Experimental Windshields", Eighth Stapp Car Crash Conference Proceedings, pp. 147-166, October 21-23, 1964, Wayne State University, Detroit Michigan.
9. V. R. Hodgson and L. M. Thomas – "Breaking Strength of the Human Skull vs. Impact Surface Curvature", Final Report to Department of Transportation, Washington D.C., Contract No. FH-11-7609, June 1971, DOT HS-800 583.
10. L. M. Patrick and K. R. Trosien – "Volunteer, Anthropometric Dummy and Cadaver Responses with Three and Four Point Restraints", Automotive Engineering Congress, Paper No. 710079, January 11-15, 1971, Detroit Michigan.

DISCUSSION

J. A. Bartz (Cornell Aeronautical Laboratories)

In your abstract you concluded that the muscle forces exerted by the human affect the measured responses significantly at low accelerations but they have little effect at high accelerations. How did you arrive at that conclusion?

L. M. Patrick

By some of the volunteer work we've done. For example, we found that at low accelerations in neck flexion studies the muscles could control the complete motion of the head; but when we got high enough in the severity we found that they could not. The head flexed, and the results were about the same as if the muscles were not present. Certainly if we extrapolated that higher, I'm sure the muscles would have had no affect.

H. E. VonGierke (Aerospace Medical Research Laboratory)

I wonder if you shouldn't add to your conclusions as a first requirement that one clearly decides what one wants to duplicate and what one wants the dummy to represent. I think there is a great difference if you want to have equivalent severity indices, equivalent dynamics of the body parts, or equivalence in accelerometer readings. I think one should make this decision before one goes a step further and freezes on the number of channels of instrumentation. I think if you want to take the severity index, or something similar to this, as an indicator; then the title of your paper should have been different. It should not be the *Dynamic Response* of the Body Compared to Dummies, but the *Severity Index* of the Body. . .; Primarily one should decide on the final method we use for assessing the "severity" of the impact, the severity index, any modification of it or any other integrated measure.

L.M. Patrick

What we should use as the severity index is currently the subject of broad discussion. However, I don't quite understand what you're driving at. The dynamic response, as I mentioned, was both macro- and microscopic; and certainly if the dynamic response during the impact is the same, then we're going to have the same severity index. If we can exactly duplicate the dynamic response, I don't care what severity index you use. They are going to compare identically.

One other point is that it may not be necessary to compare the exact values. Maybe we can have a correlation of two to one and so on. We might not even need the same test device for all conditions. Maybe we need one for forward force, one for the flexion and one for the extension studies. I think we'd like to have the same one though, if at all possible.

THE REPEATABILITY OF DUMMY PERFORMANCE

H. T. McADAMS

Cornell Aeronautical Laboratory, Inc., Buffalo, New York

ABSTRACT

Anthropometric test devices (dummies) provide a means for evaluating vehicle/-occupant restraint systems for compliance with safety standards. Inasmuch as all tests cannot be performed with the same dummy, it is important that all individuals of the dummy population exhibit as nearly the same response as possible when tested under identical conditions. Moreover, a particular dummy should also exhibit highly repeatable performance when subjected to the same impact environment. This paper describes efforts to measure (1) performance differences attributable to test-to-test variability of a particular dummy, (2) dummy-to-dummy variability of two or more dummies of the same manufacture, and (3) variability from one make of dummy to another. In recognition of the fact that repeatability depends on the restraint environment, comparisons were conducted under a number of restraint conditions.

Two makes of dummies were tested: Sierra and Alderson. Two dummies of each make were employed, and each was subjected to three replicate tests. For each run, head and chest accelerations and femur loads were recorded as functions of time. Two aspects of the time histories were isolated as being of interest in the repeatability comparisons: (1) general shape of the response pulse, and (2) high-frequency perturbations about this general shape. The time history records were digitized for computer analysis and appropriate numerical measures extracted for statistical analysis. This analysis consisted of partitioning the total variance of the observations into test-to-test and dummy-to-dummy components.

With regard to qualification of dummies for use in compliance testing, it was concluded that dummies should be qualified not on the basis of response repeatability alone but rather with due regard to their ability to discriminate between severity levels of the environment to which they are exposed. Thus, sensitivity and repeatability must both be considered in judging the efficacy of a dummy for safety compliance testing. Lack of repeatability of dummy response when subjected to supposedly identical conditions can arise because of either (a) difference in the response characteristics of the dummies or (b) differences in the test environment. Moreover, various quantities can be viewed as measures of repeatability. Some of these quantities evidence less variability than others because they, in effect, emphasize different aspects of the dummy response. Repeatability of response also depends on the nature of the test environment. In some instances, minor variations in the test environment may induce sizable changes in dummy response; in other instances, dummy response may be relatively insensitive to such variations. Finally, it was concluded that for statistically valid conclusions a larger number of tests may be required than is often realized.

INTRODUCTION

A principal use of anthropomorphic tests devices (dummies) is to provide a means for evaluating vehicle restraint systems for compliance with safety standards, specifically Motor Vehicle Safety Standard No. 208. It is important, therefore, that the response of the dummy be reproducible when subjected to repeated testing under the same impact conditions and with the same restraint system. Inasmuch as different dummies will be used by different test facilities, it is equally important that dummies produced to the same written physical specifications exhibit the same response when tested under identical conditions. It was for this reason that a program was undertaken to measure (1) the run-to-run repeatability of a particular dummy, (2) the dummy-to-dummy repeatability of two or more dummies of the same manufacture. In the event that dummies are of different makes, it must also be assured that the two makes respond similarly under the same test.

Reproducibility alone, however, is not a sufficient requirement for the use of dummies in a viable program of safety compliance testing. The object of testing is to discriminate between safe and unsafe systems. Consequently, dummies must have a high degree of *sensitivity* to different levels of severity of the test and restraint environment. This requirement is complicated by the fact that dummies are used in tests of several different forms of restraint or protective systems. Each of these systems gives rise to a different type of dummy loading and to a different emphasis on the various aspects of dummy response. In view of this fact, it is conceivable that dummies may exhibit highly repeatable performance in one test environment and highly variable performance in another.

To a certain extent, sensitivity and repeatability are conflicting requirements of any test, and dummy impact testing is no exception. If the dummy is made sufficiently unresponsive to variables in the test, it will by definition give reproducible results; however, the dummy would provide little or no discrimination between a good restraint system and a poor one, or between a severe impact and one which is substantially negligible as far as human injury is concerned. The need to arrive at some acceptable compromise between the opposing demands of sensitivity and repeatability in a dummy, therefore, must not be underestimated. It must, of course, also be borne in mind that the ability of a dummy to yield reproducible response depends strongly on the degree to which the exposure environment in the test can be repeated.

The interrelation of sensitivity and repeatability can be better appreciated by considering the schematic of Fig. 1.

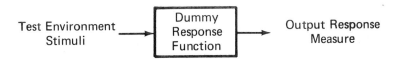

Fig. 1. Dummy stimulus-reponse relationship.

In reality, the dummy exhibits no response without a stimulus; for example, if the vehicle moves at zero velocity, the output measure would be constant and thus highly reproducible. So long as the dummy response function remains constant, variations in the output response measure can occur only by being induced by variations in the test environment stimuli.

Variations in stimuli can be regarded as of two types. In one set are those variations in the test and restraint environment which cannot be avoided; in the other are those variations which can be deliberately brought to bear on the test by design modifications of the restraint system or by intentional changes in the exposure environment. Ideally, the dummy should be unresponsive to minor variations in severity of exposure which occur because of lack of complete control over the testing environment, yet be fully responsive to variations in severity which are large enough to be of importance in the design of restraint systems. For a meaningful evaluation of test devices, therefore, the testing program should be structured so as to measure responsiveness of the dummy to significant changes in severity levels as well as to measure its repeatability at a given severity level. It is to be noted that under sufficiently severe conditions, uniform "response" may be realized because the dummy is "saturated" or overstressed beyond its range of endurance.

In view of the indicated relationship between repeatability of dummy response and exposure conditions, a test program was conceived in which several restraint configurations were employed. Four configurations were selected with a view toward spanning a range of exposure severity and, at the same time, representing plausible options for viable restraint systems. The severity of the exposure environment was controlled so that dummy response would be roughly in the range of MVSS 208 injury criteria. This test program, aimed at statistical assessment of the repeatability of dummies in the four selected configurations, is the subject of this paper.

EXPERIMENTAL CONDITIONS AND PROCEDURE

Tests simulating a 30 mph barrier crash were performed with the HYGE sled. This sled is of the reverse acceleration type — that is, the sled is accelerated from zero velocity, so that initial conditions (representing conditions at the time of impact) can be made highly reproducible. Dummies were instrumented to measure triaxial head and chest accelerations as functions of time and to measure femur-load time histories. The raw data were recorded on magnetic tape as analog signals by means of Sangamo Model 3500 tape recorders. After being digitized at a scanning rate of 9600 samples per second, the data in each channel was numerically filtered in accordance with MVSS 208 and written on magnetic tape for computation and manipulation as desired. The data in this form, along with appropriate control information, was also written on a separate magnetic tape and subsequently used as input to drive a CALCOMP plotter, which plotted each measured response parameter as a time history.

Four 50th-percentile anthropometric test devices were employed: two Sierra 292-1050 models and two Alderson VIP 50A models. Three replicate tests were conducted on each of the four dummies in each of the four test configurations. The four test configurations were:

1. Lap and torso belt restraint.

2. Lap belt and head target.

3. Energy absorbing steering wheel column and knee target.

4. Pre-inflated air bag.

Thus 48 tests were conducted in all.

A photograph of Configuration 1 (lap and torso belt restraint) is presented in Fig. 2. The shoulder belt is attached to the basic test structure over and behind the left

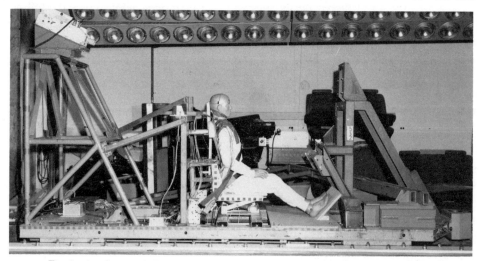

Fig. 2. Accelerator sled test configuration number 1 (three point restraint system).

shoulder and below and behind the right of the seat as is typical for the driver position of current automobiles. The lap belt is anchored symmetrically behind and aft of the seat on both sides in a typical manner. As can be seen, there are no contact targets. A new set of commerically available replacement belts was used for each test run. The belts were both adjusted to provide five pounds of pre-load at each anchor location. Care was taken always to place the dummy in the same position by using marks on the floor and seat. As per MVSS 208 and prior to final detail positioning, the spine, buttocks, etc. of the dummies were flexed and positioned by bending the torso over a 6-inch roller pulled aft against the abdomen with a 50-pound force. The same positioning procedure was used in each of the four configurations.

Configuration 2 (lap belt and head target) is shown in the photograph of Fig. 3. The lap belt is located and preloaded as in Configuration 1. The head target consists of four 2-inch thick layers of plastic foam of 1.0 lb/ft^3 density (Styro-Foam "Dorvar FR-100") bonded together on a 1/4-inch metal support plate which is attached to the basic structure as shown in the photograph. The lap belt and target foam were replaced for each sled test.

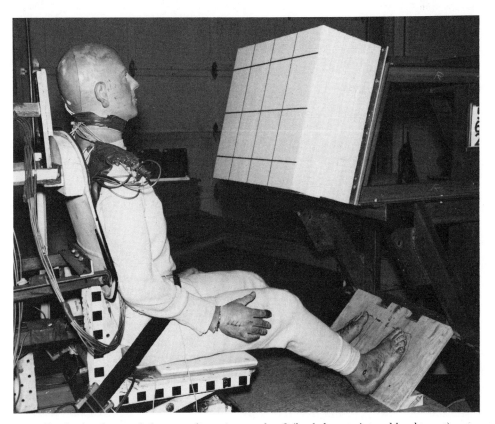

Fig. 3. Accelerator sled test configuration number 2 (lap belt restraint and head target).

The test setup for Configuration 3 (energy absorbing steering wheel column and knee target) is shown in the photograph of Fig. 4. The simulated steering wheel was made by bonding two layers of 2-inch plastic foam of 1.0 lb/ft^3 density to a 1/4-inch

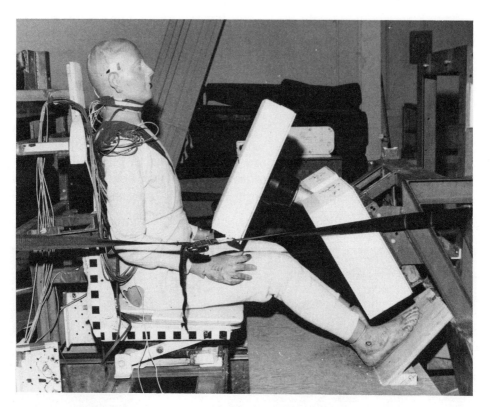

Fig. 4. Accelerator sled test configuration number 3 (energy absorbing steering column and knee target).

rectangular metal plate 16 inches wide by 18 inches high. The design of the simulated steering wheel was chosen in the interest of repeatability. The steering column consists of a 2-3/8 inch diameter steel pipe attached to the basic structure by transverse flexures which allow column axial loads to be reacted and measured by a load cell installed at the lower end. A Mercedes-Benz inver-tube is installed between the simulated steering wheel and steering column to absorb energy of the impacting dummy. The knee target was constructed of three layers of 2-inch plastic foam bonded to a 1/4-inch metal plate and attached to the basic structure as shown. The density of the outer layer of plastic foam on the knee target was 1.8 lbs/ft^3 (Styro-Foam "FR") and the density of each of the two inner layers was 3.0 lbs/ft^3 (Styro-Foam "HD-300"). The dummy was positioned as described for Configuration 1 and the simulated steering wheel, the inver-tube, and the knee target were replaced for each sled run. Because the dummy is unrestrained in this configuration, it tends to be pitched off the sled during the rebound portion of the event. To prevent this occurrence and subsequent damage to the dummy, a non-interferring belt barrier was installed on the sled on either side of the dummy as can be seen in the photograph.

The sled test setup for Configuration 4 (pre-inflated air bag and knee target) is shown in the photograph of Fig. 5. A simulated windshield and instrument panel were

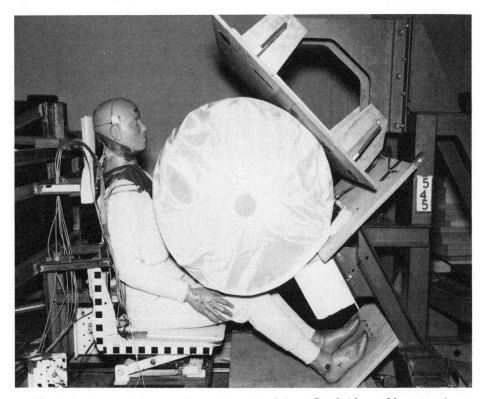

Fig. 5. Accelerator sled test configuration number 4 (pre-inflated airbag and knee target).

added to the basic structure to provide a reaction surface for the air bag. The knee targets used were the same as those described for Configuration 3 except for the vertical dimension, which was reduced to accommodate the air bag. In the interest of simplicity and repeatability, the air bag was made in the shape of a cylinder 28 inches in diameter and 30 inches long. It was fabricated from 5-ounce coated nylon fabric and lined with a very light-weight weather balloon to make the bag airtight. The bag was provided with two 3-inch vents (one on each end) and pre-inflated to a pressure equivalent to a one inch column of water. On impact of the dummy against the bag, the weather balloon rubber lines stretched through the vents and ruptured. This arrangement provided a repeatable performance of the bag. The dummy was positioned as described for Configuration 1 and the air bag and knee target were replaced for each sled run.

DATA ANALYSIS

Results from the test program are of primary interest in evaluating: (1) the repeatability of the response of a particular dummy when subjected to replicate (repeat) tests; (2) the repeatability of the response of two dummies of the same make; and (3) differences between makes with regard to (1) and (2). It was essential, therefore, to develop a rationale for data analysis which would delineate these aspects of repeatability.

What is the criterion by which two tests of a particular dummy or of two different dummies are to be adjudged "repeatable"? Are two tests to be considered repeatable only if the time histories are identical? Or is there some less exacting characteristic or feature of the time history curves which provides adequate information for purposes of the analysis? For example, it can be shown that certain response measures can be "repeatable" even though the time histories in the strict sense of time functions are not. Conceivably, there is an entire spectrum or hierarchy of possibilities, and one must choose that criterion of repeatability which is appropriate to the purpose at hand.

An example will suffice to illustrate this point.

Consider the Severity Index (SI) for the head, defined as a time integration of a function of a(t), the acceleration time history:

$$SI = \int [a(t)]^{2.5} dt \tag{1}$$

In practice, SI is approximated by numerical integration. The time axis is divided into increments and the ordinate at the midpoint of each time interval, after being raised to the 2.5 power, is multiplied by the width of the interval. These products are then summed to yield SI. Note that if the composite function

$$f [a(t)] = [a(t)]^{2.5} \tag{2}$$

is plotted instead of a(t), the procedure amounts simply to obtaining the area under the f [a(t)] curve.

Because of variables beyond control, such as small variations in initial conditions, repeat dummy runs will not produce identical time histories. For example, the instant at which the peak force or acceleration occurs may vary slightly from fun to run, but such variation may have little effect on the computed value of SI. Indeed, it can be shown that many different acceleration time histories can give rise to the same value of the Severity Index. Consequently, the apparent diversity of time histories may convey a misleading and exaggerated notion of the variability of the SI. If we are to study repeatability of the SI *as calculated by standard procedure*, therefore, we must not concern ourselves with "repeatability of time histories" in the large — only with those aspects of the time history which are germane to the calculation.

In the case of the head Severity Index, it is evident that sequential aspects of the time history can be ignored and that we need concern ourselves only with the *amplitude vs time duration* aspects of the acceleration time history. Consider Fig. 6, which is taken to represent an acceleration time history.

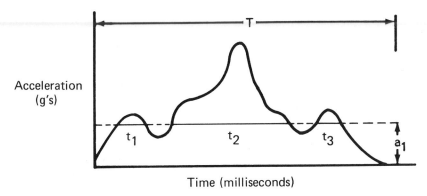

Fig. 6. Hypothetical acceleration time history.

The combined length of the line segments t_1, t_2 and t_3 gives the total time during which the acceleration equals or exceeds the value a_1, and if T denotes the total time duration, then $\frac{t_i}{T}$ gives the *fraction* of the total time which experiences accelerations equal or greater in magnitude than a_1. If the horizontal slice is made at various levels and the results plotted, one obtains a monotonic decreasing function f(a), the cumulative curve. All waveforms which reduce to the same cumulative curve will yield exactly the same value of Severity Index, regardless of the shape of the time history or the sequence in which accelerations of various magnitudes occur. From an engineering standpoint, this simplication is not particularly unreasonable; if the total integral is such as to exceed the critical value of SI = 1000, it may matter little whether the significant contributions ("the straws that breaks the subject's head") come early or late in the time history. Also, with a bit of reflection, it is clear that the area under the curve a(t) or the composite curve $f[a(t)] = [a(t)]^{2.5}$ can be as rigorously determined by slicing the curve into horizontal layers as by slicing it into vertical segments.

It is evident, therefore, that two classes of measures of the repeatability of acceleration time histories can be recognized: time-dependent measures and time-independent measures. In the first, points along the time axis must be identified and referenced with respect to zero time for the event — that is, one must deal in "clock time" as measured from a reference point. In the second, only time *duration* of the phenomenon is of interest. Under the first criterion, two acceleration time histories will be said to agree only if, for example, the acceleration was 5.1 g's at 50 ms after zero time for both. Under the second criterion, it is sufficent to know only that a given level of acceleration — say 5.1 g's — was exceeded a total of 10 ms in each, regardless of the clock time at which these exceedances occurred. In our study, attention was given to both time-dependent and time-independent measures of repeatability, but primary emphasis was on those measures of performance specified in MVSS 208.

In regard to time-dependent measures of repeatability, two aspects of the time histories were isolated as being of interest: (1) the general shape of the response pulse, and (2) the high-frequency perturbation about this general shape. Though an element of subjectivity is involved in this distinction, it will become apparent that the distinction is a useful one, even though it is to some extent arbitrary. Moreover, the criteria for the distinction are subject to constraints which do not admit much latitude in the choice of what is to be regarded as the general shape of the response pulse. Again, head acceleration data can be used to illustrate this point.

Head acceleration time history as digitized for analysis purposes consisted of resultant head accelerations recorded at time intervals of approximately 1×10^{-4} seconds. The two features of interest were isolated by observing a "running average" and "running variance" for a small increment of the time history, as shown in Fig. 7.

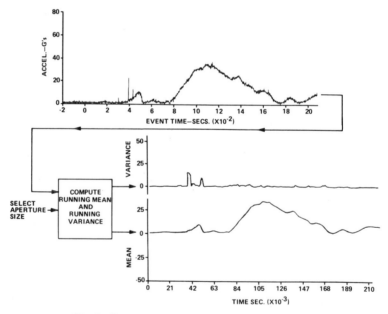

Fig. 7. Computation of running mean and variance.

The interval was chosen large enough so that several adjacent time points could be averaged to smooth out random, high-frequency variations, but not so large as to obscure the major features of the pulse. Also, the "window" or aperture was large enough to provide a "local" measure of variance or "noise" but not large enough to incur appreciable variance arising from steep up or down trends in the running average. (Detrending can, of course, be employed to eliminate such artifacts in the event that they occur). To select a window of appropriate size to delineate the two features of interest, several preliminary calculations were made in which the running average and variance were computed with various numbers of points in the averaging "window".

A 30-point window was decided upon. Thus averaging was performed over a time interval or approximately 3 milliseconds. It will be seen that such averaging is equivalent to the operation of a low-pass digital filter, which has negligible attenuation for frequencies below about 100 hertz (amplitude is down 3 dB at 200 hertz). In view of the fact that the total time duration of the test-event pulse is of the order of tenths of a second, the averaging process would not be expected to obscure major features of dummy response.

Let us direct our attention to techniques for statistical analysis for the running mean. For each dummy, three replicate runs were made and hence three curves were available. By superimposing these three curves in the same plot, one can observe qualitatively the extent of agreement of the replicated time histories. Fig. 8 illustrates

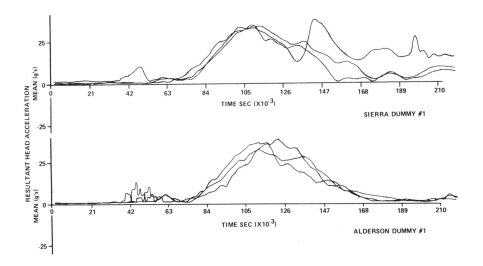

Fig. 8. Superposition of running means for three replications (NBS Strap Test).

this point for resultant head accelerations for three replications of one of the Sierra dummies and three replications of one of the Alderson dummies. The restraint system in this illustrative case was a system of straps as employed by the National Bureau of Standards and referred to as NBS Strap Tests. It will be noted that in the portion of the time histories which precede the peak values, the Sierra and Alderson dummies exhibit approximately the same degree of the agreement between replications. In the later portions of the time history, however, the Sierra dummy exhibits greater dispersion among the three replications.

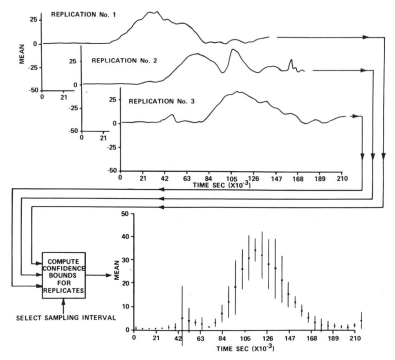

Fig. 9. Computation of confidence bounds.

In order to reduce this qualitative observation to a more quantitative basis, each of the curves was sampled at 35 equally spaced points in time. This sampling was considered adequate to retain the general shape of the curve and to provide a definitive picture of how replication repeatability varies with time, as shown in Fig. 9. The three runs for each dummy were averaged at each sample time point and the standard deviation for that time point estimated by the formula

$$\hat{\sigma} = \sqrt{1/2 \sum_{i=1}^{3} (x_i - \bar{x})^2} \tag{3}$$

where x_i denotes individual runs at a given time point and $\bar{x}$ denotes the mean of the three runs. The results are plotted as line segments centered at $\bar{x}$ and extending from $\bar{x} - \hat{\sigma}$ to $\bar{x} + \hat{\sigma}$. Comparison of a Sierra dummy with an Alderson dummy is shown in Fig. 10. In agreement with the observation in Fig. 8, it is noted that the bounds are somewhat wider for the Sierra dummies, particularly in that portion of the time history which occurs after the peak acceleration is reached.

Further insight into the factors affecting the repeatability of dummy performance is provided by analysis of variance, performed as shown in Fig. 11. The inputs to this model can be any performance measurement of interest — e.g., head severity index, maximum chest acceleration or femur load. Whatever the measurement of interest,

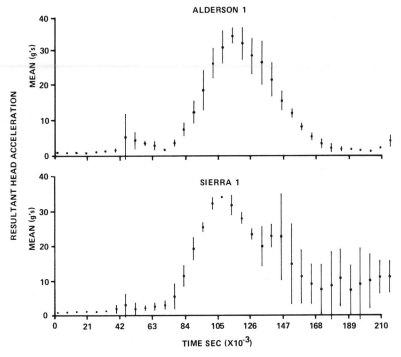

Fig. 10. Comparison of confidence bounds (NBS Strap Test).

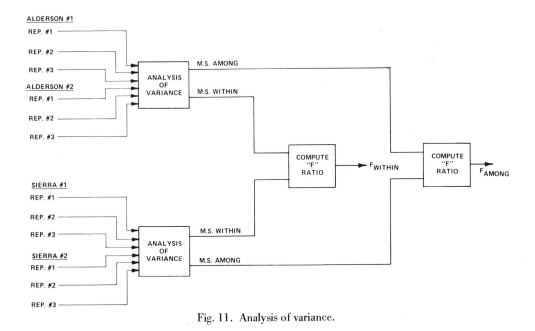

Fig. 11. Analysis of variance.

these inputs originate in the three replicate runs for each of two dummies and can be represented as:

$$y_{ij} = \mu + d_i + \beta_{j(i)} \tag{4}$$

$$i = 1, 2$$
$$j = 1, 2, 3$$

where y_{ij} is the observation for the jth replication of the ith dummy. The quantity μ is the average for all six runs. The quantity d_i denotes an excursion from this mean value occasioned by a contribution peculiar to the ith dummy. Similarly, $\beta_{j(i)}$ denotes a perturbation introduced by factors peculiar to the jth run of the ith dummy. The total variance which would be observed if many dummies were each tested once and their results analyzed is considered to be:

$$\sigma^2 = \sigma_d^2 + \sigma_\beta^2 \tag{5}$$

where σ_d^2 denotes "variance between (among) dummies of the same manufacturer" and σ_β^2 denotes "variance within repeated runs of a single dummy of the given make." These quantities are derived from two related quantities called "mean squares between" and "mean squares within" obtained directly from the analysis of variance.

Now it is an unfortunate but inescapable consequence of the limited number of tests available in this investigation that the mean squares (variance) are subject to a considerable amount of variation. If one wishes to compare the two makes of dummies with respect to repeatability, therefore, this statistical variation must be taken into account and some measure of statistical significance provided. A quantity called the "F-ratio" is used for this purpose. The mean squares within for Alderson and Sierra dummies are compared and the larger of the two is divided by the smaller. If the resulting ratio exceeds 6.39, the two makes of dummies can be considered to exhibit differences in their run-to-run *repeatability* at the 5% level of significance. In other words, if there were indeed *no* difference between the two makes, a ratio as large as 6.39 or larger would be obtained only 5% of the time. The corresponding ratio for comparing the mean squares (variance) among is 161. The ratio is much larger for the "among" comparison because only two dummies of a given make were tested, thereby providing only one degree of freedom for each dummy make. In the case of the within mean squares comparison, the two sets of three replicate runs provide four degrees of freedom for each make of dummy.

It will suffice to illustrate the analysis-of-variance approach by considering the data of Fig. 10. At each point in time, an analysis of variance can be performed to isolate the component of variance which is associated with run-to-run (replicate) variability and the component of variance associated with dummy-to-dummy variability. The analyses for these individual points in time can then be combined, as a weighted average, to provide an overall measure of run-to-run and dummy-to-dummy variability for the entire time history. Both individual time point data and an equally weighted average tends to confirm the differences observed in the confidence-bound plots. It is noted, however, that for some purposes it may be more appropriate to place heavier emphasis on some parts of the time history than on others. This can be accommodated in the analysis by changing the time weighting factors. For example, if one were to suppose that the early portion of the time history is more important than the later portion, the weighting factor could be adjusted to allow for such a comparison.

Table 1 shows the results of combining the 35 sampled points on an equally weighted basis. The quantity "MS within" denotes "mean squares within dummies," and is a measure of the degree of repeatability of the acceleration time histories among replicates for dummies of either the Alderson or Sierra type. It is obtained by pooling or combining the run-to-run variance for the two dummies of each make. The quantity "MS among" denotes "mean squares among dummies" and provides information concerning differences in the response of the two dummies of the same make.

TABLE 1

Repeatability of Tim History for Head Accelerations

	Alderson	Sierra
MS Within	6.64	18.38
Standard Deviation	2.58	4.29
MS Among	5.96	7.93
Standard Deviation	2.44	2.82

According to Table 1, the response exhibited by the Sierra dummies appears subject to somewhat greater variability than that exhibited by the Alderson dummies. A word of caution is in order, however, in drawing conclusions from such an observation. This caution derives from the statistical variation in results which must necessarily arise because of the relatively limited amount of testing conducted on the two types of dummies. First, it must be recalled that only *two* dummies of each make were tested; such a small sample of the two dummy populations severely limits the confidence which one can place in any generalizations made with regard to the two dummy populations as a whole. Similarly, only three replications were performed on each of the dummies. In judging whether the variation in the three tests for one dummy is different from the variation in the three tests for another dummy, one must ask what comparison might have resulted if two groups of three tests had been performed on the *same* dummy. A statistically significant difference can be decreed only if we can be reasonably convinced that the observed difference between two dummies is larger than that which might be expected from further testing of a single dummy. It is for this reason that one must employ the "F-ratio" or variance-ratio test alluded to above. One computes the ratio of the larger to the smaller variance and compares this value with a corresponding value obtained from tables of the F distribution. Entries in the table reflect the fact that for small amounts of data (such as replications on 2 dummies), relatively large differences must be observed between two measures of variability (such as those for Sierra and Alderson dummies) in order for them to be declared significantly different. In Table 1, the ratio of the within mean squares for Sierra to the within mean squares for Alderson is 18.38/6.64 = 2.77. Since a ratio of 6.39 is the tabled value of F for this comparison at the 0.05 significance level, the observed difference fails to reach the critical value and hence cannot be considered statistically significant on the basis of the limited amount of data available. Quite clearly, the mean squares among also fails to reach the 0.05 significance level. It should be borne in mind, however, that these conclusions apply only to the particular criterion discussed above. Conclusions with respect to other measures of performance are discussed in the following section, Results.

The running variance curves can be subjected to a similar analysis with a view toward determining whether dummies differ with respect to the amount of variability they exhibit within the averaging window. For some of the running variance curves it was noted that there are bursts of relatively high-frequency perturbations conducive to high levels of variance but that these occurred only over a small fraction of the time base. Because of the erratic nature of these occurrances, no detailed analysis was attempted, inasmuch as there did not appear to be any evident differences between the two makes of dummies in this respect. In view of these considerations, it was decided to concentrate attention on the running-mean time history. Though it is acknowledged that differences between dummies may be such as to produce differences in the variance curves, it was believed that these differences were of less immediate interest than reproducibility of the major trends of the time histories. Moreover, it was concluded that by comparing the smoothed, running-mean curve with the original time-history (see Fig. 12), one could obtain a qualitative assessment of high-frequency contributions.

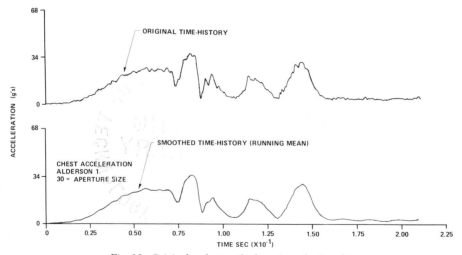

Fig. 12. Original and smoothed version of a time-history.

In general, the measures of performance specified in MVSS 208 can be regarded as time-independent. They are:

Head Severity Index*
Chest acceleration (peak)
Right femur load (peak)
Left femur load (peak)

These measures are independent of time in the sense that the clock time at which maximum accelerations or forces occur does not influence the interpretation of the values obtained.

In the following section, Results, these measures are analyzed with respect to variability which arises from lack of repeatability of the same dummy, differences between two dummies of the same make, and differences between dummy makes.

RESULTS

The inability of a single test event to discriminate between two dummies with respect to all the physical properties which significantly influence its dynamic response is evident in the test results which have been obtained. For example, consider Fig. 13. The envelope of the six time histories (3 repeats for each of two

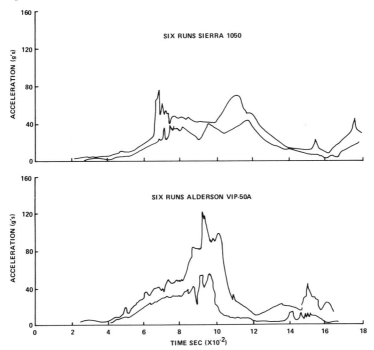

Fig. 13. Envelope of head resultant acceleration responses (lap and shoulder restraint configuration).

The analysis was conducted prior to the recent revision of MVSS 208 in which Head Severity Index has been replaced by HIC (Head Injury Criteria) number.

dummies have been pooled together) of the head resultant acceleration response in the lap and shoulder belt configuration is presented separately for the Sierra 1050 and for the Alderson VIP-50A. It is clearly evident that there is a significant difference in the basic character of the responses of these two dummies as judged by this measure (resultant head acceleration); thus these two dummies would be judged as different based on these results. The resultant accelerations of the Sierra and Alderson dummies for this same set of runs are presented in the same manner in Fig. 14. Here, however, it is observed that the chest acceleration responses are not significantly different and thus based on this measure, for this event, it would be more difficult to discriminate between the two dummies. If a similar view of the data is taken for the lap belt and head target configuration it is found that, in this event, the resultant head accelerations for the two dummies are not significantly different but the resultant chest accelerations are significantly different. Thus the response of the dummies is sensitive to different sets of its physical parameters for each event.

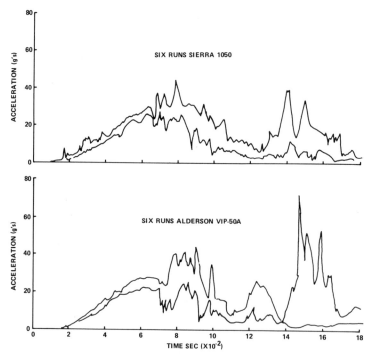

Fig. 14. Envelope of chest resultant acceleration responses (lap and shoulder restraint configuration).

As previously indicated, the selection of quantities (measures characterizing a given event) to be subjected to statistical analysis was conditioned to a considerable extent by the injury criteria as specified in MVSS 208: head severity index not to exceed 1000, chest accelerations not to exceed 60g and femur forces not to exceed 1400 pounds. For purposes of our analysis, it was not the above limiting values that were of primary interest but rather the nature of these values. For example, in all cases except the head severity index, emphasis is on the maximum values obtained during the acceleration or force time histories. In the case of the head severity index, it has been shown that the criterion value partakes of the nature of an average value.

TABLE 2

Head Severity Index

		Mean	Standard Deviation	Coefficient of Variation	Analysis of Variance	
					MS Between	MS Within
Configuration 1 — Lap and Shoulder Restraints — No Targets						
Sierra	894.54 574.02					
	830.25 761.59					
	770.98 548.10	729.913	163.123	0.223	62436.238	8695.461
Alderson	828.07 609.98					
	728.31 979.25					
	749.34 681.11	762.677	117.415	0.154	208.624	20575.000
F-Ratio					299.276(S)	2.366(A)
Configuration 2 — Lap Restraint and Head Target						
Sierra	2910.83 2504.17					
	3361.22 3080.26					
	3038.17 2359.73	2875.730	412.408	0.143	311019.938	99611.125
Alderson	2694.22 2016.66					
	2881.27 2565.26					
	2422.23 3142.41	2620.342	357.134	0.136	12457.012	185088.375
F-Ratio					24.967(S)	1.858(A)
Configuration 3 — Unrestrained — Steering Wheel and Knee Targets						
Sierra	1732.92 1562.95					
	1865.60 1375.11					
	1712.86 1384.64	1605.680	245.628	0.153	162914.688	9042.086
Alderson	1861.24 2522.20					
	2374.27 2545.87					
	2403.19 2220.81	2321.263	255.771	0.110	70455.625	62900.512
F-Ratio					2.312(S)	6.956(A)
Configuration 4 — Pre-inflated Airbag and Knee Target						
Sierra	848.93 614.42					
	675.97 619.73					
	702.75 667.48	688.213	95.281	0.138	17714.80	4760.379
Alderson	428.54 503.03					
	388.10 485.05					
	373.73 478.21	442.777	67.480	0.152	12688.641	486.010
F-Ratio					1.396(S)	9.795(S)

The results of the analysis of variance for the four measures — head severity index, chest maximum acceleration, left femur maximum load, right femur maximum load — are given in Tables 2 through 5, respectively. Tabulated in these tables are: the six input values for each make of dummy, the mean of these six values, the standard

TABLE 3

Chest — Maximum Acceleration — G

			Mean	Standard Deviation	Coefficient of Variation	Analysis of Variance	
						MS Between	MS Within
Configuration 1 — Lap and Shoulder Restraints — No Targets							
Sierra	31.48	40.42					
	30.22	30.34					
	28.56	37.12	33.022	5.177	0.157	51.797	14.297
Alderson	34.86	38.17					
	33.07	91.54					
	33.45	36.48	44.594	23.673	0.531	699.948	490.644
F-Ratio						13.513(A)	34.318(A)
Configuration 2 — Lap Restraint and Head Target							
Sierra	56.31	65.04					
	54.02	54.50					
	54.17	67.00	58.508	6.527	0.112	80.902	23.443
Alderson	95.86	90.75					
	112.32	109.00					
	117.82	84.33	101.679	13.990	0.138	292.769	147.200
F-Ratio						3.619(A)	6.279(A)
Configuration 3 — Unrestrained — Steering Wheel and Knee Targets							
Sierra	73.39	84.29					
	72.41	79.47					
	80.24	81.98	78.629	5.437	0.069	64.675	12.002
Alderson	69.21	80.21					
	76.45	72.47					
	80.26	71.65	75.041	4.251	0.057	0.419	26.903
F-Ratio						154.356(S)	2.242(A)
Configuration 4 — Pre-inflated Airbag and Knee Target							
Sierra	65.59	54.21					
	58.81	50.88					
	53.36	59.99	57.139	5.343	0.094	26.839	29.404
Alderson	72.80	71.40					
	77.53	62.87					
	87.59	72.99	74.196	9.007	0.121	156.703	43.342
F-Ratio						5.839(A)	1.474(A)

deviations, the coefficient of variation, and the mean squares within and among. In the interest of checking computational accuracy, the tables retain more than the usual number of decimal digits and should not be interpreted as having the indicated number of significant places.

TABLE 4

Left Femur - Maximum Load (lb.)

		Mean	Standard Deviation	Coefficient of Variation	Analysis of Variance	
					MS Between	MS Within
Configuration 1 — Lap and Shoulde Restraints — No Targets						
Sierra	392.59 402.44					
	332.71 402.26					
	319.79 346.07	365.979	38.411	0.105	1861.518	1282.391
Alderson	388.65 414.27					
	409.24 389.71					
	364.74 390.01	392.772	16.915	0.043	163.762	347.318
F-Ratio					11.367(S)	3.692(S)
Configuration 2 — Lap Restraint and Head Target						
Sierra	450.47 422.89					
	426.00 407.33					
	381.49 407.64	415.969	21.364	0.051	67.355	650.982
Alderson	485.20 505.13					
	490.90 465.67					
	491.45 534.56	495.484	21.947	0.044	238.455	603.278
F-Ratio					3.540(A)	1.079(S)
Configuration 3 — Unrestrained — Steering Wheel and Knee Targets						
Sierra	1123.43 1435.46					
	1292.29 1356.17					
	1223.21 1375.82	1301.066	135.976	0.105	46556.445	4455.922
Alderson	2029.96 2166.45					
	2011.47 2033.98					
	2017.86 1992.37	2042.016	61.437	0.030	2970.687	4176.336
F-Ratio					15.672(S)	1.067(S)
Configuration 4 — Pre-inflated Airbag and Knee Target						
Sierra	1345.77 1453.70					
	1402.44 1448.60					
	1429.79 1406.63	1414.486	42.272	0.030	2857.285	1251.785
Alderson	1831.80 1778.27					
	1923.56 1937.15					
	1945.70 1759.35	1862.640	85.084	0.046	8534.523	6591.551
F-Ratio					2.987(A)	5.266(A)

A word of explanation and caution is in order with regard to the standard deviations as presented in these tables. The values as given have been adjusted to represent the combined effects of among and within variance on a one-for-one basis. Since there are only two dummies of each make but three repeat runs on each, it is

TABLE 5

Right Femur — Maximum Load (lb.)

			Mean	Standard Deviation	Coefficient of Variation	Analysis of Variance	
						MS Between	MS Within

Configuration 1 — Lap and Shoulder Restraints — No Targets

			Mean	Standard Deviation	Coefficient of Variation	MS Between	MS Within
Sierra	295.36	395.02					
	234.17	347.83					
	269.48	294.90	306.128	65.709	0.215	9500.656	1726.275
Alderson	356.96	436.62					
	316.42	315.56					
	333.69	399.45	359.784	50.802	0.141	3483.125	2129.680
F-Ratio						2.728(S)	1.234(A)

Configuration 2 — Lap Restraint and Head Target

			Mean	Standard Deviation	Coefficient of Variation	MS Between	MS Within
Sierra	340.02	382.34					
	302.56	368.25					
	293.35	438.42	354.158	64.971	0.183	10674.492	994.591
Alderson	504.69	384.99					
	503.05	433.35					
	489.02	459.69	462.466	56.222	0.122	7973.289	754.773
F-Ratio						1.339(S)	1.318(S)

Configuration 3 — Unrestrained — Steering Wheel and Knee Targets

			Mean	Standard Deviation	Coefficient of Variation	MS Between	MS Within
Sierra	1422.84	1301.44					
	1523.21	1197.65					
	1433.43	1322.02	1366.766	140.771	0.103	51964.328	3742.502
Alderson	2146.83	2139.33					
	1993.74	2144.29					
	2087.19	1948.20	2076.595	78.440	0.038	2.746	9227.863
F-Ratio						18923.648(S)	2.466(A)

Configuration 4 — Pre-inflated Airbag and Knee Target

			Mean	Standard Deviation	Coefficient of Variation	MS Between	MS Within
Sierra	1308.94	1444.94					
	1453.43	1440.43					
	1370.88	1345.64	1394.045	57.713	0.041	1592.673	4199.852
Alderson	2024.05	2019.42					
	2044.12	1963.33					
	2053.07	2166.85	2044.974	61.319	0.030	124.716	5577.656
F-Ratio						12.770(S)	1.328(A)

evident that the six data values provide more opportunity for run-to-run variability than for dummy-to-dummy variability. Consider, on the other hand, a situation in which six dummies were available and each was tested once. The six data values deriving from the tests would provide equal opportunity for the between-dummy and within-dummy components of variability. The standard deviations given in Tables 2 through 5 are estimates of the standard deviations which might be expected from tests structured in this "equal-opportunity" manner. The coefficient of variation is the standard deviation expressed as a fraction of the mean for the six data values.

With the exception of Configuration 1, lap and shoulder restraints — no targets, the event duration was 220 milliseconds. In Configuration 1, the event duration was reduced from 220 to 140 milliseconds to eliminate that portion of the data corresponding to the rebound impact of the dummies against the rigid seat back. This secondary impact was not intended to be a part of the dynamic performance test event; therefore, no attempt was made in the set-up to make the force-deflection characteristics of the head rest and seat back representative. Furthermore, there is no justification for combining the primary and secondary parts of the event when evaluating the head severity index.

F-ratio significance tests for the among and within mean squares are summarized in Tables 6 and 7, respectively. The four measures, for head, chest, left femur, and right femur, constitute one axis of the table; the five restraint configurations provide the other. An entry in Table 6 (A for Alderson or S for Sierra) implies that the variance observed among two dummies of the indicated manufacturer was *significantly* greater than that observed among the two dummies of the other manufacturer. Likewise, an entry in Table 7 implies that the observed run-to-run variance for the dummies of the indicated manufacturer were *significantly* greater than that observed for the dummies of the other manufacturer. The indicated differences were established at the 5% level

TABLE 6

Summary of Significance Tests for Between Mean Squares
5% Level of Significance

	Configuration			
	1	2	3	4
Head Acceleration Severity Index	S			
Chest Acceleration G				
Left Femur Force Max Load				
Right Femur Force Max Load			S	

of significance. It will be noted that most of the cells in the tables are blank, indicating that in most cases it was not possible to establish signifcant (i.e., meaningful) difference between the observed variances for the two makes of dummies with the limited amount of data available.

TABLE 7

Summary of Significance Tests for Within Mean Squares
5% Level of Significance

	Configuration			
	1	2	3	4
Head Acceleration Severity Index			A	S
Chest Acceleration Max G		A		
Left Femur Force Max Load				
Right Femur Force Max Load				

A third comparison, and one which is of considerable interest, is the comparison between the means of the six runs for Sierra dummies and the corresponding six runs for the Alderson dummies. A significance test for such a comparison must be constructed with care. The intent of the significance test here is to make an inference regarding the hypothesis:

$$H_0 : \mu \text{ Alderson} = \mu \text{ Sierra}$$

In words, it is postulated that the population mean for Alderson dummies is equal to the mean for Sierra dummies. This postulation is called the "null hypothesis" and is denoted H_0. Its acceptance or rejection is presumed to apply to the populations of all Alderson and all Sierra dummies.

It has been noted previously that two dummies from each population provides a scant basis for inferences with regard to overall production of the two makes of test devices. We prefer, therefore, to confine our attention *to the four dummies under test* and refrain from making inferences beyond the two dummies of each type. In effect, this approach is analogous to considering the two Sierra and the two Alderson dummies as exhausting the two *entire* populations of interest. Within this restricted basis of interest, one can compare performance for the Alderson dummies and the Sierra dummies with respect to whether the observed differences in means could reasonably have occurred as the result of testing variations or are more likely to reflect a between-make difference. In this context, only the within-dummy variance need be considered; the between-dummy variance is a constant by virtue of having restricted our attention to a finite population of two individuals. The proposed comparison is summarized in Table 8.

TABLE 8

Comparison of Alderson and Sierra Dummies

Configuration	SIERRA		ALDERSON		Difference	Least Significant
	Mean	Std. Dev.	Mean	Std. Dev.	Sierra-Alderson	Diff. (5% Level)
Head Severity Index						
1	729.9	93.2	762.7	143.4	32.8	273.8
2	2875.7	315.6	2620.3	430.2	255.4	853.8
3	1605.7	95.1	2321.3	250.8	715.6	429.1*(A)
4	688.2	69.0	442.8	22.0	245.4	115.8*(S)
Chest Max. Accelerations G						
1	33.0	3.8	44.6	22.2	11.6	22.5
2	58.5	4.8	101.7	12.1	43.2	20.8*(A)
3	78.6	3.5	75.0	5.2	3.6	10.1
4	57.1	5.4	74.2	6.6	17.1	13.6*(A)
Left Femur Max. Load (lbs)						
1	366.0	35.8	392.8	18.6	26.8	64.5
2	416.0	25.5	495.5	24.6	79.5	56.6*(A)
3	1301.0	66.8	2042.0	64.6	741.0	148.8*(A)
4	1414.5	35.4	1862.6	81.2	448.1	141.8*(A)
Right Femur Max. Load (lbs)						
1	306.1	41.5	359.8	46.1	53.7	99.2
2	354.2	31.5	462.5	27.5	108.3	66.9*(A)
3	1366.8	61.2	2076.6	96.1	709.8	182.2*(A)
4	1394.0	64.8	2045.0	74.7	651.0	158.2*(A)

*   Denotes significant difference at 5% significance level.
(A)   Denotes Alderson › Sierra
(S)   Denotes Sierra › Alderson

The least significant difference is computed in the following manner. First, it will be assumed that the within-dummy mean squares is comparable for both makes of dummies. The standard error of the difference between the means for Alderson and Sierra dummies is thus given by:

$$\sigma_{diff} = \sqrt{1/3\,\sigma_1{}^2 + 1/3\,\sigma_2{}^2} \qquad (6)$$

where:

σ_1 = standard deviation for Alderson dummies

σ_2 = standard deviation for Sierra dummies

To determine the smallest difference between means which would be significant at the 5% level of significance, one must multiply by the appropriate value of t from

Student's t distribution. For means computed from three observations each (giving 4 degrees of freedom), the appropriate factor is 2.776. Therefore,

$$\text{Least significant difference} = \frac{2.776}{\sqrt{3}} \quad \sqrt{\sigma_1{}^2 + \sigma_1{}^2}$$

$$= 1.6 \quad \sqrt{\sigma_1{}^2 + \sigma_1{}^2} \tag{7}$$

Significance tests for the differences between means are summarized in Table 9. An entry in Table 9 (A for Alderson or S for Sierra) implies that the mean observed for the two dummies of the indicated manufacturer was *significantly* greater than for the other manufacturer. It is noted that significant differences are observed most prominently for configurations (2) lap restraint and head target, (3) unrestrained − steering wheel and knee targets, and (4) pre-inflated air bag and knee targets. The fact that left and right femur loads show similar trends lends further credibility to the analysis. Note that for Configuration (1) lap and shoulder restraints − no targets, significant differences were not observed for any of the injury criteria. These facts in combination indicate that establishing agreement between two dummies or makes of dummies in one constraint configuration does not insure agreement in a different constraint configuration.

TABLE 9

Summary of Significance Tests For Differences Between Means
5% Level of Significance

	Configuration			
	1	2	3	4
Head Acceleration Severity Index			A	S
Chest Acceleration Max G		A		A
Left Femur Force Max Load		A	A	A
Right Femur Force Max Load		A	A	A

DISCUSSION

The repeatability of dummy performance has been examined from a number of points of view. First, it has been noted that repeatability depends on the aspect of dummy performance which is being emphasized − i.e., whether head acceleration response, chest acceleration response, femur load response, or some other aspect of dummy behavior not focused upon in this study. Second, it has been shown that, inasmuch as dummy response is a time-varying function, many options are available as to what features of the time history are really of interest in the repeatability evaluation. Finally, it has been noted that repeatability alone is not a sufficient criterion for evaluating the usefulness of dummies as the means for evaluating passenger restraint devices or other safety-oriented systems; rather, one must balance repeatability against the sensitivity of the dummy to respond to changes in the test environment, especially those changes induced by modifications of the restraint system. A more revealing observation made evident by the study is the fact that repeatability depends strongly on the nature of the test environment. For example, Configuration 1, lap and shoulder restraints — no targets, was relatively insensitive to design differences between Sierra and Alderson dummies, whereas the other restraint configurations revealed significantly different levels of response for the two makes of dummies, in one or more of the response measures of interest.

From the standpoint of what the dummy senses, the various restraint systems can be viewed as different levels of exposure severity. It is instructive to consider that each particular dummy has its own individual response vs severity curve — that is, its own "transfer function". Unless this transfer function changes from one test run to another because of age degradation, ambient environmental conditions or the like, lack of repeatability can occur only because of inability to produce the same impact exposure conditions in successive runs. If more than one dummy is involved in a testing program, however, as must inevitably be the case, there will be superimposed on this run-to-run variability a dummy-to-dummy variability residing in differences in dummy transfer functions. Note that these differences are the only differences which can be truly leveled against the dummies themselves and that they affect the entire severity vs response curve. The prospect of "qualifying" dummies on the basis of a single test circumstance, therefore, appears questionable. Rather, the dummies should be qualified by subjecting each dummy to a graded set of severity conditions so as to establish a *calibration curve* for its response. By this means it would not be necessary to insist on absolute repeatability of dummies nor to demand that a particular dummy retain the same response throughout its response lifetime. As the dummy "ages", it could be periodically recalibrated to compensate for any changes which might occur in its transfer function. Thus each dummy could be individually "qualified" for use in vehicle compliance tests. In reality, of course, such a calibration would be difficult to implement because of difficulties involved in defining a graded set of severity conditions and the costliness of such an extensive calibration program. The alternative, however, that of producing all dummies to give uniform response, may offer little succor.

A word of apology may be appropriate with regard to the statistical approach employed in this analysis. In some respects, it differs from what might be deemed the "classical" approach to statistical inference, instance our hesitation to regard two dummies of each make as representative of the corresponding dummy population. In view of the above concern for the true nature of dummy response, however, we consider such conservatism justified until a more extensive data base is available on a larger number of dummies. The value of the initial work presented herein lies primarily in defining the nature of the repeatability problem and in providing limited but definitive data on its magnitude and seriousness.

The statistically astute may wish to point out a possible discrepancy which arises in comparing Tables 7, 8 and 9. In the computation of least significant differences between performance levels for Alderson and Sierra dummies, it was assumed that the within-dummy variance was comparable for the two makes, yet this hypothesis is rejected in Table 7 in three instances. Quite clearly, the troublesome Behrens-Fisher problem of statistical inference raises its ugly head here, and we have chosen to ignore it by treating the two significance tests — that for variances and that for means — independently. In the three cases in question, the observed differences between means so far exceeds the computed least significant difference that there seems to be little question of their statistical significance.

SUMMARY AND CONCLUSIONS

In summary, the following observations can be made with regard to dummy repeatability and to the qualification of dummies for use in safety compliance testing.

1. Lack of repeatability of dummy reponse when subjected to supposedly identical conditions can arise because of either (1) difference in the response characteristics of the dummies or (b) differences in the test environment.

2. Various quantities can be viewed as a measure of repeatability. Some of these quantities may evidence less variability than others because they, in effect, emphasize different aspects of the dummy response.

3. Repeatability of response will depend on the nature of the test environment. In some testing situations, minor variations in test control may induce sizable changes in output response; for other situations, response may be relatively insensitive to lack of test control. Also, ability to repeat a test environment will depend on its nature: a difficult-to-repeat experimental setup can induce relatively large variation in the output for a given response function.

4. For statistically valid conclusions, in view of the indicated variability among test results, a relatively large number of tests is required; in particular, these tests should include an adequate sampling of the dummy population in order to assess dummy-to-dummy variability in response.

5. Further study is required to properly relate the role of variation in dummy response function to variation in input stimuli, as these combine to produce variation in the output response of dummies.

6. Dummies should be qualified for use in compliance tests not on the basis of the repeatability of their response alone, but rather with due regard to their ability

to discriminate between severity levels of the test environment to which they are exposed. Thus, sensitivity and repeatability must both be considered in judging the efficacy of a dummy for safety complaince testing. Qualification on the basis of a single, standardized exposure environment is not adequate for qualifying a dummy for use over a range of severity conditons.

The need for a more sophisticated calibration or standarization procedure for dummies for use in safety compliance testing is suggested.

ACKNOWLEDGMENTS

The work reported in this paper was performed under Contract No. DOT-HS-053-1-129 for the U.S. Department of Transportation, National Highway Safety Administration, Office of Crashworthiness, Washington, D.C. The cooperation and encouragement of Mr. Stanley Backaitis, Project Officer, is acknowledged with thanks.

The author wishes to extend thanks to Mr. R. R. McHenry, Mr. R. A. Piziali, Mr. P. D. Mollemet, and Mr. P. A. Reese of Cornell Aeronautical Laboratory, Inc., for their important roles in the work, and to the many technicians and assistants without whose efforts this paper could not have come into being.

DISCUSSION

J. N. Silver (GM Proving Ground)

I have a two part question; first, would you care to make any estimates of what percentage contribution of the variability your work has indicated in dummy response is related to (1) the set-up techniques and the external stimuli, and (2) the dummy itself and its construction? Secondly, would you care to put some numbers on the quantity of tests required to, in your mind, statistically adequately generate either a mean or some statistical basis for saying things are the same or different?

H. T. McAdams

First question: You asked to what extent is the variation due to control of the test environment, and to what extent is it determined by the difference between dummies. Well, of course, this is the question we are trying to address. The only problem is that all one has here is the output. You can look at the variation in output and break it down into repeatability within the dummy. This, in some ways, reflects the variation in the test environment; but you have no independent way of measuring how closely you did actually control the conditions, other than by taking all the pains that you can to see that it's done. It's a very difficult question to answer. I think we need more work along that line.

Now on the other question as to how much data do you need, I haven't looked into that because I though it would be a bit frightening. I suspect that it would be at least an order of magnitude greater than the number we tested. Perhaps more than that.

W. Goldsmith *(University of California)*

I should like to add a comment regarding the possibility of change in the test configuration itself—in other words in the dummy—as a result of the test. Under accelerations from noncolision condition, changes are not likely to be as severe as under collision conditions; but even then some of the viscoelastic materials may not have had time to recover their original configuration. I submit that some tests might have to be made to establish that the dummy that is used repeatedly is indeed the same configuration and material that it was originally.

H. T. McAdams

If I understand you correctly, you're suggesting that the dummy itself may not be the same after several tests as it was initially. That is a point which I have addressed in the text of the paper and I may just say a word about it here.

It seems to me that this is a very real problem. We should not restrict our analyses to differences from one dummy to another. We also must talk about differences in dummy 1970 and dummy 1971 after it's been used a while. I suggest in the paper that there might be an approach in which we use a sort of "internal standardization".

If we could subject dummies to an ordered range of severity conditions we could establish a performance curve for how the dummy responds to different levels of severity. Each individual dummy could have his own response curve, which could be determined at the outset. In fact, it could be redetermined as the dummy ages. Perhaps some scheme could be worked out where the results could be adjusted to some standard curve on the basis of the curve which the dummy itself individually exhibits. This is not unlike the use of calibration procedures for any other instrumentation or mechanical system.

R. C. Haeusler *(Chrysler Corporation)*

There is one more point being added here — that the dummies response one hour after previous use might be substantially different than that same dummie's response after a period of a day or two. It's a matter of slow recovery of properties.

H. T. McAdams

Right. I was addressing more the degrading type of use, but the other point is important also.

R. A. Potter *(GM Engineering Staff)*

I noticed in your slides that your measure of chest response was maximum acceleration whereas MVSS 208 talks about the cumulative time of exposure above

60 g's. If you had used cumulative time as your variable as opposed to the maximum acceleration, would your statistics or conclusion have been any different?

H. T. McAdams

I don't think they would have been materially different because of the fact that we are performing this analysis on the running average, and therefore the acceleration profiles have already been smoothed to some extent.

L. M. Patrick *(Wayne State University)*

The Bureau of Standards test where they belt the dummy down in all directions is essentially just proving Newton's Law, which most of us didn't question anyway. You're coming a little closer to the real problem, and I think you've found far greater variations than the Bureau of Standards test. Suppose we went one step farther and actually impacted a windshield or something that is considerably stiffer than the devices that you used. Would you expect even greater variability than you measured here?

H. T. McAdams

I think this question is a little difficult to answer. What we may see reminds me of a photographic emulsion response. If the test conditions are not severe, we may see very little variation in response. Likewise if they are extremely severe so that the dummy is overwhelmed, we may again see very little variation. Somewhere in the responsive portion of the dummie's response curves we might expect to see greater variation. In other words, I think the dummy can be overwhelmed or underwhelmed.

H. E. VonGierke *(Aerospace Medical Research Laboratory)*

Does your separation of the variance into environment and dummy neglect a possible interaction between dummy and environment? I primarily refer to the initial conditions of the test, which might be very critical for the dynamics of the dummy. In other words, it could be more difficult to assure that one particular dummy is in exactly the proper position prior to the test than it is to assure this for another dummy. You could have dynamically ideal, perfect, 100% reproducable dummies and yet a slight variation in initial conditions would still give you differences in the dynamic response. I wonder if this is considered in your analysis?

H. T. McAdams

If I understand you correctly, the answer is yes. I think the interaction is to a certain extent included. However, our analysis of variance does not show an interaction term like one normally sees. We have handled this in a little bit different way. Actually when we see the difference between the repeatability, say in configurations 1, 2, 3 and 4, that is really of the nature of an interaction term. It's saying that the repeatability of response is affected by the initial conditions and the general level of exposure conditions. In a slightly different analysis this could be expressed as an interaction term. I have a feeling I've not answered your question. Do you want to try again?

H. E. VonGierke

I assume you have some measure of the variations in your environment which you already defined by the accelerations of the sled and the measurements on the support to the dummy. The interaction term I refer to would deal with slight differences in the seating of the dummy and changes in the cg locations of various parts of the dummy, which might be very hard to define just by optical observation.

H. T. McAdams

I think that is a problem. It is very difficult to quantify these slight differences. You presume to be positioning the dummy in exactly the same way but this isn't necessarily valid. In fact, you must not be doing it the same way, otherwise you wouldn't get the variations that you see on the output response. Some of the measurable quantities such as slight variations in the acceleration pulse, could be taken into account, however, we didn't do this.

H. E. VonGierke

Did you try, for example, to repeat the tests with a completely hard seat? We had similar difficulties in vibration testing with human subjects. The only way we could achieve reproducibility was to standardize on a completely hard seat. Then you know that you can exclude to a large extent differences in the seating of the subject.

H. T. McAdams

You're getting into some questions here that I'm not all that familiar with. Perhaps some of my colleagues back there would care to talk about that. Norm, do you know if anything like that was done?

N. DeLeys *(Cornell Aeronautical Laboratories)*

The seat that we used was, in a sense, a hard seat; that is to say it did not have any padding. However, it did have a vertical degree of freedom in order to more nearly simulate those conditions that would be experienced in an automobile where the subject does move in a vertical direction. The seat was actually a Bostrom truck seat, which has a torsion bar suspension, with the seat cushion replaced by a sheet of teflon to minimize the variability in the tests that might otherwise result from deterioration of the seat through repeated use.

J. H. McElhaney *(HSRI, University of Michigan)*

What aperture size was used for computing the mean; and since this is tantamount to filtering, what is this in terms of frequency response? Did you look at the effect of varying the aperture size?

H. T. McAdams

That's a very good question. The answer is yes, we did vary the aperture size; and having varied it and observing what it did to the response, we selected the final aperture size for the comparison. We used an aperture size consisting of 30 points. The points were spaced approximately a tenth of a millisecond apart. We did this because it seemed to get rid of the extraneous spikes which we didn't think were really relevant to repeatability in the sense we were talking about and yet retained the essential features of what we considered to be the deterministic part of the time history. Also, the aperture is still small enough relative to the total pulse duration that it does not degrade the information in the total time history. As far as the frequency is concerned, I think this amounts to something like about a 100-Hertz cut-off. Based on calculations, the signal is down 3 db at about 200 Hertz. It's a gradual fall-off.

DUMMY PERFORMANCE IN CRASH SIMULATIONS ENVIRONMENTS

A. M. THOMAS

National Highway Traffic Safety Administration, Washington, D. C.

ABSTRACT

A test procedure is presented whereby dummies can be subjected to a controlled and reproducible environment which is independent of the sled facility used. Essentially, a set of straps with slack in them couple the sled to the dummy. The deceleration phase of the sled occurs during the time that the slack is taken up, and a difference in velocity is established between sled and dummy. The straps then decelerate the dummy in a specific fashion independent of the shape of the deceleration time pulse of the sled. Results of measurements on four dummies subjected to such a test are presented. Both side facing and forward impacts were simulated, each at two different energy levels.

Maximum accelerations were about 50 g's in the head and about 40 g's in the chest. The standard deviation of 10 runs for both head and chest measurements averaged about 6%.

A procedure is described whereby a calibration number can be assigned to individual dummies so that results obtained in compliance testing can be more uniform.*

INTRODUCTION

In September of 1970, NHTSA's Safety Systems Laboratory, which was then part of the National Bureau of Standards, undertook the task of developing a test method to evaluate the dynamic performance of anthropomorphic test devices prior to their use in compliance testing.

The opinions, findings and conclusions expressed in this publication are those of the author and not necessarily those of the National Highway Traffic Safety Administration.

References p. 79

Such a test method had to be designed so that laboratories throughout industry and government could perform the required tests. It was recognized that the different types of test sleds at various facilities produce different deceleration pulse shapes and therefore different dummy responses for otherwise identical tests. A method had to be devised to overcome this problem.

Two other requirements had to be met. Dummy response variability had to be measured and therefore the test environment should be as reproducible as possible. Also, an effective restraint system was to be simulated and therefore hard impacts were to be avoided.

To meet these requirements, a method of decelerating the dummy through a system of straps was devised. These straps were to have slack built in between the dummy and the sled so that the deceleration phase of the sled motion would be complete before the straps started to load up. Under these conditions, the shape of the sled deceleration pulse would not affect the response of the test device being used. Instead the shape of the pulse experienced by the dummy is determined by the nature of the webbing, for example an elastic webbing produces a half-sine pulse whereas a force limiting webbing may produce a modified square wave.

To evaluate this test procedure, four commerically available dummies were tested in identical dynamic environments which simulated an air bag type restraint system. Each dummy was tested ten times in two configurations and at two energy levels in each configuration.

The results of this program showed that the standard deviation of the resultant maximum acceleration in the head and chest averaged about 6 percent. The variation of measurements in the pelvis had an average of about 10 percent.

Statistically significant differences could be detected between the response of different dummies in some cases. A procedure was devised whereby a calibration number based on the results of such a test as the one described here is assigned to each dummy so that results of compliance testing using this procedure will be more uniform.

PROCEDURE

Four SAE J963 type dummies were used in these tests, two Alderson VIP-50A dummies, a Sierra 292-1050, and a Sierra 292-850 modified dummy. The Sierra 292-850 modifications consisted of a replacement of the jointed neck with a rubber one and the installation of a more human-like pelvic structure.

During the series of tests a variety of repairs had to be made on the dummies. Most were considered minor, such as a tear in the skin, loosening of a bolt and the like.

Each dummy was tested in the two different configurations, side facing to simulate a collision from the side and forward facing to simulate a head on collision. The kinetic energy levels chosen for the forward facing tests were equivalent to velocities of 31.1 and 44 ft/sec. This latter speed, 30 miles per hour, is generally the level of tests conducted at this particular time in an effort to achieve occupant protection in a crash. 31.1 ft/sec was chosen since it is equivalent to one-half of the energy level at 44 ft/sec and it was felt that this spread in energy levels would serve to uncover any unusual performance of the dummies at different energy levels.

The side facing tests were conducted at 22 and 31.1 ft/sec. The value of 31.1 ft/sec was chosen to make a comparison of the dummy's dynamic reaction in side and forward facing shots. 22 ft/sec produces one-half the energy of the 31.1 fps runs and it would be possible to determine what the dynamic reaction of the dummy in the side facing mode at different energy levels would be.

The first parameter which had to be fixed was that of the location and the type of straps to be used in the forward and side facing experiments. The location of the center of gravity of various segments were determined during anthropometric measurements of the dummy. The dummy was set in position and the location of the straps marked so that lines of action would pass through the centers of gravity of the segments.

The force that should be produced by the webbings at each location was calculated by determining the kinetic energy change expected in each segment of the dummy and equating that to the elastic energy expected in the webbing for the desired deflection. In the forward facing configuration, there was an effort to obtain a deflection of 15 inches.

The side facing strap forces were calculated in a similar manner using a nominal deflection of 3.5 inches. When initial tests were conducted using straps in accordance with these calculations, the dummy moved forward in a manner as if it were virtually jointless, therefore, some of the lengths of webbing were adjusted to increase the flexure of the dummy in a more realistic way.

Fig. 1 illustrates the strap configuration used for forward facing 31.1 fps runs. Whereas the discussion has concerned accelerator type sleds, the sled at the Safety Systems Laboratory (1) simulates a crash situation by accelerating rather than decelerating which explains why the dummy is not facing in the direction of the sled motion. This sketch shows the appearance of the straps at the onset of strap loading after the sled has attained its test velocity.

The straps around the chest had an additional five inches of slack which allowed the chest to move forward to produce the articulation which had been observed in motion pictures of tests in which air bags were used (2). The number of straps was doubled for the 44 fps runs so the deflection was nominally the same.

In the side facing configuration, additional slack was allowed in the head strap as shown in Fig. 2. This produced the articulation expected in the neck when the

shoulder is restrained by the door. The specifications for the length and number of
straps at each location is included in a report to the NHTSA and included in Docket
No. 69-7 (3). The characteristics of the webbing used to make the straps is shown in
Fig. 3. The webbing in available from the Burlington Ribbon Mill at South Hill,
Virginia. Since we wish to simulate a half-sine pulse of acceleration, we chose a
webbing which was essentially elastic in the range of our tests.

The webbing used was chosen with a low enough spring constant that several straps
had to be used in most locations. This allowed a test procedure in which the same
webbing material was used throughout, and the different forces obtained on different

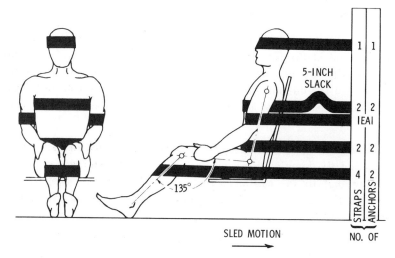

Fig. 1. Strap requirements (forward facing, 31.1 fps).

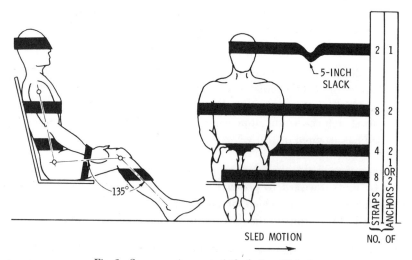

Fig. 2. Strap requirements (side facing, 22 fps).

body segments was controlled by the number and length of straps used rather than by changing the webbing material used.

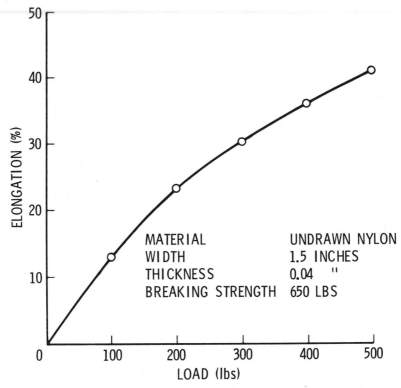

Fig. 3. Webbing characteristics.

The test setup used by anyone duplicating this procedure should also include a method of catching and restraining the rebound of the dummy due to the energy stored in the restraint straps. We found that three strategically placed restraining belts with automatic locking retractors which would restrain the dummy at the conclusion of the test was a satisfactory procedure.

Preparation of the dummy consisted of adjusting each of the joints to the one g setting, e.g., the shoulder joint was adjusted so as to just support the extended arm. When placing the dummy in the seat for test, a six inch diameter cylinder was laid horizontally across the lap of the dummy and used to push the dummy back into the seat with fifty pounds force.

Performance of the dummies was measured by triaxial accelerometer packs in the head, chest and pelvis and load transducers in the femurs. Load data in the supporting straps and sled performance data were obtained to assist in the analysis of the sled tests. Both high speed motion pictures and time-lapse Polaroid photographs were

References p. 79

taken of each run. Each of the four test devices used was tested 10 times in each configuration and at each energy level in order to produce sufficient data for meaningful statistical analysis.

DATA ACQUISITION

For each sled run, accelerometer outputs, load cell outputs, sled position markers and timing lines were recorded on strip chart recorders.

Dummy Accelerations — The nine full bridge, piezoresistive accelerometers mounted inside the dummy were calibrated prior to each run by switching a calibration resistor in parallel with one leg of the accelerometer bridge and noting the deflection on the oscillograph. Prior to each group of ten runs, a plus and minus one g calibration was performed on each of the accelerometers. To perform this test, the output of the accelerometers was amplified to give a reasonable deflection on the strip chart recorder. As a further check, some of the accelerometers were calibrated at the National Bureau of Standards.

Femur Loads — In the forward facing runs, axial loads were detected with force transducers in the dummy's left and right femur, and recorded. These force transducers were calibrated using a proving ring.

Strap Tension — The straps which accelerate the dummy are fastened by anchorages. The forces in these anchorages are detected by load cells and the outputs of these load cells recorded on the oscillograph.

Sled Velocity — Sled velocity was determined by measuring the space between position markers on the strip chart record. The position markers were produced by two magnetic transducers fixed 12 inches apart along the track and as the sled passed by, three projections on the sled activated the transducers. Since a time reference was also on the record, the average velocity of the sled between successive markers could be determined. Five average velocities over a five foot travel were thus measured.

Sled Acceleration — Two accelerometers were fastened rigidly to the sled and their outputs were recorded on the strip chart recorder with opposite polarity. The area under these curves during a given time interval is a measure of the velocity change of the sled. Careful measurements with a planimeter of the area under the sled acceleration traces up to the times of the sled position markers resulted in sled velocity measurements which agreed within 2% of the velocity measurements described before.

Sixty Cycle Time Base — As a check of the time base, i.e., the speed of the recorder paper during the test, a sixty cycle signal from the power line was applied to one channel of the records.

Timing Wheel — In order to correlate the data on the charts with high speed film taken during the test, a disc marked with alternate black and white areas rotates in the field of view at 600 rpm and three projections on the disc actuate a magnetic transducer with outputs recorded on the strip charts.

DATA REDUCTION AND PRESENTATION

Time lines were superimposed on the oscillograph records corresponding to 10 msec intervals and aligned so that the 90 msec line coincided with a particular sled position marker recorded on the trace. Deflections of the traces were measured and recorded at each 10 msec interval from the zero time line until after the strap loads had returned to zero, usually about 100 to 150 msec, depending on the energy level and configuration for the run. From these measurements uncorrected resultant accelerations were calculated for the head, chest, and pelvis.

Correction Factors — Correction factors were applied to the accelerations experienced by the dummy. As an example, one correction factor was applied to compensate for the fact that actual sled velocity was slightly different from the nominal test velocity.

Corrections from all sources ranged from about two percent to minus 17 percent. The correction factors were calculated for each series of runs and applied to the average resultants. A typical plot of corrected, average, resultant accelerations for the four dummies is shown in Fig. 4. This example is for a 31.1 fps series of runs and shows head accelerations for the side-facing configuration.

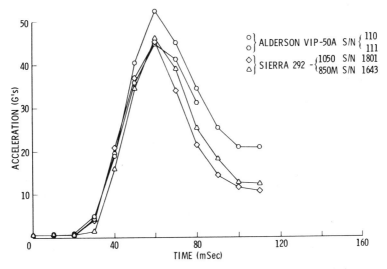

Fig. 4. SSL dummy tests (head acceleration, side facing, 31.1 fps).

CONCLUSIONS

The results of this program showed that the accelerations experienced in the head and chest of J963 type dummies when subjected to the environment which has been

References p. 79

described, have a standard deviation of about 6 percent. The coefficient of variation for each of these measurements is presented in Fig. 5. Ninety-five percent confidence intervals have also been calculated for each of the dummy's responses, both in the head and in the chest for each configuration and energy level. These 95 percent confidence intervals are presented in Figure 6. Accelerations varied from about 20 to 50 g's in the head and about 20 to 40 g's in the chest. In each group, the order of presentation is the two Alderson VIP-50A dummies with Serial Nos. 110 and 111, the Sierra Model 292-1051 with Serial No. 1801 and finally the modified Sierra Model 292-850, Serial No. 1643. Comparison between any two dummies in any group can be performed approximately in this figure to determine if there is a statistically significant difference between their responses. The criterion for a significant difference would be that the average acceleration of both dummies falls outside the range of the 95 percent C.I. of the other. For example, the average head accelerations of each of the Alderson dummies in the side facing 22 fps tests is outside the range of the 95 percent C.I. of the other, therefore, the indication is that there is a statistically significant difference between these averages. More precise calculations of the comparison of these averages at the 95 percent confidence level confirmed this conclusion.

$$\left[100 \cdot \frac{S_x}{\overline{X}} \right]$$

			SIDE FACING			FORWARD FACING			
		22 fps		31.1 fps		31.1 fps		44 fps	
		HEAD	CHEST	HEAD	CHEST	HEAD	CHEST	HEAD	CHEST
ALDERSON VIP-50A S/N { 110		7	4	6	5	9	4	5	5
111		5	3	6	3	5	6	5	2
SIERRA 292- { 1050 S/N 1801		7	19	3	4	4	10	6	14
850M S/N 1643		12	6	5	4	4	2	8	6
		7.7	8.0	5.0	4.0	5.5	5.5	6.0	6.8

Fig. 5. Coefficient of variation of maximum accelerations.

The standard deviation of the accelerometer measurements in the pelvis averaged about 10 percent and were in general higher than in the head and chest measurements. The standard deviations of the axial loads in the femurs have been calculated and the variability in these measurements is quite high with an average of about 33 percent and, in one case, as high as 99 percent. Similar lack of repeatability in femur load data has been reported by D. J. Evanoff (4). In our tests, the lower legs were free to swing out and it is felt that this led to the wild variations noted in the femur loads.

Proposed Calibration Procedure — From the results of the program conducted in our laboratory, it appears as if a calibration procedure of the type described for each dummy before its use in compliance testing is a reasonable approach. Certain modifications of the procedure are certainly in order. Ten runs for each dummy would not be necessary after having established the precision to be expected from the tests. The environment experienced in our tests was essentially a half-sine pulse of acceleration, and it may be desirable to change this to some other pulse shape; a square wave for instance could be produced with force limiting webbing or hardware. Some method of restraining the lower legs could reduce the large variability of the femur loads. These and other changes in the details of the procedure may be desirable and could be implemented without changing the essential features of the test method. These features are: first, capability of any sled facility to perform the tests; second, repeatability of the environment; and third, simulation of a working restraint system.

The calibration procedure recommended is best described by an example taken from Fig. 6. Consider the chest accelerations for the forward facing 44 fps runs. A nominal sensitivity for accelerations under these conditions may be chosen arbitrarily. Looking at the data presented in Fig. 7, 38 g's should not be unreasonable number, the average of these measurements in our tests being 37.95 g's. Having chosen a nominal sensitivity, a calibration number can then be assigned to each dummy which is equal to the ratio of the nominal sensitivity to the measured response for that particular dummy.

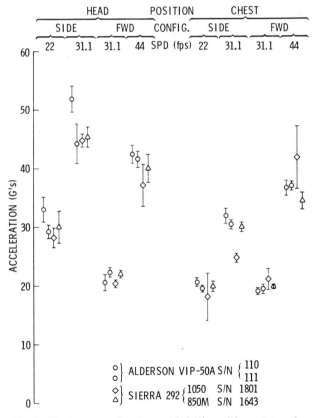

Fig. 6. Maximum accelerations with 95% confidence intervals.

References p. 79

	DUMMY	AVERAGE MAXIMUM ACCELERATION	CALIBRATION FACTORS
ALDERSON VIP-50A S/N { 110		37.1	1.024
111		37.4	1.016
SIERRA 292 - { 1050 S/N 1801		42.4	0.896
850M S/N 1643		34.9	1.089
AVERAGE		37.95	

ASSIGNED NOMINAL SENSITIVITY = 38

Fig. 7. Calibration factors (forward facing, 44 fps, chest).

Future measurements of dummy response to a specific environment using this dummy must then include a correction to the chest acceleration measurements equal to the calibration number just described.

For example, the Alderson dummy with Serial No. 111 had a response in the chest of 37.4 g's and the Sierra with Serial No. 1643 had a response of 34.9 g's. Correcting to a nominal response of 38 g's, calibration factors of 1.016 and 1.089 would be assigned to this Alderson and this Sierra dummy respectively.

In future compliance testing using these dummies, chest accelerometer measurements would then be multiplied by these calibration numbers.

SUMMARY

Dummies which are to be used to make measurements of crash severity must be subjected to a calibration technique if the measurements are to be at all meaningful. A test procedure has been presented which can be used to determine the magnitude and precision of the dummy response when subjected to a specified dynamic environment. This environment was chosen with the following requirements in mind. First, the tests should be designed so that other laboratories can repeat the tests regardless of the type of sled available, i.e., swing seats, accelerators or decelerators could all be used in these tests. Second, the environment had to be repeatable as it is dummy performance which is being studied. Third since the dummies are to be used in compliance testing, the environment should represent an effective restraint system so that hard impacts should be avoided. Again, the environment is defined by the system of straps employed and the differential velocity established between sled and dummy.

The average maximum resultant accelerations in the head and chest had a standard deviation of about 6 percent for 10 runs in the tests which we performed and statistically significant differences between responses for different dummies were detected in some cases.

Finally, it is proposed that dummies prior to use in compliance tests be subjected to a calibration procedure such as the one described here to determine their sensitivity and repeatability. Calibration factors based on such a procedure and assigned to the dummies should lead to more reproducible measurements.

REFERENCES

1. R. W. Armstrong, "NBS Dynamic Seat Belt Tester", Tenth Stapp Car Crash Conference, SAE, 1967.
2. "Studies of Inflating Restraint Systems", Highway Safety Research Institute, University of Michigan, DOT Report HS-800497.
3. "Test Procedures and Performance of Anthropomorphic Test Devices", Safety Systems Laboratory, NHTSA, Docket No. 69-7 of NHTSA.
4. D. J. Evanoff, "Evaluation of Anthropomorphic Test Devices Response to Head, Chest, and Femur Impacts – Sierra 1050 vs. Alderson VIP-50A, 50th Percentile Male Dummies", Ford Motor Company.

DISSCUSSION

J. N. Silver (GM Proving Ground)

Did you make any attempts to calibrate your test procedure by using a nonarticulated dummy or something else that would determine what the variability of your test is, and try to separate this from the variability of the dummy itself?

A. M. Thomas

We did not do that specifically. The procedure we went through for setting up the dummy at the beginning of the test was to carefully control the strap characteristics and sled velocity for the test. I think your recommendation to make that measurement is a good one and it probably should be done in the future.

H. T. McAdams (Cornell Aeronautical Laboratories)

I find that your approach to a calibration number is similar to the one I was suggesting in my talk. However, I wonder if one set of test conditions is adequate. In the calibration number, as you've shown there, you have adjusted individual dummies according to their response. One has a factor of .89 and another one has a factor of 1.05 and so forth. It seems to me that this amounts to ordering these several dummies in relation to their responsiveness to a specific set of circumstances. We do not know if we subject those same dummies to a different set of circumstances that they would be ordered in the same way. Perhaps the one that has the .89 factor now has a factor of 1.05 and the one that was 1.05 is now .7. Will you address that question please?

A. M. Thomas

Yes. This was a development program. In our particular case, we were trying to simulate an effective airbag system. We believe that if these dummies, after having been subjected to this test, were then subjected to an environment where an effective airbag system had been used, the ordering of these calibration numbers would not have changed appreciably. If we wanted then to do other tests, perhaps where the environment would have been somewhat different, we would have to change our environment in the calibration procedure to try to simulate that new environment.

H. T. McAdams

In other words, you would have to work in analogues?

A. M. Thomas

Yes.

R. C. Haeusler *(Chrysler Corporation)*

Would you expect, if you use that procedure, that you then would be able to compare dissimilar restraints; for example, belts versus airbags?

A. M. Thomas

I would guess that if the restraint system were effective, so that a soft impact were experienced, then intercomparisons could be made; however, this is just a guess and further testing would be necessary to determine the dummies responses to environments other than the one presented here.

S. W. Alderson *(Alderson Biotechnology Corporation)*

I would like to address this question to both of the preceding authors who have been involved in measuring the repeatability of dummies. Has any work been done in establishing the source of nonrepeatability in dummies? In other words, what are the principle criteria in the design and construction of dummies to be obeyed if repeatability is to be improved.

A. M. Thomas

We certainly think that in setting up the test, the setting of the joint is one of the big variables. Again, this has more importance at certain energy levels than at others.

In the design of the dummy, I think Professor Patrick's suggestion of less articulated dummies may be the direction to go to attain repeatability.

H. T. McAdams

I would just like to reiterate my previous comment that striving for repeatability without regard for the sensitivity of the dummy to respond to those things which are important in injury assessment is a dangerous thing. We should not look at repeatability for repeatability's sake. I concede that you could probably design a dummy by locking up all the joints and certain other things, which would be repeatable, but would not be very useful in evaluating potentially injurious situations.

V. L. Roberts *(HSRI, University of Michigan)*

I'm slightly confused by your calibration factor. Would you suggest that this be applied throughout the pulse or just in the peaks. It seems to me that when you're dealing with a computed resultant things add together differently. Also, is it not, in fact, possible that the calibration factor could change throughout the pulse?

A. M. Thomas

Our work was strictly on the maximum accelerations (peaks) in this case. I think we would have to look into that much deeper; especially from the standpoint of what kind of calibration factor one would use for the various severity indices. These would be different; ours was a straight linear type of calibration factor.

V. L. Roberts

I was also interested by your comment that the femur load cells didn't seem to contribute very much and that they perhaps ought to be neglected. Did the femurs hit anything?

A. M. Thomas

In our case, no. We had a large variability in this and others have also noted this variability in femur loads in various tests. Ours, I think, arose from the kicking-out of the lower legs. We could get quite a bit of variation in the loads due to the way in which the lower legs kicked-out.

L. M. Patrick *(Wayne State University)*

As far as joints are concerned, I think we've all had the same experience. We set them for 1g, move them and we find that they're no longer at 1g. We can't put the dummy in a car, for example, without moving it at times; so this initial joint setting is probably not very realistic. Going back to your short film, it didn't look to me like there was sufficient movement to allow the joints to bottom out. If the joints don't bottom out, they offer very little resistance, so I question whether the slack in the belt really did much good. Also, no matter how you try to limit the use of your calibration factor, once its published it's going to be used for everything, every kind of impact you can think of. For example, our tolerance curve has been applied to the chest and I suppose by some people even the toe. It was originally meant only to be used for frontal-force forehead impact to hard, flat surfaces. Lets be a little careful how we apply calibration factors.

A. M. Thomas

As concerns the film strip, the articulation experienced by the dummy did match fairly well that shown in the films of air bag tests which we had studied. Also the one film shown was for a low energy run and more articulation was evident in the high energy runs.

There still seems to be confusion about the purpose for the slack in the belts and I would like to try to clear this up. It was stated earlier that results of tests on dummies depend on the sled facility used because of the different pulse shapes produced by different sleds. It was for this reason that our method was developed, that is, using the method described, a given test environment can be reproduced regardless of the type of sled used. This is accomplished by using the slack in the belts as a kind of time delay so that the acceleration phase of the sled is completed before the dummy is subjected to the crash environment and this environment is totally defined by the restraining system and the differential velocity between the sled and the dummy.

SESSION II

IMPACT RESPONSE AND APPRAISAL CRITERIA
HEAD AND THORAX I

Session Chairman
L. M. PATRICK

Wayne State University
Detroit, Michigan

BIOMECHANICAL ASPECTS OF HEAD INJURY

J. H. McELHANEY, R. L. STALNAKER and V. L. ROBERTS

University of Michigan, Ann Arbor, Michigan

INTRODUCTION

With the advent of high speed air and land transportation, engineers have become increasingly aware of the mechanical frangibility of the human body. Thus, we have seen the evolution of various isolating and load distributing devices ranging from seat belts and padded sun visors, to ejection seats, crash helmets, and acceleration couches. While there is a large amount of information available regarding the response of inanimate systems to vibration and impact, there is a comparable dearth of knowledge pertaining to the mechanical responses of biological systems. Therefore, the design of much supporting and protective equipment is often based on intuition because of the lack of information available about the mechanical behavior of the human body. In addition, such knowledge would be helpful in the treatment of injury by serving to identify the mechanism of trauma. Thus, both a rational design procedure for impact protection and a rational therapy for treatment of trauma cannot be developed until a quantitive description of the mechanical responses of the human body is obtained.

To position this research program in its proper place in the spectrum of serious problems facing our modern society, consider the following statistics from the National Safety Council. Accidents are the fourth leading cause of death in this country, following heart disease, cancer, and stroke. In the group aged one to forty, accidents are the most common cause of death. Based on recently published estimates, there were about 53 million accidents in 1971 resulting in 85,000 fatalities and 2,000,000 disabling injuries with overall costs of about $14.3 billion in terms of wages lost, medical expenses, and insurance administration, but not counting property damage. Of this terrible toll, it is estimated that head injuries account for more than 66%.

References pp. 109-110

In order to properly design devices aimed at minimizing head injury in the automotive crash environment, engineers require a means of predicting potential injury or a so-called Head Injury Criteria. This criteria might be used in real accident reconstructions, car crash and sled test experiments or mathematical simulations.

The automotive crash environment encompasses a wide range of impulse durations and directions. Thus, a viable head injury criteria must provide appropriate mechanisms that realistically account for the frequently observed, but poorly documented, relations of head impact tolerance and impulse duration and direction. In addition, two distinct types of loading are observed:

1. An impact or blow involving a collision of the head with another solid object at an appreciable velocity. This situation is generally characterized by large linear accelerations and small angular accelerations during the impact phase.

2. An impulsive loading including a sudden head motion without direct contact. The load is generally transmitted through the head-neck junction upon sudden changes in the motion of the torso and is associated with large angular accelerations of the head.

It has been shown (1) that in the moderate but survivable automotive crash environment (30 mps barrier equivalent) no significant head injuries occur when the fully-belted occupant rides the crash down without head-to-vehicle contact. This does not mean that the Type Two loading described above can not produce injury, but only that the levels of linear and angular acceleration required to produce head injury without contact do not occur in moderate crashes. If, however, vehicles are stiffened to provide less compartment intrusion at higher velocities, injuries of the type described by Ommaya and Hirsch (2) might become commonplace.

Thus a rational head injury criteria for current automotive design may be concentrated on the first type of loading. There is, of course, a possible defect in this rationale that concerns the possible increased potential for head injury involved with a combination of the first and second type of head loading. An attempt to reconcile this problem has lead to the development of the Mean Strain Criterion. This head injury criteria considers the total linear acceleration history of the head but assumes a single injury mechanism.

MEAN STRAIN CRITERION (MSC)

The dynamic structural characteristics of monkey skull and brain were determined over a wide frequency range of Stalnaker and McElhaney in 1971. These results, reported as the change of mechanical impedance (force/velocity) with frequency, allowed the conceptual characterization of the head as two masses coupled by a spring and dashpot. The mathematically predicted dynamic response of the model agreed well with the experimental data.

Experimental impacts delivered to the heads of various size primates showed that the dynamic model postulated on the basis of vibration studies accurately predicted head-impact injuries. It was further found that for head impacts of a known magnitude, the resulting injuries could be grouped by comparing the mean strain as predicted by the theoretical model with injury levels, (where mean stain is defined as the displacement of one side of the head relative to the other, divided by the distance across the cranium). This experimentally derived head injury data for living primates formed the basis for a Mean Strain Criterion (MSC) for head injury to humans. Using the value of predicted strain in the Rhesus monkey head as a criterion of injury, a tolerance curve was derived which related average acceleration and time for a constant level of mean strain.

The derived tolerance curve for the subhuman primate was validated by plotting the experimental data points necessary to produce minor, but identifiable, brain injury in the living test subject for a wide range of pulse durations (Fig. 1).

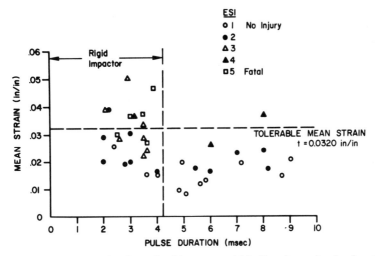

Fig. 1. MSC strain levels for rhesus head impacts variable direction and pulse duration.

It was determined that the heads of several species of subhuman primates, squirrel monkeys, Rhesus monkeys, the chimpanzee and the fresh human cadaver had mechanical impedance characteristics over a broad frequency band (5 to 5000 Hertz) which were similar in shape but varied in the mechanical characteristics of mass, stiffness and damping. Using the maximum predicted strain as the basis for injury, experimental head impacts in the laboratory documented the validity of the theory and formed the basis for establishing modeling relationships through which extrapolation of the MSC to other size heads may be made. This approach was tested by comparing the MSC to human volunteer and fresh human cadaver head impacts (Fig. 2.). The mean tolerable head strain for humans (0.006 in./in.) was calculated from the mathematical model and the scaling technique referenced above.

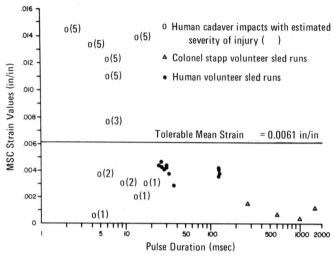

Fig. 2. Mean strain criterion for humans vs pulse duration.

A study was undertaken to document the validity of the MSC for arbitrarily directed head impacts. Thirty carefully selected Rhesus monkeys were impacted at increasing levels in various directions (front, side, back top and mid-front). The impact level was increased until autopsy studies indicated that an Estimated Severity of Injury (ESI) of a moderate but reversible type (Class 3) was obtained. Only closed brain injuries were considered.

The mechanical impedance was determined for five monkeys of approximately the same weight as the ones used in the impact study. These impedance curves were obtained for the top, side, rear, and front of the head (Fig. 3).

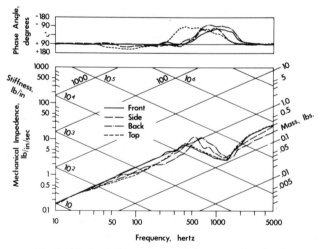

Fig. 3. Mechanical impedance of rhesus monkey head.

The values of tolerable acceleration and impedance data were then input to the MSC model. The predicted mean strain values for each direction varied less than 7.5%. This indicates that, while widely varying accelerations are required to produce an injury level of 3 in the Rhesus monkey, the corresponding strains are approximately equal. The results of this study are given in the form of an acceleration surface for a constant strain level of 0.032 in./in. for Rhesus monkeys subjected to rigid striker impacts (Fig. 4).

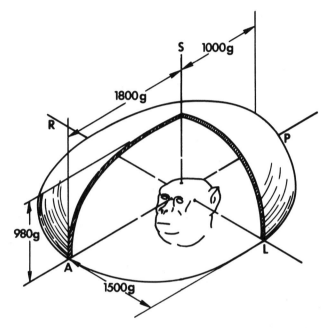

Fig. 4. Critical acceleration surface for rhesus monkey, injury index = 3.

TABLE 1

Results of Rhesus Monkey Head Impacts and Impedance Tests

Direction of Head Impact	Acceleration (G's)	Pulse Duration (msec)	Model Constants				Mean Strain ϵ
			w_1	k	c	w_2	
Front	1800	3.6	.051	39,000	1.6	1.1	0.032
Side	1500	2.8	.040	33,000	2.1	1.0	0.032
Top	980	7.0	.030	18,000	1.2	0.9	0.032
Back	1000	3.4	.035	20,000	2.9	1.1	0.032

Preliminary studies on the fresh intact cadaver indicate that a tolerable mean strain level of 0.0061 as predicted by the MSC model may be used for arbitrary impact directions with model constants appropriate for that direction. However, a sufficient number of impact tests and driving point impedance measurements have not yet been made to verify a direct extrapolation of the Rhesus monkey data. Figs. 5 and 6 show the currently available impedance data for the fresh human cadaver contrasted with Sierra and Alderson Dummy Heads. Fig. 7 shows predictions and measured values of the MSC for human head impact in the sagittal plane.

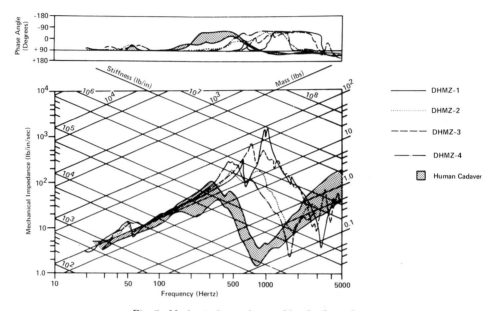

Fig. 5. Mechanical impedance of heads (frontal).

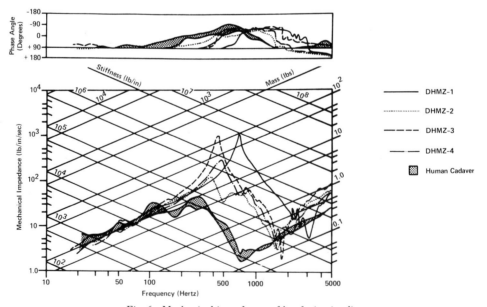

Fig. 6. Mechanical impedance of heads (parietal).

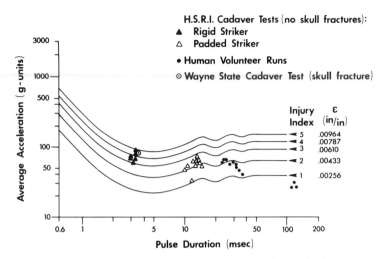

Fig. 7. Mean strain criterion for humans, sagittal plane loading.

COMPARISON OF HEAD INJURY CRITERIA

The preceding discussion involved the most recent work on the Mean Strain Criterion for head impact. The following section contains the results of a series of analysis aimed at comparing various head injury criteria.*

Values of various head injury indices were computed for two crash simulations.

1. Dummy resultant head accelerations in frontal automotive crash environments were used. The dummy was unbelted, and struck the windshield a clean blow. The windshield was not penetrated except a small tear in the laminate was allowed (Class A), or several small tears (Class B).

2. Resultant head accelerations from a recent series of high speed human volunteer and dummy tests at Holloman Air Force Base using airbag restraints.

Wayne State Tolerance Curve (WSTC) — The Wayne State Tolerance Curve was introduced by Lissner in 1960. Originally this curve was developed from data obtained by dropping embalmed cadaver heads onto unyielding flat surfaces. Linear skull fracture was used as the criterion of injury. In 1962 Gurdjian published the Wayne State Tolerance Curve as it appears today. This curve was devloped by combining a wide variety of pulse shapes, animal types and injury mechanisms. The failure criteria used was generally skull fracture and/or concussion, except for long pulse duration from human volunteers with no discernible injury. In 1963 Patrick proposed that the original horizontal asymptote of 45 G's be raised to 80 G's to adjust for additional data from tests against yielding surfaces.

---

*This work was undertaken at the suggestion of Jules Brinn acting as Chairman of the SAE Human Tolerance Subcommittee.

References pp. 109-110

The injury assessment is based on the average acceleration and pulse duration. A given average acceleration at a particular pulse duration which lies below the WSTC is considered to cause at most cerebral concussion without permanent after-effects, while any point which lies above the curve is considered to be dangerous to life. For single head impacts into a rigid flat surface the average acceleration and pulse duration is quite easy to determine, but for sled testing where multiple impacts are quite common the "effective" pulse, that is, the part of the pulse upon which the average is based, is not well defined. In spite of the many interpretive difficulties associated with this curve, it has been the principal source for head injury tolerance information used by the automotive safety community.

Gadd Severity Index (GSI) — Because of various interpretive difficulties associated with the use of the WST curve, C. W. Gadd introduced the Gadd Severity Index as a generalization of the Wayne curve. More recently the GSI has been extended for long pulse duration by means of the Eiband tolerance data and other primate sled runs. The severity index equation has the following form:

$$G.S.I. = \int_0^\tau a^n dt \tag{1}$$

where

    a = head acceleration response function
    n = weight factor, general 2.5
    τ = pulse duration
    t = integral parameter of time

The head injury threshold severity index number was determined from comparison with the WST curve and the number 1000 recommended.

The assessment of injury hazard is obtained performing the above calculation. If this number (GSI) is greater than 1000, the acceleration pulse is considered to be dangerous to life. If this index is less than 1000, the acceleration pulse is then considered not to be life threatening.

Head Injury Criterion (HIC) — The Head Injury Criterion was first proposed by J. Versace and then modified by NHTSA. This criterion is based on a new interpretation of the Gadd Severity Index.

Versace pointed out that because the WST curve was plotted for average acceleration, any comparison to the WST curve should be made using the average acceleration of the pulse of interest.

The question of long pulse head accelerations has posed some problems when using the S.I. to predict head injuries. In order to provide a better comparison with human volunteer tests, a head criteria has been proposed as

$$HIC = [\frac{\int_{t_1}^{t_2} a\,dt}{t_2 - t_1}]^{2 \cdot 5} (t_2 - t_1) \qquad (2)$$

where

t_1 = an arbitrary time in the pulse

t_2 = for a given t_1, a time in the pulse which maximizes the HIC

a   = resultant acceleration at the head center of gravity

If this index is less than 1000, the situation is considered not to be life-threatening.

Vienna Institute Index (JIT) — The Vienna Institute Index was introduced by A. Slattenschek and is based on a single degree-of-freedom vibration model.

With the damping assumed to be critical $\beta = 1$, and two triangle acceleration pulses determined from the WST curve, the natural angular frequency value of $\omega = 635$ rad/sec and a maximum tolerance displacement $x_{to\omega} = 0.092(2.35mm)$ inches was obtained from the following equation:

$$\ddot{x} + 2\beta_\omega \dot{x} + \omega^2 x = \ddot{y}(t) \qquad (3)$$

where:

x = relative displacement of brain mass to skull

$\dot{x}, \ddot{x}$ = ralative velocity and relative acceleration

ω = natural angular frequency of vibration 635 rad/sec

β = viscous damping coefficient of 1.0

$\ddot{y}(t)$ = acceleration pulse measured at the head

The maximum deviation between the model and the WST curve is -4 percent. To access an impact, the amplitude x_{max} corresponding to the acceleration pulse to be analyzed is determined from the model and compared to the tolerable amplitude $x_{Tolr} = 0.092$ inches. A J tolerance index is then defined by:

$$J = \frac{x_{max}}{x_{Tolr}} \qquad (4)$$

where:

x_{max} = maximum x generated by the model for a given acceleration pulse

x_{Tolr} = tolerable amplitude from the Wayne State Tolerance Curve 0.092 inches (2.35mm)

According to Slattenschek, impacts with a J tolerance of $J_2 = 1$ just reach the threshold of human tolerance; values $J < 1$ at worst cause cerebral concussion without permanent after-effects, while values $J < 1$ are considered to be hazardous to life.

Effective Displacement Index (EDI) — The effective displacement index was introduced by J. Brinn and is similar to the Vienna Institute model with changed

References pp. 109–110

damping and the angular frequency. New angular frequency and damping values for the Slattenschek model were determined by matching the model to the WST curve. The emphasis was placed on matching for short duration events (3-5mscs pulse duration). The best fit of the model to this portion of the WST curve was found by using the model parameters ω = 482 rad/sec and β = 0.707. These model parameters and the Slattenschek model were exercised for points on the WST curve and a displacement value of 0.15 inches was determined as a "Design Bogie" for human AP head impacts. Because of the unhuman like response of dummy heads, the Design Bogie is raised to a dummy Test Limit of 0.17 inches in the AP direction. When the resultant acceleration is used, a Design Bogie of 0.18 inches and a dummy Test Limit of 0.20 inches is recommended.

Revised Brain Model (RBM) – The revised brain model was introduced by W. R. S. Fan and is a modification of the Vienna Institute model. Like the JTI, the RBM is a single degree-of-freedom mass-spring-dashpot model of the brain. The viscous damping coefficient for this model was estimated from published values of brain material properties. With an estimated damping coefficient of 0.4 data from the WSTC for long duration inputs, a natural angular frequency of 175 rad/sec and the theoretical tolerable brain deformation (S_d) of 1.25 inches was estimated. A tolerable brain velocity S_V was then calculated from the WST curve for short pulse duration and was found to be 135.3 in/sec.

The recommended measure of brain injury potential is $\dot{x} \langle S_V$ for impact pulse durations less than 20 msec and $x \langle S_d$ for pulse durations greater than 20 msec, as calculated from the differential equation of the Slattenschek model with revised coefficients.

A summary of head injury criteria is shown in Fig. 8.

RESULTS OF COMPARISON

The results of the computations of the various head injury criteria are presented in Table 2. Table 3 gives normalized values of the head injury indices obtained by

SI	HIC	JTI	RBM	EDI	MSC
SEVERITY INDEX (GADD)	HEAD INJURY CRITERION (VERSACE & NHTSA)	J-TOLERANCE INDEX (SLATTENSCHEK)	REVISED BRAIN MODEL (FAN)	EFFECTIVE DISPLACEMENT INDEX (BRINN)	MAXIMUM STRAIN CRITERION (STALNAKER)
Weighted Impulse of a(t) $SI = \int_0^T [a(t)]^{2.5} dt$ Time in seconds Acc. in g-units	Weighted Impulse of a(t) Let $\bar{a}_{12} = \frac{\int_{t_1}^{t_2} a(t)dt}{(t_2 - t_1)}$ $HIC = \left\{ \begin{matrix} \max \\ \text{all} \\ t_1, t_2 \\ 0 \leq t_1 < t_2 \leq T \end{matrix} \left[\bar{a}_{12}^{2.5} (t_2 - t_1) \right] \right\}$	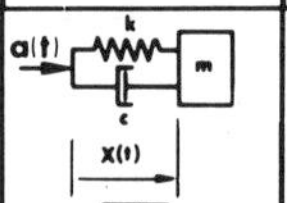 $\omega_n = \sqrt{k/m}$ (rad/sec) $\beta = c/c_c$ $\omega_n = 635.$ $\beta = 1.0$	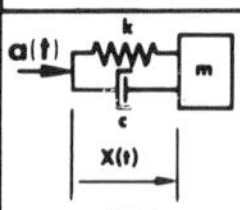 $\omega_n = 175.$ $\beta = 0.4$	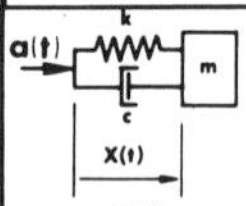 $\omega_n = 482.$ $\beta = 0.707$	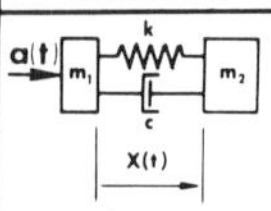 $m_1 = 0.6$ (lbs) $m_2 = 10.0$ (lbs) $c = 2.0$ (lb-sec/in) $k = 50000.$ (lb/in)
$SI_{tol} = 1500$	$HIC_{tol} = 1000$	$J = \frac{X_{max}}{0.0925}$, $\frac{in}{in}$ $J_{tol} = 1.0$	T< 20ms: $\dot{x}_{tol} = 135.3 \frac{in}{sec}$ T >20ms: $x_{tol} = 1.25$ in	X_{tol} = A-P / RES. HUMAN: 0.15 in / 0.18 in DUMMY: 0.17 in / 0.2 in	$\varepsilon = X_{max}/L$ HUMAN: L = 5.75 in (A-P) $\varepsilon_{tol} = 0.0061$ in/in

Fig. 8. Summary of head injury criteria.

dividing the particular computed index by the appropriate cut off value (i.e. 1500 for the SI and 1000 for the HIC). Fig. 9 shows the results of Table 3 in more graphic form.

TABLE 2

Summary of Head Injury Indice Comparisons

	Pulse I.D.	Acceleration Pulse		GSI	HIC		HIC	JTI	RBM	EDI	MSC
		Dura-tion (msec)	Peak (g's)		Dura-tion (msec)	Aver. Accel. (g's)					
	Sine	10	100	458	7.8	83	415	0.853	–	0.146	.0039
	Triang	10	100	286	5.7	72	247	0.691	–	0.118	.0045
	Square	10	100	1000	10.0	100	1000	1.026	–	0.172	.0061
Dummy Windshield Class A	22	128	162	1170	40	49	680	0.749	0.909	0.120	.0037
	23	105	207	1609*	4	125	702	1.087*	0.905	0.180	.0053
	24	187	144	922	33	54	704	0.718	0.950	0.123	.0027
	25	188	109	717	28	52	555	0.683	0.937	0.111	.0039
	26	182	111	825	43	45	597	0.764	0.934	0.122	.0028
	41	187	248	2080*	38	61	1082*	1.117*	0.890	0.182	.0056
	42	211	290	3066*	2	254	2057*	1.608*	1.283*	0.269*	.0065*
	43	202	150	917	46	47	716	0.668	0.741	0.109	.0041
	44	250	117	1154	32	59	841	0.839	0.863	0.137	.0033
	45	250	111	825	18	66	644	0.901	1.098	0.147	.0037
Dummy Windshield Class B	11	300	418	2229*	36	75	1273*	1.018*	1.373*	0.164	.0046
	12	155	151	1020	50	46	701	0.721	0.909	0.115	.0030
	13	153	174	1194	24	68	903	0.812	1.087	0.130	.0028
	21	105	150	1275	15	61	438	0.922	0.940	0.153	.0038
	31	88	85	395	43	31	232	0.603	0.685	0.100	.0018
	32	62	58	400	31	42	355	0.527	0.733	0.088	.0024
	47	250	98	1577*	34	69	1360*	0.913	0.116	0.146	.0060
Holloman Human Air Bag Tests	51	162	80	1246	26	60	718	0.795	1.051	0.130	.0047
	53	167	66	915	12	49	544	0.666	0.858	0.107	.0038
	54	189	56	697	36	41	392	0.568	0.782	0.092	.0029
	55	169	75	1249	30	55	682	0.759	0.980	0.122	.0043
	57	154	76	1324	26	61	765	0.764	0.997	0.122	.0043
	58	173	79	1212	24	63	763	0.798	1.059	0.128	.0044
	59	187	78	1224	28	59	751	0.783	1.031	0.126	.0041
	5B	159	78	1446	30	61	875	0.754	1.011	0.120	.0044
	5C	154	67	1077	126	29	563	0.666	0.866	0.109	.0033
Dummy Air Bag	5A	147	78	1305	124	34	848	0.683	0.892	0.110	.0042
	52	158	65	987	122	29	546	0.628	0.819	0.102	.0036
	56	144	71	1394	124	34	832	0.716	0.975	0.116	.0041

Exceeds tolerable value of applicable criterion.

References pp. 109-110

TABLE 3

Comparison of Normalized Head Injury Indices

	Test I.D.	SI	HIC	JTI	RBM	EDI	MSC
Dummy Wind-shield Tests	22	.780	0.680	0.749	0.727	0.600	0.607
	23	1.073	0.702	1.087	0.724	0.900	0.869
	24	.614	0.704	1.718	0.760	0.615	0.443
	25	.478	0.555	0.683	0.750	0.555	0.639
	26	.550	0.597	0.764	0.747	0.610	0.459
Class 'A'	41	1.386	1.082	1.117	0.712	0.910	0.918
	42	2.043	2.057	1.608	1.026	1.345	1.066
	43	.611	0.716	0.668	0.593	0.545	0.672
	44	.769	0.841	0.839	0.690	0.685	0.541
	45	.550	0.644	0.901	0.878	0.735	0.607
Dummy Wind-shield Tests	11	1.485	1.273	1.018	1.098	0.820	0.754
	12	.680	0.701	0.721	0.727	0.575	0.492
	13	.746	0.903	0.812	0.870	0.650	0.459
	21	.850	0.438	0.922	0.752	0.765	0.623
	31	.263	0.232	0.603	0.548	0.500	0.295
Class 'B'	32	.267	0.355	0.527	0.586	0.440	0.393
	47	1.051	1.360	0.913	0.893	0.811	0.983
	51	.831	0.718	0.795	0.841	0.722	0.770
Hollo-man Air Bag Tests	53	.610	0.544	0.666	0.686	0.594	0.623
	54	.465	0.392	0.568	0.627	0.511	0.475
	55	.832	0.682	0.759	0.784	0.677	0.705
	57	.882	0.765	0.764	0.798	0.677	0.705
	58	.808	0.763	0.798	0.847	0.711	0.721
	59	.816	0.751	0.783	0.825	0.700	0.672
Human	5B	.964	0.875	0.754	0.809	0.666	0.721
	5C	.718	0.563	0.666	0.693	0.605	0.623
Hollo-man Air Bag: Dummy	52	.658	0.546	0.628	0.655	0.566	0.590
	56	.929	0.832	0.716	0.780	0.644	0.672
	5A	.870	0.848	0.683	0.714	0.611	0.689

Based on accident statistics, it is felt that head impacts of this type into the HPR windshield seldom involve serious head injury and an appropriate head injury criteria should so indicate. Study of Table 2 shows the GSI to be quite conservative in this situation. The HIC is less conservative, but still indicates four life-threatening situations in the windshield tests. The RBM, EDI and MSC all predict essentially the same injury levels for both series.

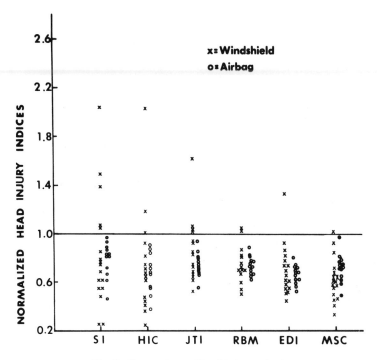

Fig. 9. Comparison of head injury criteria.

IMPACT PERFORMANCE CRITERIA FOR CRASH TEST DEVICE HEADS

In order to apply head injury criteria developed from cadaver and human volunteers to crash test devices, the impact response of the head of the crash test device should be similar to that of the human for a variety of impulse magnitudes. This section contains the results of a series of tests aimed at establishing force-time and acceleration time histories for fresh cadaver heads to allow the specification of the appropriate impact responses of a new generation of crash test device heads.

Preparation of the Cadaver — The specimens used in this study are fresh, unembalmed cadavers obtained from the Anatomy Department of The University of Michigan Medical School (Fig. 10). They are stored at $37°F$ for two of the three days between time of death and impact. The cadavers are allowed to reach room temperature before testing. These procedures insure that the effects of rigor mortis have disappeared and the blood is again fluid.

Frontal (A-P) Head Impacts — The axis of the impactor is in the mid-sagittal plane and is normal to the surface of the skull at the point of initial impact. The cadaver is positioned so that the inferior aspect of the four-inch diameter impactor is adjacent to the glabella and therefore no part of the impactor comes in contact with the subject's nose. The point of impact is located two inches superior to the glabella and in the mid-sagittal plane.

References pp. 109-110

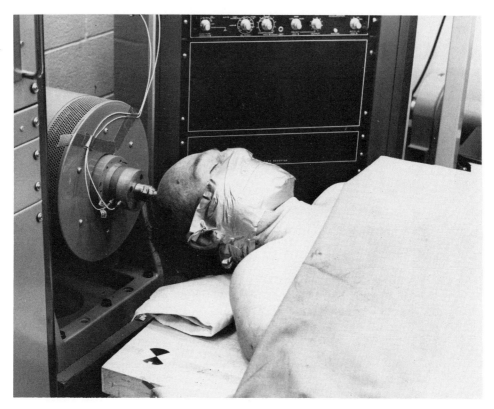

Fig. 10. Typical driving point impedance setup.

The targets used for photographic analysis are in a plane extending in the L-R direction, common to the axis of impact. They are located two inches posterior and one inch anterior to the intersection of a line perpendicular to the axis of impact passing through the external acoustic meatus, with the above described plane. A biaxial accelerometer pack is mounted at a point on the occupant, directly opposite the point of impact. Class 1000 filters were used in all channels as per J211.

After being targeted and equipped with accelerometers, the cadaver is placed in a specially designed chair. All surfaces against which the cadaver might come in contact with in its post-impact movements are thickly padded with styrofoam to prevent damage to the cadaver.

The cadaver is carefully positioned so that its head is in the correct position relative to the impactor and at the same time the whole cadaver acts as a relatively free body. The head is suspended and held in place by four strands of 000 thickness surgical thread. This thread supports only the weight of the head, and breaks easily on impact (Fig. 11).

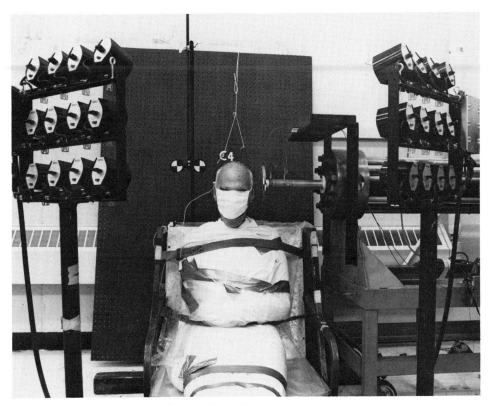

Fig. 11. Typical head impact setup.

TABLE 4

Description of Cadavers

No.	Sex	Age	Height	Body Wt. lbs.	Head Wt. lbs.	W in.	L in.	H in.	Circ. in.	Brain Wt. lbs.	Chest in.	Days Dead	Cause of Death	Comments
C1	F	48	5'2"	129	10.6	6.2	8.0	5.2	23.0	3.25	32.1	3	Cancer of Abdomen	
C2	M	66	5'3"	79.2	9.7	6.0	6.8	5.7	21.0	2.75	29.7	6	Cancer of Prostate	
C3	M	50	5'8"	176	10.3	6.0	8.1	5.9	23.0	2.93	35.0	8	Cardiac Arrest	Left Cranium Hydrocephalic
C4	M	78	5'9"	158.8	11.1	6.1	7.7	6.3	23.0	3.15	34.1	5	Multiple Myeloma	
C5	M	84	5'6"	92.8	8.9	5.7	7.5	5.9	21.5	3.02	29.3	8	Pneumonia	
C6	M	72	5'8"	126.5	10.3	5.8	7.8	5.7	22.5	2.96	32.1	3	Congestive Heart Failure	

References pp. 109–110

TABLE 4 (Continued)
Description of Cadavers

No.	Sex	Age	Height	Body Wt. lbs.	Head Wt. lbs.	Head W in.	Head L in.	Head H in.	Head Circ. in.	Brain Wt. lbs.	Chest in.	Days Dead	Cause of Death	Comments
C7	M	64	5'7"	112.0	9.7	6.2	8.1	5.8	23.0	2.93	32.0	7	Metastatic Nemanoma	
C8	M	72	5'7"	112.4	10.1	6.2	7.8	5.8	22.5	2.83	31.7	5	Asphyxiation	
C9	M	99	5'8.5"	148.5	10.8	6.8	7.7	6.3	23.0	2.97	32.7	1	Cerebral Vascular Accident	
C10	M	70	5'3"	92.4	9.3	6.0	7.5	6.8	21.6	3.07	30.0	5	Coronary Occlusion	
C11	M	70	5'6"	123.8	10.7	5.9	6.8	5.4	21.0	3.09	34.1	2	Acute Coronary	
C12	M	72	5'2"	84.7	7.9	5.6	7.3	5.3	20.5	3.01	29.3	2	Lobar Pneumonia Bilateral	Legs and Arms Bent Up
C13	M	85	5'7"	135.7	10.7	6.0	7.7	5.5	22.3	2.74	33.1	9	Pneumonitis	
C14	M	73	–	–	10.8	6.0	7.8	4.9	22.0	3.00	31.3	3	Unknown	Double Amputee
C15	M	65	5'2"	77	8.3	4.5	6.3	4.4	20.0	2.94	30.1	3	Bilateral Extensive Broncho-pneumonia, Fibrosis, Emphysema	
C16	M	88	5'8"	149.4	10.9	6.3	8.4	5.5	23.6	3.11	37.1	4	Unknown	
C17	M	49	5'11"	155.0	11.3	6.2	7.3	5.9	22.3	2.81	38.2	2	Acute Myocardial Infarction	
C18	F	65	5'3.5"	100.0	9.6	5.4	6.8	4.8	20.3	2.97	30.8	3	Generalized Carcino-matosis	
C19	M	52	5'10"	203.0	11.7	6.3	7.9	5.5	22.5	3.12	42.0	3	Coronary Heart Disease	
C20	F	75	4'8"	88.0	9.3	6.2	8.1	5.6	20.8	3.11	28.5	5	Coronary Thrombosis	
C21	M	62	6'0"	113.0	10.5	6.1	7.7	4.7	21.5	2.89	31.7	2	Coronary Heart Disease	
C22	M	63	5'7"	128.0	11.0	6.1	7.4	5.3	22.0	2.84	33.3	1	Cardiosenic Shock	
C23	M	58	5'10"	155.0	12.1	6.1	8.0	6.3	23.0	3.11	39.5	3	Aspiration of Pneumonia	

Two kinds of impacting surfaces are used. The first surface is a four-inch diameter rigid metal disc. The second surface consists of the same metal disc with a three-inch thick pad of polystyrene foam (density = 1.78 lb/ft^3), fixed to the contacting surface of the disc. The weight of the impactor in both cases was 22 pounds.

Side (L-R) Head Impacts — The point of impact is the left temporal region, two inches superior to the external acoustic meatus. The photographic targets for L-R head impacts are located on the supra-orbital ridge, two inches on either side of the glabella (i.e., two inches either side of the mid-sagittal plane). Other details of the test setup are the same as for the A-P tests.

Static Load Deflection Tests — The static load deflection characteristics of the head were measured in an Instron Testing Machine. The cadaver was positioned so that its head rested between two six-inch diameter steel platens. Tests were performed with the load axis in the A-P and L-R directions. Figs. 12 and 13 show the results of these tests. The static stiffness of the group of cadaver heads studied varied from 4,000 to 10,000 pounds per inch in the L-R direction and 8,000 to 20,000 pounds per inch in the A-P direction.

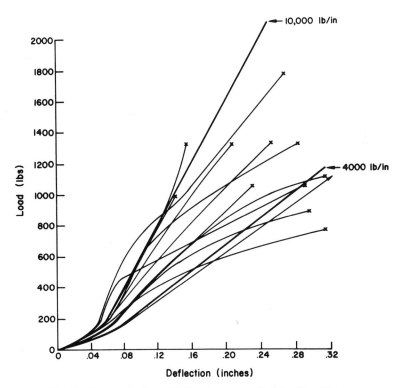

Fig. 12. Human skull load-deflection curves, L-R loading, 12 tests.

References pp. 109-110

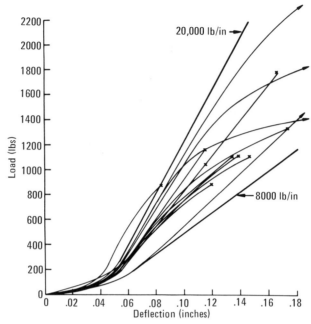

Fig. 13. Human skull load-deflection curves, A-P loading, 12 tests.

Impact Test Results — A series of head impacts was performed to establish the mechanical impact characteristics of the cadaver head under survivable non-skull fracturing conditions. Figs. 14 through 25 show the results of these tests with a rigid or deformable striker impacting either the front or the side of the head. Impact durations varied from 3 to 18 milliseconds, and peak loads were as high as 2,300 pounds. The test results are shown as composite sets of curves, so that an envelope of responses may be constructed.

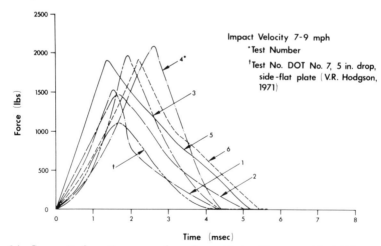

Fig. 14. Composite force-time curves for cadaver side head impacts with rigid impactors.

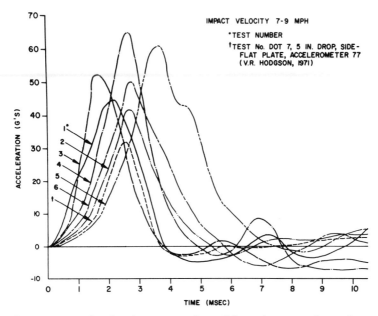

Fig. 15. Composite acceleration-time curves from S-I accelerometer for cadaver side head impacts with rigid impactor.

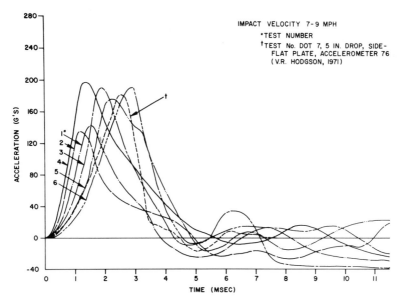

Fig. 16. Composite acceleration-time curves from L-R accelerometer for cadaver side head impacts with rigid impactor.

References pp. 109–110

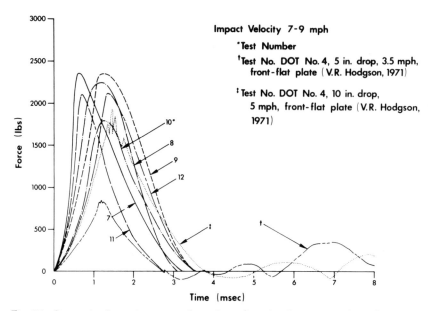

Fig. 17. Composite force-time curves for cadaver front head impacts with rigid impactor.

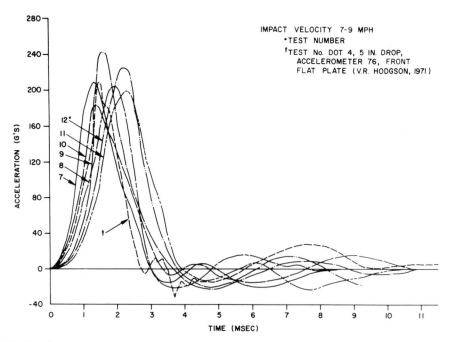

Fig. 18. Composite acceleration-time curves from A-P accelerometer for cadaver front head impacts with rigid impactor.

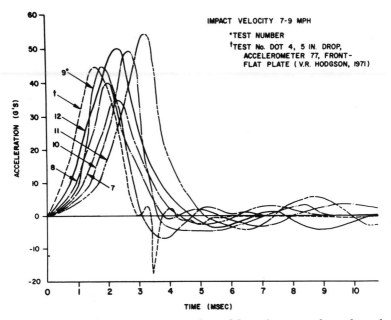

Fig. 19. Composite acceleration-time curves from S-I accelerometer for cadaver front head impacts with rigid impactor.

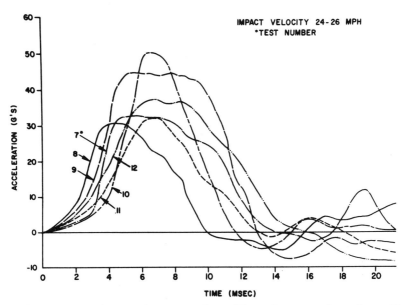

Fig. 20. Composite acceleration-time curves from S-I accelerometer for cadaver side head impacts with padded impactor.

References pp. 109–110

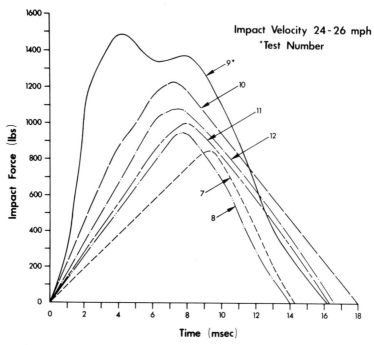

Fig. 21. Composite force-time curves for cadaver side head impacts with padded impactor.

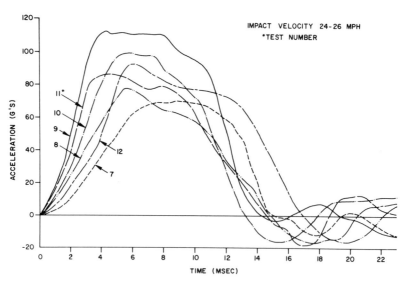

Fig. 22. Composite acceleration-time curves from L-R accelerometer for cadaver side head impacts with padded impactor.

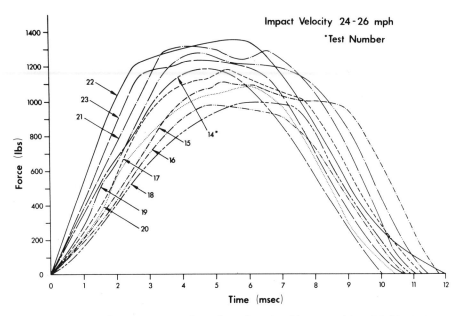

Fig. 23. Composite force-time curves for cadaver front head impacts with padded impactor.

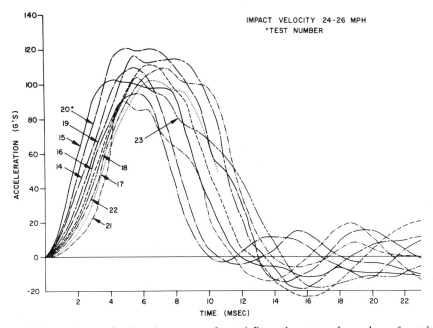

Fig. 24. Composite acceleration-time curves from A-P accelerometer for cadaver front head impacts with padded impactor.

References pp. 109–110

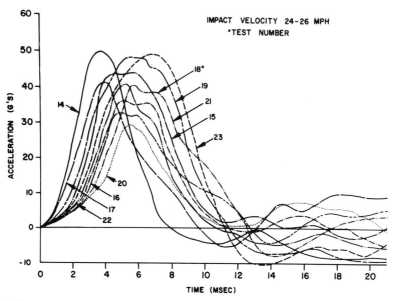

Fig. 25. Composite acceleration-time curves from S-I accelerometer for cadaver front head impacts with padded impactor.

DISCUSSION

The driving point impedance studies indicate that the mechanical response of the primate head may be approximated as a two-mass system. Injury levels for blunt impacts have been related to the compression of a spring in the simple two-mass model. The MSC model indicates decreasing tolerance with impulse duration for pulses of approximately 10 times the resonance period or less. For pulses longer than 10 times the resonance period, a quasi-static response is indicated that is unaffected by further increases in pulse duration. The prediction is quite similar to that produced by EDI, JTI and the RBM, but considerably different from the SI and HIC, which indicate that a quasi-static response is never obtained. Obviously, as with all simple models of complex phenomena, extrapolation of model predictions beyond the range of validation or to new situations is dangerous and should be done with caution. The MSC model has been developed for blunt impacts where the amount of bone and scalp in contact with the impactor approximates that loaded by the coupling clamp during the impedance tests. When the loads are applied to large sections of the head or through the neck, many of the arguments used in the model development do not apply. In addition, it is probable that the injury mechanisms change considerably with these different types of loading, and a single mechanism model, as are all the ones discussed in this paper, would be inadequate.

The impact data provides bands of responses. The variation in these bands is due primarily to the normal inherent biological differences in the cadavers. It is intended that this information form part of a crash test device performance specification that will result in a human analogue with human-like impact responses. The fresh cadaver has no muscle tone, however, and the responses reported here represent a bound on human behavior with joint stiffnesses somewhat lower than in the anesthetized human.

REFERENCES

1. J. H. McElhaney, V. L. Roberts, and J. W. Melvin, "Biomechanics of Seat Belt Design," 16th Stapp Car Crash Conference Proceedings, 1972.

2. A. K. Ommaya and A. E. Hirsch, "Protection of the Brain From Injury During Impact: Experimental Studies in the Biomechanics of Head Injury," AGARD Conference Pre-Print No. 88 on Linear Acceleration (Impact Type), 1969.

3. S. W. Alderson, "The Development of Anthropomorphic Test Dummies to Match Specific Human Responses to Accelerations and Impacts," 11th Stapp Car Crash and Field Demonstration Conference Proceedings, pp. 62-67, October 10-11, 1967.

4. J. Brinn and S. E. Staffeld, "Evaluation of Impact Test Accelerations: A Damage Index for the Head and Torso," 14th Stapp Car Crash and Field Demonstration Conference Proceedings, November 17-18, 1970, Paper 700902, pp. 188-202. New York: Society of Automotive Engineers, Inc., 1970.

5. Department of Transportation National Highway Traffic Safety Administration (49 CFR Part 571) [Docket No. 69-7; Notice 17], "Occupant Crash Protection Head Injury Criterion."

6. C. W. Gadd, "Use of a Weighted-Impulse Criterion for Estimating Injury Hazard," 10th Stapp Car Crash and Field Demonstration Conference Proceedings, Paper 660793, New York: Society of Automotive Engineers, Inc., 1966.

7. A. G. Gross, "A New Theory on the Dynamics of Brain Concussion and Brain Injury," Journal of Neurosurgery, 15, pp. 548-561, 1958.

8. E. S. Gurdjian, V. R. Hodgson, L. M. Thomas, and L. M. Patrick, "Significance of Relative Movements of Scalp, Skull, and Intracranial Contents During Impact Injury of the Head," Journal of Neurosurgery, Volume 29, pp. 70-72, 1968.

9. T. Hayashi, "Study of Intracranial Pressure Caused by Head Impact," (2nd Report), Journal of the Faculty of Engineering, University of Tokyo, Vol. XXX, No. 2, 1969.

10. V. R. Hodgson and L. M. Patrick, "Dynamic Response of the Human Cadaver Head Compared to a Simple Mathematical Model," 12th Stapp Car Crash and Field Demonstration Conference Proceedings, pp. 280-301, 1968, Society of Automotive Engineers, Inc., 1968.

11. V. R. Hodgson and L. M. Thomas, "Testing the Validity and Limitations of the Severity Index," 14th Stapp Car Crash and Field Demonstration Conference Proceedings, November 17-18, 1970, Paper 700901, pp. 169-187. New York: Society of Automotive Engineers, Inc., 1970.

12. A. H. S. Holbourn, "Mechanics of Head Injuries," Lancet, Vol. 2, pp. 438-441, 1943.

13. J. Kulowski, Crash Injuries, The Integrated Medical Aspects of Automobile Injuries and Deaths, C. C. Thomas, Springfield, Illinois, 1960.

14. S. O. Lindgren, "Studies in Head Injuries: Intracranial Pressure Pattern During Impact," Lancet, Vol. 1, pp. 1251-1253, 1964.

15. J. H. McElhaney, R. L. Stalnaker, V. L. Roberts and R. G. Snyder, "Door Crashworthiness Criteria," 15th Stapp Car Crash and Field Demonstration Conference Proceedings, Paper 710864, p. 39, New York: Society of Automotive Engineers, Inc., 1971.

16. V. L. Roberts V. R. Hodgson, and L. M. Thomas, "Fluid Pressure Gradients Caused by Impact to the Human Skull," Proceedings of the Human Factors Conference, ASME, Paper No. 66-HUF-1, 1966.

17. K. Sano, N. Nakamura, K. Hirakawa, H. Hashizume, T. Hayashi, and S. Fujii, "Mechanism and Dynamics of Closed Head Injuries," Neurologia Medico-Chirurgica, Vol. 9, pp. 21-33, 1967.

18. A. Slattenschek, and W. Tauffkirchen, "Critical Evaluation of Assessment Methods for Head Impact Applied in Appraisal of Brain Injury Hazard, In Particular in Head Impact on Windshields," International Automobile Safety Conference Compendium, 1970, Paper 700426, pp. 280-301, New York: Society of Automotive Engineers, Inc., 1968.

19. G. R. Smith, S. S. Hurite, A. J. Yanik, and C. R. Greer, "Human Volunteer Testing of GM Air Cushions," Second International Conference on Passive Restraints, SAE, Paper 720443, 1972.

20. R. L. Stalnaker, J. L. Fogle, and J. H. McElhaney, "Driving Point Impedance Characteristics of the Head," Journal of Biomechanics Vol. 4, No. 2, pp. 127-139, March 1971.

21. J. P. Stapp, "Voluntary Human Tolerance Levels," Impact Injury and Crash Protection, p. 308, Charles C. Thomas, Springfield, Ill. 1970.

22. F. Unterharnscheidt, and K. Sellier, "Mechanics and Pathomorphology of Closed Brain Injuries," Chapter 26 in Head Injury pp. 321-341. (Edited by Wm. F. Caveness, M.D., and A. Earl Walker, M.D.), J.B. Lippincott Co., Philadelphia, 1966, 589.

23. J. Versace, "A Review of the Severity Index," Ford Technical Report, No. S-71-43, November 12, 1971.

24. K. U. Zulch, "Medical Causation," The Late Effects of Head Injury, C. C. Thomas, Springfield, Illinois, 1969.

DISCUSSION

L. M. Patrick *(Wayne State University)*

The Wayne Tolerance Curve is based on uniaxial acceleration, so when one goes to a triaxial resultant acceleration, immediately the tolerance curve is too conservative.

W. Goldsmith *(University of California)*

That's an excellent piece of work, gentlemen, but I do have one question. The establishment of the maximum strain criterion and the driving-point impedance data would seem to indicate that you consider the head system as a whole, whereas if you try to compare this with some kind of pathological damage and relate it to an estimated severity index, this would seem to be a local phenomenon. The question then is does this mean that the properties of the head are reasonably uniform? In other words, can you expect damage that is fatal pretty much throughout the head if you have an index of 5 or beyond this threshold within the band that you've shown?

J. H. McElhaney

I think the answer to that question is yes, with certain qualifications. The qualifications are that the model we propose, which as we see predicts essentially the same things as many other models, is based on information from essentially rigid or almost hard impacts. Under these conditions we get a certain type of injury. We have certain injury mechanisms involved. We're starting to believe, more and more, that this mechanism is cavitation whereas at one time we thought it might be just a gross deformation in the brain; hence the development of this model. We have found, however, that this model we're talking about, and other models, do predict a number that is sort of proportional to the average strain in the brain; and we've related this in pretty much a curve matching way to our laboratory experiments. We're taking a different tack now in our modeling. We're still staying with a lumped-parameter model, but we're introducing many more parameters in an attempt to build into it the ability to consider different mechanisms of injury. We recognize particularly that as the pulse duration increases, there is a different mechanism of injury; and in fact, in the rigid striker impacts we see predominately contracoup type injuries. When we pad this striker we don't see contracoup injuries anymore. We see minute hemorrhaging in much larger areas in the brain. Both of these injury types causes some difficulty in assessment. If the injuries occur in one region of the brain our pathologist tells us it's of no consequence; but if they occur in another region of the brain or tear a major artery, for example, it's very serious. Thus looking at different directions gets to be rather complex. When we go to long pulse durations where we have a lot of angular displacement of the head or high angular velocity, we see a completely different type of injury mechanism. We see subdural hematomas and brain stem involvement. What I think we will wind up with in the future is either several injury criteria, each relating to a specific injury depending on the mechanism, or possibly one injury criterion that covers all of the injury mechanisms. We're looking at both approaches. This is a difficult problem, and its not going to be solved right away; but at the same time the automotive community is also faced with a difficult problem that they have to solve right away, and I don't have much more advice then to go slow until we learn more.

J. N. Silver *(GM Proving Ground)*

You mentioned that most of your information on response of the human head was based on the minor injury level, and that you didn't want to get into the severe injury responses. Do you assume that this mechanical response extends linearly into the injury range and then you make your cut-off where injury begins?

J. H. McElhaney

It depends on the kind of injury. If we're talking about skull fracture, then it's different all together. The response is all together different. We don't generally go beyond severe brain damage without skull fracture. Under these circumstances, if we sacrifice the animal immediately after impact, we don't find very much change. You have to wait some time for bleeding to develop. We think that there is not much structural change from that point of view.

Y. K. Liu *(Tulane University)*

In your particular model how did you arrive at the system constants; and especially, what do you mean by strain. Your model obviously can only take force and displacement, so what do you mean by strain? It is not inherent in the model itself.

J. H. McElhaney

I have to apologize for not describing these models in more detail. There are papers written on each of the models and we didn't have time to go into that; I just wanted to compare the results. The MSC model was originally developed to match driving-point impedance characteristics of the side of the head over a frequency range of nominally zero to 5,000 Hertz; and it does that. We were embarrassed naturally, at the time, to find that such a simple-minded model would match the impedance characteristics; but it does, and the constants were evaluated from the driving point impedance. We then wanted to use this model to predict injury, and we selected the displacement of the spring to represent nominally the average strain that might occur across the driving point. We looked at some data to see whether this could be used to make such a prediction. Under certain circumstances, particularly rigid impacts to Rhesus monkeys, we have a lot of information to indicate that this can be used to predict injury. We've added now, in this presentation, more information looking at different directions; and the model still seems to predict that a certain compression of this spring is associated with a certain level of injury.

HEAD MODEL FOR IMPACT TOLERANCE

V. R. HODGSON

Wayne State University, Detroit, Michigan

ABSTRACT

A human head model has been developed which may be useful in the study of impact attenuation of motor vehicle components and for possible use on an anthropomorphic dummy in vehicle crash studies. The model has a 1 piece self skinning urethane foam skull which is cast from a stiff rubber mold made by a slightly modified human skull, a silicon gel filled cranial cavity and a silicon rubber coated skin simulating cover. A solid silicon rubber rudimentary neck is held onto the head by means of a steel strand cable attached at the foramen magnum.

A triaxial accelerometer is mounted near the CG in a recess into the cranial cavity accessible through the underside of the mandible. The triaxial signals are amplified and then summed into a resultant by special circuitry for display or mathematical operation. It has not yet been shown to behave quantitatively and qualitatively as a frangible model in regard to its lacerative and skull fracture properties but preliminary results show that human linear acceleration cerebral concussion injury indices can be applied directly to model response.

INTRODUCTION

A head model is in the process of development to replace a rigid metal head form which has been used for years to evaluate the impact attenuation capabilities of protective head gear. Manufacturers of most football helmets have long been convinced that the game required a resilient liner, at least in part, for this multiple impact sport. A product liability case was brought against one of the major firms following a head injury by a player wearing one of their helmets. (1) One of the charges brought against the defendant was that they had not adopted a non-resilient liner which had been found to be superior to resilient liners by investigators using a rigid metal head form whose tests were designed primarily for racing helmets (one or at most 2 successive massive impacts) (2). Consultation experiments by the manufacturer during the trial showed that the metal head form was very different than a human in its response to impact even when protected by a helmet. They also found that the rigid form distorted the relative protective effect of resilient and non-resilient padding as compared to the human (3). Not long after the trial, several football groups banded together to fund research of which to study simulated football impacts with cadavers and to develop a head model to have a shape and acceleration response similar to a representative human head.*

MECHANICAL CONSTRUCTION

In the prototype model the design aim was for a size 7-1/4 head, which represents the most important size among adult football helmet wearers. Anthropometry of a group of 13 cadavers with near size 7-1/4 head was obtained and averaged as shown in Table 1. The head which most nearly fit the average was chosen as the pattern, after

TABLE 1

Comparison of Cadaver Head Weights and Measures with Model

	cad.	w-lb (kg)	circ-in (cm)	a-p -in- (cm)	lat -in- (cm)	I_{a-p} lb-in-sec (kg-cm-sec^2)	a-p/lat.
1.	1717	10.0 (4.6)	23.0 (58.4)	7.9 (20)	6.1 (15)	.20 (.23)	1.3
2.	1745	11.5 (5.2)	22.8 (58)	7.8 (20)	6.5 (17)	.23 (.27)	1.2
3.	1699	10.0 (4.6)	22.5 (57.2)	8.0 (20)	6.3 (16)	.20 (.23)	1.27
4.	1819	10.0 (4.6)	23.0 (58)	7.9 (20)	6.4 (16)	.20 (.23)	1.23
5.	1829	9.3 (4.2)	23.0 (58)	7.5 (19)	6.4 (16)	.19 (.22)	1.17
6.	1859	9.2 (4.2)	22.5 (57.2)	7.8 (20)	6.0 (15)	.18 (.21)	1.3
7.	154	10.6 (4.8)	22.0 (56)	7.4 (19)	6.0 (15)	.20 (.23)	1.23
8.	1890	9.7 (4.4)	22.8 (58)	7.1 (18)	6.4 (16)	.20 (.23)	1.11
9.	1905	9.2 (4.2)	22.8 (58)	7.5 (19)	6.1 (15)	.19 (.22)	1.23
10.	1912	9.7 (4.4)	22.5 (57.2)	8.0 (20)	6.1 (15)	.19 (.22)	1.31
11.	1940	10.7 (4.9)	22.0 (56)	7.4 (19)	6.5 (17)	.21 (.24)	1.14
12.	1936	10.3 (4.7)	22.8 (58)	7.6 (19)	6.3 (16)	.21 (.24)	1.19
13.	1932	10.7 (4.9)	22.5 (57.2)	7.6 (19)	6.4 (16)	.21 (.24)	1.19
SUM.		131 (60)	291 (740)	100 (254)	81.5 (204)	2.61 (3.02)	15.9
AVE.		10.0 (4.6)	22.4 (57)	7.67 (19)	6.3 (16)	.20 (.23)	1.22
MODEL		10.0 (4.6)	22.5 (57.2)	7.6 (19)	6.4 (16)	.18 (.21)	1.19

*National Operating Committee on Standards for Athletic Equipment, Inc., American College Health Association, Athletic Goods Manufacturers Association, National Athletic Trainers Association, National Collegiate Athletic Association, National Federation of State High School Athletic Associations, and National Junior College Athletic Association.

being modified by thickening several wafer thin areas primarily around the temporal, orbital, nasal and palate regions. Also some smoothing and reinforcing of thin, jagged, boney edges and recesses were performed on the base of the skull. (Compare Figs. 1a, 1b, and 2).

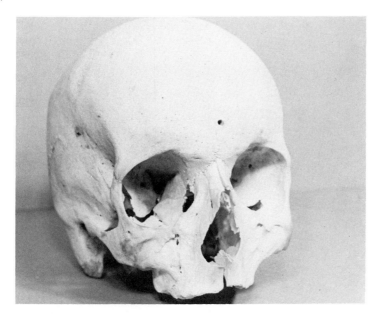

Fig. 1a. Front view of human skull used as model pattern.

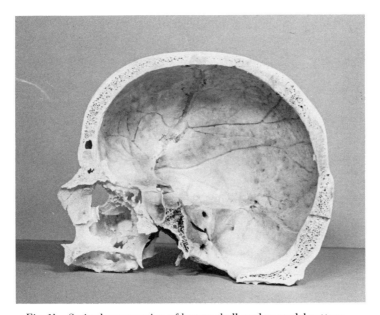

Fig. 1b. Sagittal cross-section of human skull used as model pattern.

The skull was split in the sagittal plane, built up to replace the saw cut and pinned together. A brain pattern was obtained by pouring a casting plaster into the cranial cavity and removing it after hardening (see Fig. 2). The brain pattern was modified to provide a recess at the head CG. A split silicon rubber mold was poured around the brain pattern which was supported on an armature inside a box. Similarly, silicon rubber molds were made from nasal sinus and accelerometer mounting hole patterns (not shown) as viewed in Fig. 3.

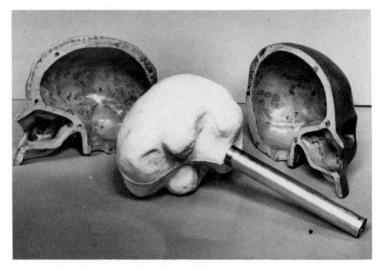

Fig. 2. Split view of human skull modified with body putty to strengthen weak spots and fill upper tooth space for use as model skull pattern. Shown also is the brain core on its armature.

Fig. 3. Brain and nasal sinus core molds.

Water soluable plaster was poured into the respective molds to form brain, sinus and accelerometer mounting hole cores mounted on armatures. The cores were suspended inside the skull mold which was housed in a structural aluminum box to minimize distortion from high casting pressures (Fig. 4). Self skinning urethane foam was poured into the void between the skull outer surface mold and the three cores. Different materials were poured into the mold and statically tested after curing for comparison with load-deflection test results on several human skulls.

Fig. 4. Model skull mold housing. Vent covers and filler holes of top.

A self-skinning urethane foam was chosen as the material which appeared to have the best handling and physical properties for the prototype head model and casting equipment. The density, strength and cross-sectional structure of the material can be varied by the amount of material poured into the mold, which is initially vented and then sealed after most of the air has escaped from the mold. The skull model is removed after a few hours and the cores are dissolved out with water and mechanical action. A view of the skull model after removal from the four piece rubber mold is shown in Fig. 5.

The modulus of elasticity of most plastics on the market, known to the author, is below that of the skull bone which is on the order of $1 - 2 \times 10^6$ (4). However it is possible to increase the density and strength of the self-skinning foam by increasing the pressure. This can be done up to a limit with a rubber mold, by increasing the amount of foam poured into the mold. It would be necessary to use external pressure in conjunction with an expensive metal mold if higher density was required than that obtained by filling the mold completely before the chemical process generates heat and causes expansion of material. Fortunately, it is possible with the available equipment to cast skulls which had load-deflection characteristics which fall within

References p. 126

the two extremes of a group of four cadavers as shown in Fig. 6. The load-deflection characteristics of both cadavers and models for lateral loading were linear up to a 1000 lbs., which is as high as the test was carried.

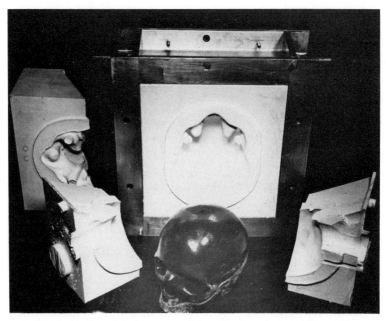

Fig. 5. Skull model after removed from 4 piece silicon rubber mold.

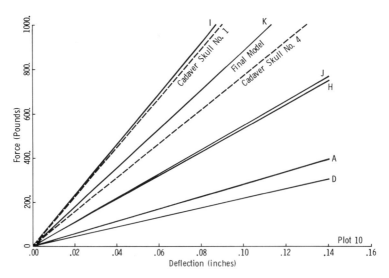

Fig. 6. Static lateral load-deflection tests of 6 skull models compared to limiting cases of 4 moist cadaver skulls.

Mandible — The cadaver mandible was used as a pattern to make a rubber mold for the model mandible similar to those shown for making the cores in Figure 3. The mandible pattern was built up with body putty as had been the cadaver skull to fill in the teeth space of the cadaver which was without teeth. Model mandibles are cast using self-skinning urethane and glued with silicon rubber cement along the bite plane and at the heads of the mandible.

Soft Tissue — The model skull and mandible assembly were coated with beeswax to dimensions obtained from the cadaver head before it was dissected. This assemblage was used as a pattern around which was poured a rubber mold of the outer surface of the head. To form a rubber surface model the beeswax pattern is replaced with a model skull-mandible assembly suspended on a core inside the head mold and silicon rubber is poured into the void between the skull-mandible model and the head mold. This forms a simulated superficial soft tissue over the head model as in a drop carriage facility in Fig. 7.

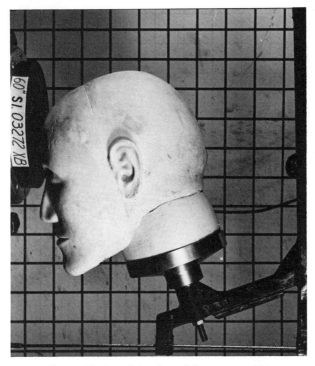

Fig. 7. Head model in frontal drop test position.

Brain — The cranial cavity of the model is filled with a silicon gel. It has the feel and slump characteristics which are somewhat similar to those observed in fresh animal brains but no physical measurements were made to simulate the brain tissue accurately.

References p. 126

Neck — The model is attached to a test carriage or body of an anthromorphic dummy through a cable attached at the foramen magnum is shown in Fig. 8. A split washer is used inside the foramen magnum to provide anchorage for the bolt swaged to the upper end of the cable and a bolt is swaged to the lower end of the cable for attachment to a test fixture. The properties of the silicon rubber can be coarsely adjusted by the amount of hardener used in vulcanizing and it was originally mixed to have a penetrometer resistence similar to the data supplied by Gadd et al (5). However it has been observed that the rubber becomes more firm with age.

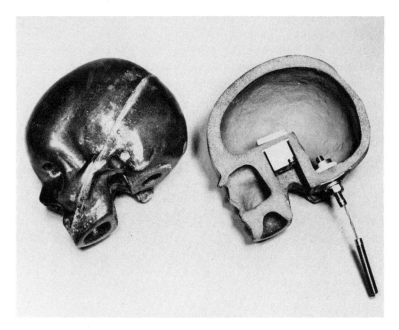

Fig. 8. Skull model and cutaway view showing method of neck attachment and triaxial accelerometer mount.

INSTRUMENTATION

Accelerometer Mount — An accelerometer is mounted on a magnesium plate cast into the skull in the recess as shown in cut away view of the skull model in Fig. 8. The cable can be brought out to an access hole on the under side of the mandible, or in cases where damage is anticipated in this area the cable can be brought out through the rear of the neck.

Circuitry — A schematic of the circuitry used to condition the signals from a dc triaxial accelerometer is shown in Fig. 9. The three signals are amplified by a dc conditioner followed by a summing amplifier which performs the function of combining the signals into the absolute value of the result ($|\sqrt{A_x^2 + A_y^2 + A_z^2}|$). The resultant signal is then stored on tape for later display or digitization for computer operation.

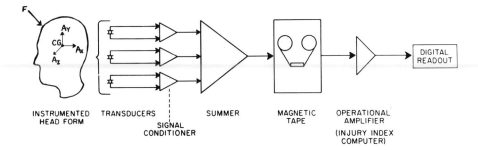

Fig. 9. Signal conditioning for measuring head model impact accelerations.

IMPACT RESPONSE

Comparison to a Rigid Head Form — A comparison was made between the rigid metal head form (MHF) presently used to test protective head gear and the head model in the series of drop tests with a group of helmets comprising the major types used in football. Most helmets had similar shells, but various types of liner materials including non-resilient, resilient, sling and hydraulic. One comparison of the oscillographic records for the MHF and the model as a result of a 48″ drop on the front of the helmet onto a simulated synthetic turf, is shown in Figs. 10 and 11. The acceleration response is shown in the upper trace and the force in the lower trace of each record. As is seen in Fig. 10, the MHF has experienced higher force and acceleration levels than those measured for the model because the MHF has bottomed the liner material. In the series of frontal impacts to compare the responses of the MHF and the model for frontal impact on a basis of Severity Index, the MHF response was higher in every case ranging from 20 to over 150% for 7 different types of helmet liners.

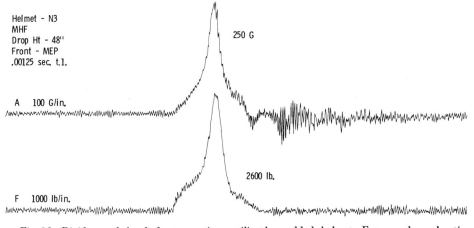

Fig. 10. Rigid metal head form wearing resiliently padded helmet. Force and acceleration response (upper trace) impact drop test oscillograms.

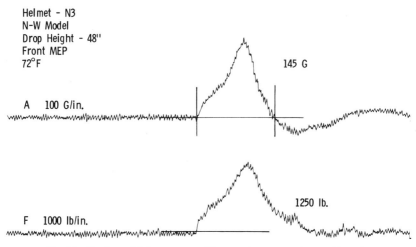

Helmet - N3
N-W Model
Drop Height - 48"
Front MEP
72°F

145 G

A   100 G/in.

1250 lb.

F   1000 lb/in.

Fig. 11. Head model drop test oscillogram for comparison to Fig. 10.

Head Model Response Compared to Cadaver — A cadaver head was decapitated and placed in a carriage as shown in Fig. 12. The head was held to the carriage by means of fibreglas tape as compared to the neck cable shown in Fig. 7 for the head model. The tests were conducted with a variety of conditions and unfortunately the only test in which the cadaver head and model can so far be compared, produced a fracture in the cadaver as shown in Fig. 13 on the acceleration trace. The oscillograph record for the model is shown in Fig. 14. The head model and carriage weight 13 lbs. as compared to 17 lbs. for the cadaver head weight and carriage. Consequently if the

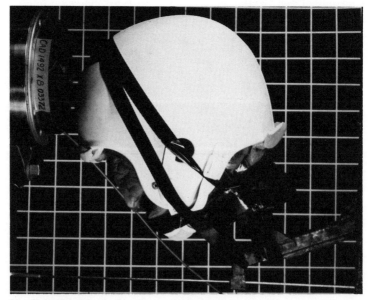

Fig. 12. Cadaver head taped to drop test carriage for comparison to model response.

cadaver head force at any time is multiplied by a factor of 13/17 it is seen that the two pulses have close correspondence, i.e. the adjusted peak value for the cadaver head would be 1450 lbs. as compared to 1480 measured for the head model. These tests were conducted with a small accelerometer (Endevco 2222B) attached to similar points on the skull of the model and cadaver on the forehead near the impact site under the helmet.

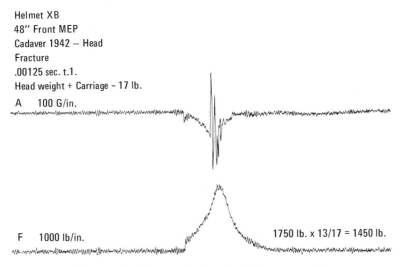

Fig. 13. Oscillogram of cadaver head drop test results.

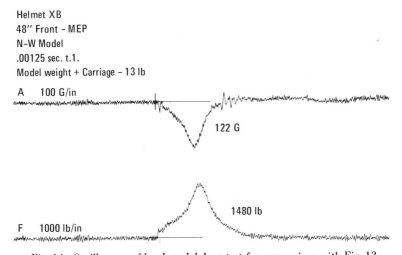

Fig. 14. Oscillogram of head model drop test for comparison with Fig. 13.

It is of particular interest to determine the relationship of response of the CG resultant linear acceleration of the model to that of a representative group of cadavers, since automotive head impact safety standards are based upon CG resultant accelerations. Some data in the cadaver has been developed at Wayne State University (6) in which CG resultant acceleration was deduced mathematically using the Mertz (7) rigid body computer program and based upon biaxial acceleration measurements made on the side of the head for plane motion, frontal impact against rigid surfaces. Other data is also being generated at HSRI as part of the AMA dummy program (8). Tests at both institutions have been conducted with intact cadavers. It is planned to compare the cadaver results to model head only and model attached to several available dummy bodies in the near future as more models and the instrumentation diagrammed in Fig. 9 becomes available.

Repeatability — To determine the repeatability and durability of the head model components, including the neck attachment, a series of 24 drop tests were conducted in a vertical guide wire drop test apparatus as partially shown in Fig. 8. The tests were conducted on the front from a height of 60" onto an artificial turf simulating surface using a hydraulic helmet to protect the model. The statistical variation in response to the helmet is shown in Table 2, showing a mean acceleration of 142 G's with a

TABLE 2

Statistical Variation in 24 Drops Using N-W Model
Helmet XB, Front, MEP — 60 Inch

Drop No.	Accel. -G-	M-G	(M-G)^2	Drop No.	Accel. -G-	M-G	(M-G)^2
1	154	12	144	14	143	1	1
2	144	2	4	15	145	3	9
3	145	3	9	16	140	2	4
4	153	11	131	17	145	3	9
5	143	1	1	18	140	2	4
6	133	9	81	19	140	2	4
7	138	4	16	20	135	7	49
8	132	10	100	21	135	7	49
9	132	10	100	22	150	8	64
10	132	10	100	23	155	13	169
11	132	10	100	24	145	3	9
12	145	3	9				
13	145	3	9	Σ	3401		1175

$$\text{Mean (M)} = \frac{3401}{} = 142$$

$$\text{Stand. Dev.} = \pm \sqrt{\frac{\Sigma \,(M\text{-}G)^2}{n\text{-}1}} \;\; = \;\; \pm \sqrt{\frac{1175}{23}} = \pm 7 \, G \;\; (\pm 5\%)$$

NOTE: For 9 frontal drops of MHF onto MEP over the period of the several days of the tests, the acceleration range was 295-315G, mean 301 G, Standard Deviation + 8 G or + 2.7% accuracy.

standard deviation plus and minus 7 G's (± 7%). The acceleration measurements were made at a point adjacent to the blow site on the front with a small accelerometer (Endevco 2222B) glued to the surface of the model skull.

Reproducability — The reproducibility of the response of several models produced by the model mold has been compared only statically. The load-deflection curve for final model K is the mean of 3 model load-deflection characteristics which were almost identical. The 3 models are being instrumented and fitted into identical drop facilities for use in round robin tests which have been devised to obtain an objective evaluation of the model by 3 different laboratories.

Mechanical Impedance Tests — Driving point mechanical impedance tests at various points on the model skull and are currently underway at WSU and by R. R. Bouche, Endevco Corporation, Pasadena, California, for comparison to cadaver and other models.

INJURY INDICES

Impact response results to date are limited to a comparatively few comparisons against human cadaver head detached from the body using accelerations on the surface of the skull adjacent to the impact site under the helmet on the front, but give encouragement that human injury indices may be applied directly to the model. A comparison must be made between resultant CG accelerations and with the model attached to an anthropomorphic dummy body before the comparison to intact cadaver data can be more conclusive. It is also not yet known how fragile the model is when used without protective head gear. With the modifications shown in Fig. 1 and 2 it is probable that model skull fractures will not be qualitatively the same as in the cadaver. Quantitative force levels to cause fracture also have not been checked out, although it has been found that whereas the cadaver head breaks down gradually and eventually fractures at threshold concussion acceleration levels even though the head is protected by a helmet, the model has been dropped over 100 times at this level without fracture. It is possible that with the use of close material controls and a metal mold, skulls of various strengths from realistically frangible to much higher strength can be produced depending upon the degree of human-like response accuracy desired.

The silicon rubber used to simulate the outer tissue can have a marked effect on the head response, particularly for more rigid impact surfaces, and it has been found to become more firm with age. This will also affect its lacerating qualities, and consequently more investigation of the synthetic soft tissue used on the model is essential if it is to be used to gage concussion and possible fracture and lacerative hazards.

CONCLUSION

It is possible to produce a head model from synthetic materials which results of preliminary frontal tests using protective head gear show to have similar to human head impact response yet which is less frangible and more repeatable than the human head. Further comparative tests against the human head and investigation of the frangible, repeatable and reproducable qualities are necessary to validate use of the model for assessing concussion hazard in automotive impact testing.

REFERENCES

1. E. V. Pelton, Rawlings Sporting Goods Company, Sacramento, California, 1/70.
2. G. D. Snively, et al, "Design of Football Helmets." National Conference on the Medical Aspects of Sport, First Proceedings, 1959.
3. V. R. Hodgson, L. M. Thomas and P. Prasad, Testing the Validity and Limitation of the Severity Index," Proceedings of Fourteenth Stapp Car Crash Conference, p. 169, paper 700901. New York: Society of Automotive Engineers, Inc., 1970.
4. F. G. Evans and H. R. Lissner: "Tensile and Compressive Strength of Human Parietal Bone." Journal of Applied Physiology, Vol. 10 (1957), pp. 493-497.
5. C. W. Gadd, et al: "Tolerance and Properties of Superficial Soft Tissues In Situ." Proceedings of Fourteenth Stapp Car Crash Conference, p. 356, paper 700910. New York: Society of Automotive Engineers, Inc., 1970.
6. V. R. Hodgson and L. M. Thomas: "Comparison of Head Acceleration Injury Indices in Cadaver Skull Fracture." Proceedings of Fifteenth Stapp Car Crash Conference p. 190, paper 710854. New York: Society of Automotive Engineers, Inc., 1970.
7. H. J. Mertz, Jr.: "The Kinematics and Kinetics of Whiplash." Ph.D. dissertation, Wayne State University, 1967.
8. Oral Report Relative to Progress on "Crash Test Device Performance Requirements." Program Sponsored by Automotive Manufacturers Association, August, 1972. Highway Safety Research Institute, University of Michigan.

ACKNOWLEDGEMENTS

The patterns, molds and casting technique employed to produce the head model are the creative work of Matthew W. Mason, M.A., Department of Neurosurgery, Wayne State University, under a contract funded by the National Operating Committee on Standards in Athletic Equipment, entitled: "Biomechanical Study of Football Head Impacts Using a Human Cadaver."

DISCUSSION

J. N. Silver (GM Proving Grounds)

I read recently in a program at the University of Michigan football game, that there is a fellow at Northwestern University that has been carrying on some volunteer studies, with respect to football type injuries, with instrumented players. From the looks of this particular write-up, there might have been some information that would be usable by yourself or by this group. Are you familiar with this? If so, what kinds of information is he developing?

V. R. Hodgson

Well, I wish I could say it was good information. Actually, I think its very premature. He has come out with a recent paper which I think indicates that the data they have developed so far is just not very good. For example, in some tests they had the player instrumented with three accelerometers and some EEG electrodes. He said that the probable human tolerance for concussion is in the range of 188 g's to 278 g's for 300 to 400 milliseconds. Even giving him the benefit that the average acceleration was half of those values, the minimum velocity change is 90 feet per second and the maximum is 180. I think its laudable work, but he should look into it a little better. It's a very difficult problem to measure impact accelerations accurately on the head of a live person.

L. M. Patrick *(Wayne State University)*

Didn't you do some of this type of work, Voigt?

V. R. Hodgson

We did and we backed off because the coaches get a little tired of us coming up with information thats not really benefiting them directly. It sort of interferes with their game. They are apathetic about developing human tolerance data. They're out to win football games.

C. L. Ewing *(Naval Medical Department)*

What was the source of the mass distribution data that were used to design the model, and do the various adjustments of the headform permit the center of gravity of the head-helmet mass to be maintained in a constant relationship to the center of the striker plate?

V. R. Hodgson

It is not intended to maintain a constant relationship between head-helmet center of gravity and the center of the striker plate for the purpose of testing helmets at various locations. However if it were so desired for other applications the line of action could be maintained through the CG of head-helmet mass.

The model was designed to have the mass distribution of the average of 13 size 7-1/4 cadaver heads which we have measured.

R. A. Wilson *(GM Proving Ground)*

The accelerometer mounting looked like just a drop-in sort of a thing. Wasn't something else done there?

V. R. Hodgson

In the movie we're just dropping them in there. In the photograph that you have in your paper, we had it mounted on a magnesium plate, as well as a backup plate on the opposite side of the skull surface and it is anchored into the plates with a couple of screws. Presently, we have this designed for an Endevco aluminum triaxial accelerometer. These could get smaller and could make the headform a little more realistic.

S. H. Backaitis (NHTSA, Department of Transportation)

I would like to know, how are you going to correlate the acceleration values measured on a cadaver and the values measured on a dummy? It looks like the two accelerometer mountings are entirely different. One is internal, the other is external.

V. R. Hodgson

The way we have done this in the case of plane motion is to enable calculation of the cg acceleration of the cadaver by using externally mounted biaxial accelerometers to define the plane motion of the head. We then use Dr. Mertz's rigid-body mechanics program to compute the cg acceleration.

H. E. VonGierke (Aerospace Medical Research Laboratory)

With all the attention paid to modeling the skull and head as such, you didn't pay much attention to the neck; or at least it was not mentioned. In your actual tests, the mechanical properties of the neck are very important, aren't they?

V. R. Hodgson

I think neck properties get more and more important as the duration of the pulse gets longer. In the past these helmets have been tested in an environment where the duration of the pulse is in the range of 7 to 10 milliseconds in which the neck rigidity appears to play a minor role. As we go along we'll attempt to evaluate the effect of the neck. For now, we feel we might want to take the approach of comparing a very rigid neck to a very flaccid neck. As necks are developed by such people as Mertz, we'll incorporate them. In the meantime, we already see a response which is similar to a decapitated head falling on the same carriage, so I think its at least an improvement on the units that already existed. I don't mean to say that we're there yet. I'm sure the neck is going to be an important factor under other conditions we intend to simulate because necks are being injured in football.

H. E. VonGierke

But for the time being you assume the neck to be completely stiff compared to the skull, so the components used in your tests are on the stiff side.

V. R. Hodgson

Yes, the neck we've used in on the stiff side.

A BASIS FOR CRASH DUMMY SKULL
AND HEAD GEOMETRY

R. P. HUBBARD and D. G. McLEOD

General Motors Research Laboratories, Warren, Michigan

ABSTRACT

As an essential step toward improving the repeatability and reproducibility of crash dummy head response to impact events, the geometric configuration of dummy heads must be more completely defined. If these dummy heads are to be used for assessment of human head injury hazard, then their geometric characteristics should be based on human anthropometry. Measurements of a large number of human heads have been previously published, but these measurements alone are not sufficient for location of landmarks and determination of contours of the human head. Also previously published are measurements of a small number of human skulls which are sufficient for location of significant skull landmarks and construction of a skull model which represents an average American male skull. The location of skull landmarks and the construction and configuration of this skull model are documented in this report. By adding reasonable skin and flesh thicknesses to the skull model, a second model was constructed which represents the external configuration of the human head and not only agrees with head dimensions from a large number of subjects but also correctly locates the features of human head. These features include anatomically correct headform coordinate axes, head-neck articulation, cranial contours, and facial structure. In using the information presented here, it is important to realize that duplication of human structural geometry in dummy head design is not sufficient to insure similarity of human and dummy responses.

INTRODUCTION

The geometric requirements for 50th percentile male crash dummy heads specified in S.A.E. Recommended Practice J963 (1) are based on data from a study of U.S. Air Force personnel by Hertzberg, et al. (2). From a second study of this data by

References p. 146

Churchill and Truett (3), it is apparent that the nominal values of the head dimensions specified by S.A.E. are the 50th percentile values for the Air Force sample and the specified tolerances insure that these dummy head dimensions are between the 25th and 75th percentile values of the Air Force sample. The Churchill and Truett study documents a generally poor correlation between the dimensions of the head and face with the correlation coefficients between head circumference and length of 0.67 classified as "high" and between circumference and breadth of 0.54 classified as "moderate" but close the "high" correlation group. These correlations were among the best in the study. The correlation coefficient between head length and breadth was 0.12, classified as "slight." The low degree of correlation between the dimensions of the head and face indicates that no single geometric shape can be constructed which is highly representative of a large segment of the human population. Yet, head structures are designed, fabricated, and placed on "50th percentile" crash test dummies.

The heads of crash dummies commonly used for automotive restraint system evaluation all comply with the S.A.E. Recommended Practice J963. Midsagittal sections of two of these heads are shown in Fig. 1 with their centers of gravity as specified by S.A.E. J963 placed coincidently and their base planes horizontal. Even

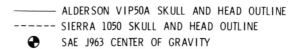

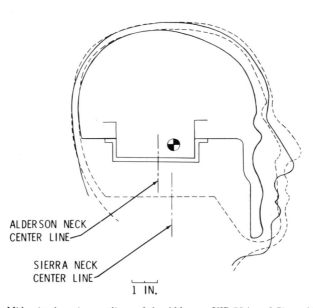

Fig. 1. Midsagittal section outlines of the Alderson VIP 50A and Sierra 1050 dummy head.

though these heads comply with the geometric requirements of S.A.E. J963, their configurations are markedly different with specific differences in facial contours and head-neck attachment location. These differences indicate that the geometric requirements specified by S.A.E. J963 are not sufficient information for crash dummy head design configurations. If crash dummy heads from the same or different builders are to give reproducible and repeatable response to impact environments, a more complete geometric description of required dummy configuration is necessary. If these dummy heads are to yield responses which are meaningful bases for human injury assessments, the geometric description for dummy heads must be based on human anthropometry. A highly detailed geometric description of human head geometry may not be academically defensible in light of the findings of Churchill and Truett (3), but such a description is required in the interest of improving dummy design and usage.

The objective of this paper is to document a study of human anthropometric data which provides a geometric foundation for 50th percentile crash dummy head design. The most complete set of external measurements of the human head taken on the largest sample currently available is documented by Hertzberg, et al. (2). Even though this set of external human head measurements is the most extensive available, it is not sufficient for the construction of a head model for which the locations of the anthropometric landmarks are completely determined. For this reason, the approach taken in this study was to use the skull measurement data given by Byars, et al. (4), to determine skull landmark locations, and construct a master skull geometry model which is representative of the 50th percentile American adult male. Using this skull model as a basis, reasonable skin and flesh thicknesses were added to construct a master head geometry model which complies with the head measurement data reported by Hertzberg, et al. (2).

GEOMETRIC DETERMINATION OF SKULL LANDMARK LOCATIONS

As part of a program directed toward the construction of a human head model, 125 measurements were taken by Byars, et al. (4) on 16 Caucasian male adult skulls from the Terry collection at the Smithsonian Institution. Reported in that paper are mean values for many of the measurements and nearly a complete set of measurements taken from skull Number 533 of the Terry collection which was selected by the investigators as the "average skull" for an adult Caucasian male. For both sets of measurements, the relative locations of cranial landmarks can be determined through graphical and analytic geometry. Definitions of head and skull landmarks and other anthropometric terms are given in the Appendix.

From the information given by Byars, et al. (4), it was possible to graphically determine the locations of the skeletal landmarks which lie on the midsaggital plane for both the Terry skull Number 533 and the average of the Caucasian skulls. This graphical construction was accomplished by scribing arcs with radii equal to the distances between the various landmarks. Many of the landmarks were redundantly determined with virtually coincident arc intersections in most cases and arc intersections within a 0.1 in. radius in all cases. For a few of the chords in both the Terry 533 and Caucasian mean measurement sets, the angles of these chords are given relative to the Frankfort plane. From these angles, the orientation of the midsagittal section and landmarks relative to the Frankfort plane can be determined. The results of these graphical constructions are shown in Fig. 2.

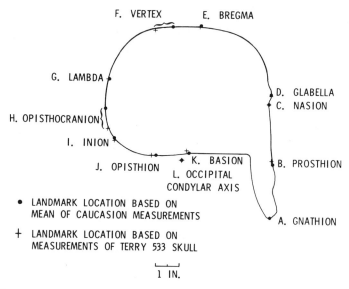

Fig. 2. Midsagittal landmark locations based on data from Byars, et al. (4).

With the midsagittal sections oriented so that their Frankfort planes are horizontal (Fig. 2), the coordinates of the skull landmarks can be determined. The coordinate system used is shown in Fig. 3 and is orthogonal with the x axis oriented horizontally and directed from posterior to anterior, the y axis oriented laterally and directed from right to left, and the z axis oriented vertically directed from inferior to superior. The origin of this coordinate system is located at the skull nasion. The coordinates of the skull landmarks lying on the midsagittal plane determined from both the Caucasian mean data and the Terry 533 data are given in Table 1. From these two sets of landmark coordinates, a set of master skull coordinates for landmarks on the midsagittal plane have been determined and are also listed in Table 1. Where the two coordinate sets differ, the Caucasian mean locations were given priority. The occipital condylar axis was not included as a landmark in the Byars study but has been located by the authors relative to the opisthion and the basion and has been included in Table 1. Except for the vertex and opisthocranion the locations of the corresponding midsagittal landmarks for the Caucasian mean and Terry 533 data are virtually coincident. The vertex and opisthocranion are points of tangency of the skull with horizontal and vertical planes, respectively, and they are difficult to locate precisely, because their locations vary greatly with small changes in skull contour and orientation.

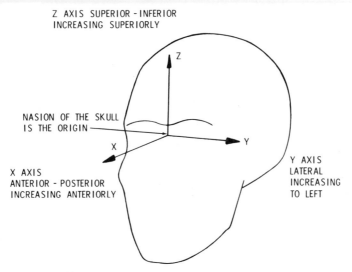

Fig. 3. Coordinate system for head geometry.

TABLE 1

Midsagittal Skull Landmark Locations

	Landmark Coordinates for Terry Skull No. 533 (in.)		Landmark Coordinates for Caucasian Mean (in.)		Master Skull Landmark Coordinates	
	x	z	x	z	x	z
A. Gnathion			0.10	-4.80	0.10	-4.80
B. Prosthion	0.15	-2.40	0.20	-2.55	0.20	-2.50
C. Nasion	0.00	0.00	0.00	0.00	0.00	0.00
D. Glabella	0.10	0.35	0.05	0.40	0.10	0.40
E. Bregma	-3.00	3.25	-2.95	3.30	-3.00	3.25
F. Vertex	-4.90	3.10	-4.15	3.25	-4.15	3.25
G. Lambda	-6.90	1.00	-6.85	1.00	-6.90	1.00
H. Opisthocranion	-6.90	-1.10	-7.00	-0.25	-7.00	-0.25
I. Inion	-6.60	-1.60	-6.60	-1.50	-6.60	-1.50
J. Opisthion	-5.00	-2.20	-4.80	-2.20	-4.90	-2.20
K. Basion	-3.50	-2.00	-3.40	-2.10	-3.45	-2.50
L. Occiputal Condylar Axis	-3.70	-2.40	-3.70	-2.40	-3.70	-2.40

TABLE 2

Off-Midsagittal Plane Landmark Locations

Landmark		Landmark Coordinates for Terry Skull No. 533						Landmark Coordinates for Caucasian Mean			Master Skull Landmark Coordinates		
		Three radii location			Two radii location			Two radii location					
		x	y	z	x	y	z	x	y	z	x	y	z
1. Gonion	left												
	average										-2.35	+1.80	-3.90
	right							-2.34	1.82	-3.92			
2. Porion	left	-3.39	2.23	-1.44	-3.39	2.38	-1.33						
	average	-3.41	+2.31	-1.36	-3.41	+2.38	-1.30				-3.40	+2.35	-1.30
	right	-3.43	-2.38	-1.27	-3.43	-2.38	-1.27						
3. Zygion	left	-2.19	2.48	-1.56	-2.16	2.66	-1.46	-1.99	2.55	-1.25			
	average	-2.14	+2.57	-1.48	-2.12	+2.66	-1.45				-2.10	+2.55	-1.35
	right	-2.08	-2.65	-1.40	-2.08	-2.66	-1.43						
4. Zygomaxillare	left	-0.83	1.32	-2.07	-0.78	-1.75	-1.88	-0.80	1.75	-2.01			
	average	-0.78	+1.35	-2.06	-0.73	+1.75	-1.88				-0.75	+1.75	-1.95
	right	-0.72	-1.38	-2.05	-0.67	-1.75	-1.88						
5. Orbitale	left	-0.46	1.20	-1.33	-0.36	1.45	-1.11						
	average	-0.45	+1.27	-1.30	-0.38	+1.45	-1.15				-0.40	+1.45	-1.30
	right	-.043	-1.33	-1.27	-0.39	-1.45	-1.19						
6. Frontomalare Orbitale	left	-0.59	1.90	-0.27	-0.59	1.89	-0.29						
	average	-0.59	+1.90	-0.24	-0.59	+1.89	-0.26		+1.89		-0.60	+1.90	-0.25
	right	-0.58	-1.90	-0.21	-0.58	-1.89	-0.23						
7. Frontotemporale	left	-0.89	1.96	0.57	-0.91	1.93	0.57						
	average	-0.89	+1.94	0.70	-0.89	+1.93	0.69		+1.88		-0.90	+1.90	0.70
	right	-0.88	-1.92	0.82	-0.88	-1.93	0.81						
8. End of the Coronal Suture	left	-1.79	2.14	0.22	-1.74	2.18	0.30	-1.55	2.18	0.59			
	average	-1.80	+2.17	0.30	-1.79	+2.18	0.33				-1.65	+2.20	0.45
	right	-1.80	-2.20	0.37	-1.83	-2.18	0.35						
9. Stephanion	left	-2.52	2.06	2.10	-2.53	2.05	2.09	-2.40	2.15	2.01			
	average	-2.45	+2.10	2.14	-2.53	+2.05	2.10				-2.50	+2.10	2.10
	right	-2.37	-2.13	2.18	-2.52	-2.05	2.11						
10. Frontal Eminence	left	-0.56	1.11	1.69	-0.59	1.27	1.62	-0.50	1.04	1.53			
	average	-0.52	+1.11	1.73	-0.54	+1.18	1.70				-0.50	+1.10	1.70
	right	-0.48	-1.11	1.77	-0.48	-1.09	1.78						
11. Euryon	left	-3.94	2.74	0.96	-3.65	2.58	0.74	-3.97	2.76	1.02			
	average	-3.96	+2.83	0.98	-3.60	+2.58	0.65				-4.00	+2.75	1.00
	right	-3.98	-2.86	1.00	-3.54	-2.58	0.56						
12. Asterion	left	-5.49	2.24	-0.92	-5.46	2.21	-0.89	-5.21	2.24	-0.99			
	average	-5.51	+2.21	-0.91	-5.48	+2.21	-0.87	-5.23	+2.24	-1.02	-5.30	+2.25	-1.00
	right	-5.53	-2.18	-0.89	-5.49	-2.21	-0.85	-5.25	-2.24	-1.05			
13. Mastoidale	left	-3.77	2.11	-2.22	-3.77	2.11	-2.22						
	average	-3.78	+2.09	-2.26	-3.80	+2.11	-2.25		+2.03		-3.80	+2.05	-2.25
	right	-3.78	-2.07	-2.29	-3.83	-2.11	-2.28						

In the Byars study (4), chord measurements were taken from skull landmarks on the midsagittal plane to landmarks off the midsagittal plane. Width measurements were also taken from landmarks off the midsagittal plane to the corresponding landmarks on the opposite sides of the heads. Where sufficient data was reported, off-midsagittal plane landmarks were located by the intersection of three spheres whose centers were the on-plane landmarks of known location and whose radii were reported chord lengths from the respective on-plane landmarks to the off-plane landmark being located. Sufficient data was available for location of all the off-plane landmarks on the Terry 533 skull, except the gonions, by this three-radii technique. The results are given in Table 2. The mean values of the Caucasian measurements were not sufficient for any of the landmarks to be located by this technique.

As a second approach, the off-plane landmark in question was located by the intersection of a plane parallel to the midsagittal plane and a circle generated by the intersection of two spheres whose centers were on-plane landmarks of known location and whose radii were distances from the respective on-plane landmarks to the off-plane landmark being located. The distances from the midsagittal plane to off-midsagittal plane landmarks were taken as half the respective widths thus assuming the skulls measured were symmetrical about the midsagittal plane. Sufficient data was reported by Byars so that this two-radii technique could be used to locate eight of the off-plane landmarks from the mean of the Caucasian measurements and all of the Terry 533 landmarks for which the three-radii technique was used. The results of the two-radii landmark location are shown in Table 2.

Also shown in Table 2 are master skull coordinates for landmarks off the midsagittal plane. These coordinates were based on the landmark coordinates for the mean values of the Caucasian skull measurements and the measurements of the Terry 533 skull. Priority for selection of the master coordinates was given to the Caucasian mean results in the few instances that significant conflicts existed between the results of the two data sets or location techniques.

CONSTRUCTION OF THE GMR 50TH PERCENTILE
SKULL GEOMETRY MODEL

Using the master skull landmark locations as a basis (Tables 1 and 2), a skull geometry model was constructed to demonstrate that these landmark locations lead to a realistic skull configuration and provide a format for addition of skull features not completely defined by the landmarks. A wooden form of roughly the correct shape served as a starting point. To this form, sculpturing clay and polyester resin putty were added resulting in a shape which complied closely with the master skull landmark locations and provided realistic detail. This form was used to cast a Silastic E silicone rubber mold. A Hysol aluminum filled expoxy skull form was then cast in this mold and the skull form was further refined in shape using only polyester resin putty. The resulting skull form was dimensionally stable, complied very closely with the master skull landmark locations and incorporated many realistic features of the human skull. This form was selected as the basic configuration for the master skull geometry model.

References p. 146

Measurements of the skull model were made by attaching it to an angle plate and orienting it so that any one of the three coordinate axes was perpendicular to the measurement table surface. A machinist's height gage was used to measure the landmark locations on the model relative to the nasion (Fig. 4). This technique was laborious but proved to be accurate and repeatable.

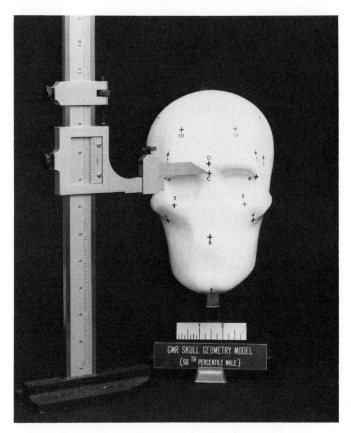

Fig. 4. Model measurement technique.

The GMR 50th percentile male skull geometry model is shown in Figs. 5, 6, and 7 with the landmark locations marked on the model surface with crossed "v" shaped grooves and labeled with letters for landmarks on the midsagittal plane and numbers for the off-midsagittal plane landmarks. These letters and numbers correspond to the landmarks listed in Tables 1 and 2, respectively, and to the definitions in the Appendix. There are two exceptions to this landmark locating scheme. The base of the occipital region of the skull model is parallel to the Frankfort plane at the level of the opisthion. The basion is located 0.15 in. upward from the marked location on the skull model base plane, and the occipital condylar axis lies 0.20 in. below the marking line on the skull model base plane.

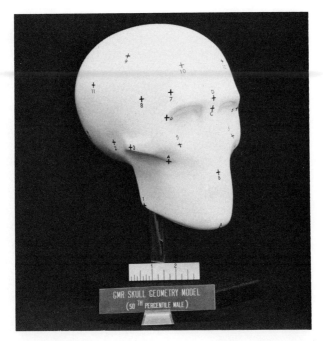

Fig. 5. GMR 50th percentile skull geometry model — oblique view.

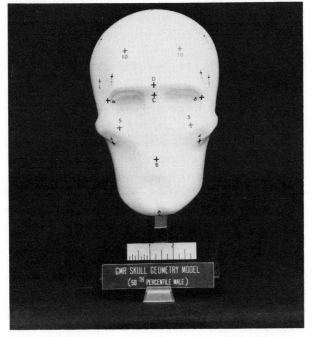

Fig. 6. GMR 50th percentile skull geometry model — front view.

References p. 146

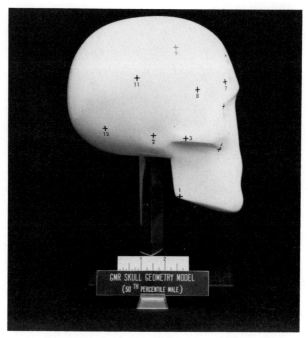

Fig. 7. GMR 50th percentile skull geometry model — side view.

The actual locations of the landmarks as they are marked on the master skull geometry model were measured and listed in Table 3. When these values are compared with the master skull landmark coordinates, it is apparent that the two sets of landmark locations are in close agreement. The skull geometry model is an accurate three-dimensional representation of the master skull landmark coordinates.

The skull model was also measured in accordance with the measurements originally taken by Byars, et al. (4). These measurements not only included the chord lengths which used for the landmark location as described above, but also included arc lengths taken between landmarks over the surface of the skulls. The agreement between the mean of the measurements on the Caucasian skulls by Byars and the corresponding measurements on the skull geometry model were within 0.05 in. in all cases. The agreement in arc lengths indicates that the master skull landmark locations is an accurate representation of the data from the Byars study. Compliance with the mean of the Caucasian arc measurements of Byars, et al. (4), between the landmarks on the calvaria of the skull model, indicates that the contours of the model over the calvaria are truly representative of human contours in this region. The details of the facial skeletal features are less well described by the landmarks and measurements between them. These facial details were added to the skull model so that the landmarks were correctly located and the contours between the landmarks were representative of human structure. The protrusions which represent the zygomatic arches, the margins of the eye orbits and the treatment of the maxilla-mandible region are the features which were subjectively considered to be important to the representation of the human skull. These features were modeled with as much accuracy to human structure as possible. The details of the eye sockets, nasal bones, and teeth were considered to be less important and were not incorporated into the model with significant fidelity.

TABLE 3

Landmark Locations on the GMR
50th Percentile Skull Geometry Model

Midsagittal Landmarks	Location (in.)	
	x	z
A. Gnathion	0.10	-4.77
B. Prosthion	0.22	-2.50
C. Nasion	0.00	0.00
D. Glabella	0.10	0.40
E. Bregma	-3.00	3.23
F. Vertex	-4.15	3.23
G. Lambda	-6.88	1.00
H. Opisthocranion	-7.00	-0.25
I. Inion	-6.59	-1.50
J. Opisthion	-4.90	-2.20
K. Basion	-3.45	(0.15 above the model base plane)
L. Occipital Condylar Axis	-3.70	(0.20 below the model base plane)

Note: The skull model base plane is parallel to the Frankfort plane and is at the level of the opisthion, i.e., z = -2.20.

Off-midsagittal Landmarks	Location (in.)			
	x	+y(left)	-y(right)	z
1. Gonions	-2.34	1.79	-1.80	-3.90
2. Porions	-3.40	2.32	-2.36	-1.30
3. Zygions	-2.10	2.53	-2.56	-1.35
4. Zygomaxillares	-0.75	1.73	-1.75	-1.95
5. Orbitales	-0.40	1.45	-1.45	-1.30
6. Frontomalare Orbitales	-0.60	1.90	-1.90	-0.25
7. Frontotemporales	-0.90	1.87	-1.95	0.70
8. Ends of the Coronal Sutures	-1.65	2.22	-2.35	0.45
9. Stephanions	-2.50	1.94	-2.22	2.10
10. Frontal Eminences	-0.50	1.10	-1.10	1.70
11. Euryons	-4.00	2.69	-2.81	1.00
12. Asterions	-5.30	2.20	-2.30	-1.00
13. Mastoidales	(not located on the skull model)			

Note: The off-midsagittal landmarks were located on the skull model within less than 0.005 in. in the x and z directions. The y coordinates on the left and right sides were then measured.

References p. 146

CONSTRUCTION OF THE GMR 50TH PERCENTILE
HEAD GEOMETRY MODEL

The GMR 50th percentile head geometry model was constructed by the addition of material representing the soft tissue to the skull geometry model. Thicknesses of soft tissues at various locations on the human head have been measured by Gadd, et al. (5) and Hodgson and Thomas (6). Hodgson and Thomas report soft tissue thicknesses ranging from 0.13 in. to 0.30 in. on the frontal regions, from 0.18 in. to 0.30 in. on the sides, and from 0.20 in. to 0.25 in. on the posterior of human calvaria. Gadd, et al., report average values of soft tissue thickness 0.16 in. on the frontal region, 0.185 in. on the side 0.220 in. on the top, 0.31 in. over the zygomatic arches, and 0.34 in. over the maxillary regions of human cadaver heads. These figures show a wide variation in the thickness of soft tissue covering the human head.

Soft tissue thicknesses of 0.25 in. over the calvarium, 0.30 in. over the zygomatic arches and orbital margins, and 0.35 in. over the maxilla-mandible regions were selected for initial head model build-up. Plastic plugs of known length were attached to the skull model so that material could be added in a controlled manner, and the location of the skull model would be known within the head model as it was being formed. These plastic plugs were located at most of the skull landmarks and at additional points between widely spaced landmarks. The tragions are very significant in the Hertzberg head measurements scheme and they were located for the head model by lateral extension of the skull model portions by an amount sufficient to comply with bi-tragomatic diameter (4). Modeling clay was added to the skull model to comply with selected mean values for external head measurements from the study by Hertzberg, et al. (2). The clay head model was then used to cast a Silastic E silicone rubber mold. In this mold was cast a Hysol head model which served as the basis for refinement of the head model to its final configuration. The soft tissue thicknesses between the skull model and head model are listed in Table 4. These thicknesses are reasonable in light of existing biomechanical data.

The GMR 50th percentile head geometry model is shown in Figs. 8, 9, and 10. The dimensions of this head model and the corresponding mean values from the Hertzberg study are listed in Table 5. In all cases, the agreement is quite close. The head geometry model not only agrees closely with the Hertzberg data, but, because it was based on a skull shape which is well defined by anthropometric data, this head model represents the human head to a greater extent than if it were based on the Hertzberg dimensions alone. The GMR 50th percentile head geometry model is as representative of the significant aspects of the 50th percentile adult American male as presently available anthropometric knowledge permits. This head configuration can be used as a basis for anthropometric crash dummy design with the assurance that the resulting dummy head configuration is more representative of the human than present dummy heads, and that it is a configuration which is close enough to the human that it may not need significant modification in light of future anthropometric data.

TABLE 4

Distances Between the GMR Head and
Skull Geometry Models

Measurement Site	Measurement Direction	Distance (in.)
Gnathion	S-I	.30
Pogion (anterior aspect of the chin)	A-P	.40
Prosthion	A-P	.60
Nasion	A-P	.30
Glabella	A-P	.30
Vertex	S-I	.40
Opisthocranion	A-P	.30
Frontal Eminences (left & right)	Normal to Surface	.40
Stephanions (left & right)	Normal to Surface	.30
Euryons (left & right)	Lateral	.30
Zygions (left & right)	Lateral	.30
Gonions (left & right)	Lateral	.35

Fig. 8. GMR 50th percentile head geometry model — oblique view.

Fig. 9. GMR 50th percentile head geometry model – front view.

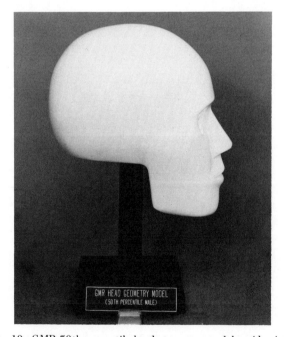

Fig. 10. GMR 50th percentile head geometry model – side view.

TABLE 5

GMR Head Geometry Model and WADC Technical Report
52-321 Head Dimension Comparison

Dimensions	WADC 52-321		GMR Head Geometry Model (in.)
	Mean (in.)	50th percentile (in.)	
Head Length	7.76	7.7	7.75
Head Breadth	6.07	6.1	6.06
Minimum Frontal Diameter	4.35	4.3	4.25
Maximum Frontal Diameter	4.71	4.7	4.52
Bizygomatic Diameter	5.55	5.5	5.55
Bigonial Diameter	4.27	4.3	4.30
Bitragion Diameter	5.60	5.6	5.40
Nose Length	2.01	2.0	2.06
Nose Breadth	1.31	1.31	1.43
Nose Protrusion	.89	.90	.70
Philtrum Length	.77	.76	.68
Menton-Subnasale Length	2.63	2.6	2.80
Lip-to-Lip Distance	.64	.63	.65
Lip Length	2.03	2.0	2.16
Head Height-Tragion to Vertex	5.11	5.1	5.12
Nasal Root to Wall	7.75	7.8	7.60
Tragion to Wall	4.30	4.0	4.04
Head Circumference	22.47	22.5	22.60
Bitragion-Coronal Arc	13.83	13.8	13.75
Minimum Frontal Arc	5.44	5.4	5.25
Bitragion-Minimum Frontal Arc	12.05	12.0	12.00
Bitragion-Menton Arc	12.78	12.8	12.55
Bitragion-Subnasale Arc	11.45	11.4	11.00

Note: The definitions of these dimensions are in References 2 and 3.

With confidence that the GMR head geometry model is representative of the 50th percentile adult American male, the locations of significant external head surface landmarks can be measured. The coordinate system for these measurements is identical to the coordinate system used for skull landmark location, including the placement of the origin at the nasion of the skull. The resulting head landmark locations are listed in Table 6. They not only indicate the relative locations of landmarks on the external surface of the head, but also the relative locations of the external surface landmarks in relation to landmarks on the skull. A particularly useful example is the location of the occipital condylar axis relative to the external surface of the head.

DISCUSSION

A set of skull landmark coordinates, a 50th percentile skull geometry model, a 50th percentile head geometry model, and a set of head landmark coordinates are

TABLE 6

GMR Head Geometry Model Landmark Locations

Landmark	Landmark Coordinates (in.)		
	x	y	z
Menton (Gnathion)	0.10	0.00	-5.10
Nasal Root (Nasion)	0.30	0.00	0.00
Vertex	-4.15	0.00	3.65
Posterior Pole (Opisthocranion)	-7.30	0.00	-0.25
Euryons	-4.00	±3.05	1.00
Tragions	-3.40	±2.70	-1.30
Zygions	-2.10	±2.75	-1.35
Gonions	-2.35	±2.15	-3.90

presented in this paper. These geometric descriptions are all representative of the average American adult male head, are compatible with each other, and are necessary additions to the body of knowledge which will lead to improvements in automotive crash dummies. In using the information presented here, it is important to realize that these landmark coordinates and models are geometrically representative of human structure and that duplicating human structural geometry in dummy head design does not insure similarity of human and dummy responses. Only if the other significant characteristics of the dummy head are like the human, can the structural geometry of the human be adopted without modification to obtain a dummy head which responds to impact like a human head. If the human head cannot be simulated in all respects, then the geometric characteristics of the human head described in this paper must be interpreted and modified for usage in dummy design.

The facial region of the GMR 50th percentile skull geometry model represents the human skull in that the landmarks in this region were correctly located and the margins of the facial features were modeled as closely as subjectively possible to agree with human structure. This model is not a replica of human structure. The zygomatic arches, maxilla, eye sockets, and nasal bones of the model were filled and blended. If the skull geometry model configuration were adopted for dummy design without modification, the facial region of the dummy skull structure would probably be much stiffer than its human counterpart. On the skull structure of a dummy head, the prominence for the facial features should probably be reduced to allow a thicker skin covering and a more compliant resonse to impact loading.

At the beginning of this paper, the midsagittal sections of two commonly used dummy heads were superimposed and compared. In Fig. 11, the skull structure outlines of these sections are superimposed with the midsagittal sections of the GMR 50th percentile skull geometry model. It is apparent that both dummy heads would require alteration to coincide with GMR model sections. A detailed comparison is left to the reader.

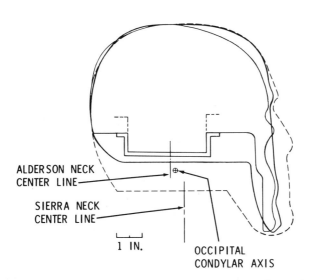

GMR 50TH PERCENTILE SKULL GEOMETRY MODEL
ALDERSON VIP50A SKULL OUTLINE
SIERRA 1050 SKULL OUTLINE

ALDERSON NECK CENTER LINE

SIERRA NECK CENTER LINE

1 IN.

OCCIPITAL CONDYLAR AXIS

Fig. 11. Midsagittal section outlines of the Alderson VIP50A and Sierra 1050 dummy skulls and the GMR 50th percentile skull geometry model.

There are many approaches to utilizing the results of this paper to effect improved crash dummy head design. One possibility is to base the head design on the skull configuration. Present dummy head designs include a skull-like structure to which the neck, accelerometers, and a skin-flesh layer are attached. Under impact, most of the head deformation occurs in the skin material and the mechanical properties and thickness of the skin are primary factors in determining the time-history of load generated between the dummy head and the impacted object. A realistic dummy head may be best obtained by closely specifying the configuration of the skull structure which would locate the head-neck attachment, the accelerometers, and the stiff structures of the head, and by covering this skull with a skin whose characteristics would be governed by the desired mechanical response of the dummy head.

A second approach would be to adopt the external head configuration of the head geometry model and define the skull structure configuration by accounting for a skin thickness which would result in the desired impact response. The geometric characteristics of a dummy head design resulting from the first approach would not necessarily be the same as a design based on the second approach. If it were possible to duplicate the mechanical properties of all significant tissues of the human head, then both the approaches outlined above would result in the same dummy head

References p. 146

configuration. At this time, it is questionable that this degree of duplication is possible.

CONCLUSION

This paper documents the development at GMR of:

1. the 50th percentile skull landmark locations,
2. the 50th percentile skull geometry model,
3. the 50th percentile head geometry model, and
4. the 50th percentile head landmark locations.

These models and landmark locations are based on anthropometry data, are compatible with each other, and are a valid basis for designing human-like crash dummy heads. The interpretation of this information in dummy head design should be considered carefully to avoid configurations which closely duplicate some aspects of human structure but do not respond to impact loading in a human-like way.

ACKNOWLEDGMENT

The authors gratefully thank Edward A. Jedrzejczak for his patient, precise, and skillful work in sculpting and molding of the skull and head models presented in this paper.

REFERENCES

1. *"Anthropomorphic Test Devices for Dynamic Testing — SAE J963," SAE Recommended Practice, Society of Automotive Engineers Handbook, 1972.*
2. *H. T. E. Hertzberg, G. S. Daniels, and E. Churchill, "Anthropometry of Flying Personnel — 1950," WADC Technical Report 52-321, Wright Air Development Center, Wright-Patterson Air Force Base, Ohio, 1953.*
3. *E. Churchill, and B. Truett, "Metrical Relations Among Dimensions of the Head and Face," WADC Technical Report 56-621, Wright Air Development Center, Wright-Patterson Air Force Base, Ohio, 1957.*
4. *E. F. Byars, D. Haynes, T. Durham, and H. Lilly, "Craniometric Measurements of Human Skulls," a paper prepared for presentation at the 1970 A.S.M.E. Winter Annual Meeting, A.S.M.E. paper No. 70-WA/BHF-8, 1970.*
5. *C. W. Gadd, A. M. Nahum, D. C. Schneider, and R. G. Madeira, "Tolerance and Properties of Superficial Soft Tissues In Situ," Proceedings of the Fourteenth Stapp Car Crash Conference, published by the Society of Automotive Engineers, 1970.*
6. *V. R. Hodgson, and L. M. Thomas, "Comparison of Head Acceleration Injury Indices in Cadaver Skull Fracture," Proceedings of the Fifteenth Stapp Car Crash Conference, published by the Society of Automotive Engineers, 1971.*

APPENDIX

Glossary of Anthropometric Terms as Shown in Figs. 12 and 13.

Midsagittal Landmarks

A. gnathion	the most inferior point on the midline of the mandible
B. prosthion	the most inferior point on the midline of the upper alveolar arch
C. nasion	the midpoint of the nasofrontal suture
D. glabella	the point midway between the superciliary arches
E. bregma	the junction of the coronal and sagittal sutures
F. vertex	the most superior point on the skull when oriented in the Frankfort plane
G. lambda	the junction of the lambdiod and sagittal sutures
H. opisthocranion	the point most distant from the glabella
I. inion	midpoint of external occiptal protuberance
J. opisthion	the most posterior point of the foramen magnum
K. basion	the most anterior point of the foramen magnum
L. occipital condylar axis	a line perpendicular to the midsagittal plane through the center of rotation of the atlantoöcipital joint

Off-madsagittal Landmarks as Shown in Figs. 12 and 13.

1. gonion	the most lateral point on the angle of the mandible
2. porion	the most superior point on the external auditory meatus
3. zygion	the most lateral point on the zygomatic arch
4. zygomaxillare	the most inferior point on the zygomaxillary suture
5. orbitale	the most inferior point on the infra-orbital margin
6. frontomalare orbitale	the junction of the frontomalare suture and the orbital margin.
7. frontotemporale	the most medial point on the superior temporal line
8. the end of the coronal suture	the inferior end of the coronal suture
9. stephanion	the intersection of the coronal suture and the superior temporal line
10. frontal eminence	the prominence of the frontal bone

11. euryon   the most lateral point on the cranium

12. asterion   the junction of the lambdoid, parieto-mastoid, and
                      occipito-mastoid sutures

13. mastoidale   the most inferior point on the mastoid process

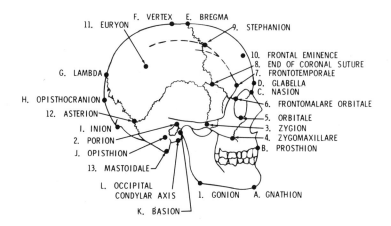

NOTE: MIDSAGITTAL LANDMARKS ARE LETTERED AND
OFF-MIDSAGITTAL LANDMARKS ARE NUMBERED

Fig. 12. Human skull landmark positions – side view.

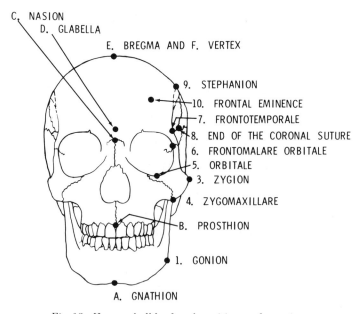

Fig. 13. Human skull landmark positions – front view.

Anatomical Planes

Frankfort plane is an anatomical plane in the head which is normally horizontal and is determined in the skull by the porions and the orbitales. The anterior-posterior and lateral directions are parallel to this plane and the superior-inferior direction is perpendicular to it.

Midsaggital plane is an anatomical plane in the head which is normally vertical and is the plane about which the head is most symmetrical. The anterior-posterior and superior-inferior directions are parallel to this plane and the lateral direction is perpendicular to it.

DISCUSSION

S. W. Alderson *(Alderson Biotechnology Corporation)*

Was any consideration given in the development of that skull to the elimination of some of those landmarks which represent random impacts and give discontinuities on the records?

R. P. Hubbard

I think it's important to realize that this study was an effort to characterize human head geometry and that imitation of human geometry in dummy head design could be misleading. The overall goal for the design and development of dummy heads is to create a useful tool to study and assess human response to impact environments. That is, dummy heads should represent as close a simulation to the significant aspects of human head response to impact as we can assure based on available biomechanical information. Dummy heads may not need to be a duplication of human heads either because of incomplete biomechanical information or because a duplication may not be appropriate for the attainment of equivalent response to various significant types of impacts. If human heads are not to be duplicated in all respects, then human structural geometry may need to be interpreted and modified for dummy heads which will respond like human heads. As a specific example, I suggest that the facial features of crash dummy heads shoud be less prominent to attain human-like facial impact response with presently accepted dummy head design concepts.

W. Goldsmith *(University of California)*

I was involved in this program to some extent but I forgot why Professor Byars and his crew used those 16 particular heads out of the hundreds that were available in the Terry collection. Do you know if there were some special criteria used? For example, did these heads come from automobile fatalities or something of that sort?

R. P. Hubbard

Their paper is discouragingly nondescriptive on some of these matters. I think the heads were selected as being representative of average size. To my knowledge there was no basis for selecting them other than subjective selection along those lines. That was one point that bothered me about the study. The fact that addition of reasonable skin and flesh thicknesses to a skull geometry that is defined by their data leads to a head geometry which agrees closely with the mean of the Air Force data, indicates that the skull size is fairly close to being an average skull. The Air Force data is accepted as being representative of the population at least for the head.

C. L. Ewing *(Naval Aerospace Medical Department)*

Concerning the anthropometric data from the Air Force, certainly that 1950 anthropometric survey was extremely well done, and I think that the 1969 study was equally well done; but it seems to me that there is the possibility that a biased sample was used in the 1950 anthropometric survey without the knowledge of the investigators. My understanding is that a large proportion of the sample used in that 1950 study was made up of fighter pilots who entered the service during 1943 and 1944. During these two years only there was an absolute limit on selection of fighter pilot trainees to 5'10" in standing height. That creates a bias in the data; not due to the investigators, but due to the fact that they were unknowingly furnished a biased sample on which to perform the study. Perhaps this would not invalidate the fifieth percentile head, but might indeed invalidate the 95th percentile. Do you have any comment on that?

R. P. Hubbard

I was aware of this bias in the data, and I think that the data is valid for use in our study. As you say, there have been subsequent anthropometric studies, but I have not received reports on these surveys. I used the phrase earlier in the talk, "generally available data", and the new Air Force data is not generally available.

The head dimensions that were measured in the Public Health Service study in the early '60's and published in 1962 agree with a tenth of an inch, or less, with the 50th percentile measurements of the 1950 Air Force study. This indicates that the measurements of the Air Force study are valid for use in our head model build-up.

In reference to your comment regarding extreme percentiles, 50th percentile values of data can be used to construct geometric models without too many statistical problems, but difficult problems arise when defining configurations to represent the extremes of the percentile scale. We did not try to build a 95th percentile head geometry model so we haven't actively dealth with these problems yet. There definitely needs to be a rational evaluation of configurations for dummies at the extreme percentiles.

D. J. Thomas *(Naval Aerospace Medical Department)*

You selected a reference origin at the nasion of your geometric study. For the study of human impact kinetics a spatial coordinate system must be established for measurement and expression of inertial properties of the human head as well as its geometric properties. Because the inertial properties such as center of gravity location and mass moments of inertia can't be measured directly on living subjects, they must be measured on cadavers and then related to the living subjects using measurements which can be made. There are probably certain anatomical simplicities in the human so that it may be possible to choose a coordinate system so that this transfer from cadavers to living subjects can be made with a minimum of error. Was this consideration included in the selection of your origin and coordinate system?

R. P. Hubbard

There were three primary reasons for the selection of the nasion of the skull as the coordinate system origin for my study. It is an easily identifiable landmark of the skull. It can be related by addition of appropriate soft tissue thickness to the nasion of the external surface of the head which is also an easily identifiable landmark. Finally, it is a point which is common to the brain case and facial regions of the head. The selection of the origin would not effect the relative placement of the skull landmarks so that it was not a critical selection decision in the context of my study.

To answer your question directly, the consideration of error in relating the inertial properties of cadaver heads to living subjects was not included explicitly in the origin selection. However the factors which make the nasion a desirable origin for my study are factors which would make it a desirable point for inclusion in a placement and orientation scheme for inertial properties of the human head.

I think that it is important to note that the coordinate system which you defined in the recently released monograph on neck mechanics (NAMRL monograph 21, USAARL 73-1) and the system which I have defined for my study differ only in their origin locations. Transfer from one system to the other involves addition of appropriate constants to the coordinate values. The coordinate axes of the two systems are parallel and in the same direction.

Authors Note:

After the formal discussion Dr. Thomas sent two papers to me which deal with the problem of establishing a meaningful biomechanical coordinate system. 1. "Specialized Anthropometry Requirements for Protective — Equipment Evaluation," D. J. Thomas, presented at the 29th meeting of the Aerospace Medical Panel of AGARD, NATO, Glasgow Scotland, Sept. 8, 1972, proceedings in press. 2. "Theoretical Mechanics for Expressing Impact Accelerative Response of Human Beings," D. J. Thomas and C. L. Ewing, AGARD Conference Proceedings no. 88 on Linear Acceleration of Impact Type, Aerospace Medical Panel Specialists, Oporto, Portugal, June 23-26, 1971.

P. G. Fouts *(Chrysler Proving Ground)*

In the discussions that we have had here already today, there seems to be a group that suggests that in order for the dummy to become a meaningful tool it must become humanlike. On the other hand, we have a group that suggests less complication of the test device. I submit that unless test procedures are much more exact than what we know them to be today, that becoming more humanlike isn't going to gain us much; so I just submit this thought for your consideration. There seems to be two schools of thought here, and somewhere along the line they're either going to have to be brought together, or else they're going to have to be carried along on parellel lines.

R. P. Hubbard

I agree that there is a need for reconciliation. Let me also say that dummy heads are designed, placed on necks, and used on ahthropometric test devices. Why not at least locate the neck attachment at the right place and the chin in the right place. In other words, the geometric constraints this study implies are the types of constraints that are already designed into the dummies. If we use biomechanical information as a basis to more completely define dummies, and many of the differences in dummy responses are due to insufficient specifications, we will move toward a more useable test device as well as one that is more meaningful for human injury assessment. The apparent conflicts which you mentioned can be reconciled by this approach.

AN ANATOMICAL SKULL FOR IMPACT TESTING

D. G. McLEOD and C. W. GADD

General Motors Research Laboratories, Warren, Michigan

ABSTRACT

Progress is described on the development of a trauma-indicating headform for assessment of localized and superficial head injury hazards. Using as a basis the facial and parietal bone impact tolerance values obtained in collaboration with Nahum, an anatomically shaped skull has been developed in which particular attention has been given to achieving humanlike fracture tolerances in the zygomatic and frontal regions.

Beginning with a polyester skullform cast from a human skull, a series of developmental steps is described which includes modifications to "idealize" geometrical symmetry, tailor the section thicknesses for average fracture force intensity in the principal regions, and select a skull structural composite material and overlying soft tissues for balance in fracture strength for both large area and localized impacts.

The frangible skull in its present state of development has been used as a research tool to indicate facial and frontal bone fracture. The accuracy with which these evaluations represent human bone fracture is not well established because of the limited biomechanical data base. As a more complete understanding of human skull fracture is developed and with further studies to control the material properties and structural geometry of the frangible head model, it may become a useful device for evaluating human skull fracture hazard in impact environments.

INTRODUCTION

Anthropomorphic dummies generally have been developed and have seen the greatest practical utility in the automotive industry as devices for studying overall car occupant trajectories and gross bodily loadings or accelerations. They have not been "trauma-indicating" in the sense of possessing anatomical fidelity or material strength

References pp. 160-161

properties analogous to their use as direct indicators of human injury hazard. For most usage, the building of such capability into a dummy as a whole would be incompatible with the needs for reasonable simplicity of construction and for strength sufficient to allow repetitive use.

On the other hand, the authors have felt there will be a need in the future, in the research and development areas, for components or test devices which will indicate fracture for particular body areas. The objective of this report is to outline progress made on a trauma-indicating head structure.

A considerable body of skull fracture tolerance data (1, 2, 3) have been gathered for the principal facial and cranial regions, and this has been used as the primary basis for the design of the headform described herein. Other recent studies (4, 5) have dealt with the crushability of the overlying soft tissues of the head, and this information has also been utilized.

It should be emphasized that the structure of the human head varies widely in its resistance to failure, as illustrated in recent studies encompassing large numbers of test specimens (6). Thus it cannot be said that a single headform design quantitatively duplicates the skull failure properties of a large segment of the population. Certain aspects of human head variability, for example, differences in strength between left and right side and effects of fluctuations in thickness over the cranium, can be minimized by "idealizing" the skull geometry to correct for such differences. This has been done in the present headform program. Variations in strength still remain, however, over a large sample of people, and a particular headform design can only be selected to represent a rough average of the car occupant population.

PRELIMINARY PROGRAM

First experimental trials were conducted utilizing a polyester skull used for medical demonstrations and manufactured by Medical Plastics Laboratories (MPL) Gatesville, Texas (Fig. 1). A small number of zygomatic, frontal, and parietal impacts were made on this demonstration skull using a one-inch-diameter free-fall impactor (1). The skull model was tested as it was received except for cementing the removable calvaria in place with an epoxy adhesive. The load distribution and cushioning effect of the superficial soft tissues was provided by placing a 1/4-inch thick Tramasaf skin and flesh simulation, as described in (7), over the contact region. The cranial vault was filled with lead wool to bring the total head weight to 10 pounds. The head was supported in an essentially freely floating state upon a thin, frangible board of styrofoam. Fracture forces read from the impactor load cell were:

Frontal bone (15 lbs. dropped from 36-in. height)   — 1075 lb.
Parietal bone (3.33 lbs. dropped from 48-in. height)   — 530 lb.
Zygoma (3.19 lbs dropped from 36-in. height)   — 440 lb.

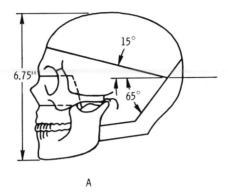

A

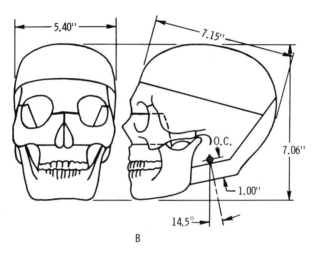

B

Fig. 1. Original MPL skull 1A, and enlarged model 1B. Parting lines between the six plastic skull sections and aluminum occiput are shown. Angles are from Frankfort plane. Neck is attached one inch below axis of occipital condyles.

While these values were within the range reported (1, 2, 6), under the same test conditions, it was evident that the commerical skull, designed only for illustration purposes, possessed a number of shortcomings as a trauma-indicating device. Dimensional variation was present, stemming chiefly from the fact that the model was a replica of a real skull, and overall size was smaller than that of a 50th percentile male. Another problem was the catastrophic nature of the crack propagation in the skull model which would have produced only a linear fracture in a real head.

The zygomatic fracture load of 440 pounds was within the range seen for humans (2), but was substantially higher than the 225 pounds reported in the same reference for first clinically significant fracture in elderly subjects. It was found possible, however, to adjust the 440-pound value downward to the latter range by increasing the size of the maxillary sinus behind the zygoma.

As a result of the encouraging results of this first test series, a decision was made to pursue several additional avenues with the aim of developing the most practically useful device. These are outlined below.

DESIGN REFINEMENT

The original MPL skull was quite small (Fig. 1). A larger skull from MPL (middle of Fig. 2) was selected as a starting point for further frangible head development. To increase the overall head size, new patterns were developed in cooperation with MPL to increase the major head dimensions.

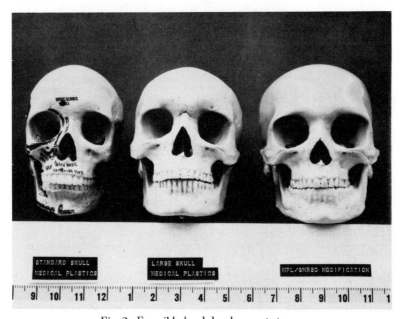

Fig. 2. Frangible head development stages.

An additional modification to the large MPL skull involved replacing the occipital region with an aluminum cap, carried integrally down under the skull where it provided a high-strength base for attachment of the head to a neck. The exposed edges of the calvaria left after removel of the rear portion were reinforced to permit the remaining skull to be screwed securely to the aluminum cap. This change was not felt to seriously compromise the value of the head as a trauma-indicating device since it was planned that the head would be used initially for frontal impacts.

As the next step in the development of the test headform, the cranium was smoothed or "idealized" (Fig. 3) in exterior curvature to eliminate slight undulations stemming from the use of a human head as a master pattern. This smoothing was accomplished by building up the flatter areas of the MPL large skull and then

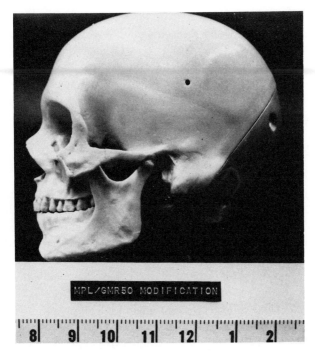

Fig. 3. Current reinforced polyester skull with aluminum occiput.

recutting the entire cranial surface. Computer controlled milling was utilized to aid in this reshaping, as well as to ensure perfect geometrical symmetry between the left and right sides of the cranium. A similarly idealized pattern was also developed to serve as a master for the interior surface of the cranium. The model incorporating these changes was then used as a master pattern by MPL to cast subsequent skulls.

Additional modifications included partially filling in the spaces between the teeth to eliminate stress concentrations which might produce unrealistic cracking of the mandible and maxilla under impact loading. The areas of the glue lines were also increased to minimize crack propagation along these lines. A brain pack was developed which was cast inside the skull. It consisted of equal parts GE-41 RTV and GE-910 diluent. A cavity within the skull located immediately above the neck attachment plate was provided for optional installation of accelerometers. A skin simulation, developed at GMR, is laid up on the inner surface of the female pattern for the exterior surface of the head. It consists of a layer of GE-154 RTV matrix with an overlay of sprayed nylon flock, followed by a second coat of the matrix, resulting in a total skin thickness of 1/16 inch. The soft tissue simulation is formulated using equal parts of GE-602 potting compound and GE-910 diluent, and is cast in place between the skin and skull surface.

Samples of this skin and flesh system were tested under impact conditions similar to those used by Gadd et al. (4, 8) for evaluating human tissue response. A skin

References pp. 160–161

simulant of 1/16 in. in thickness with a flesh simulant of 3/16 in. in thickness produced impact responses which agree reasonably well with the data obtained by Gadd et al. (4). The duplication of this type of material behavior is of primary importance if the skin-flesh simulant is to transfer the impact loading to the skull structure in a manner similar to human skin and flesh. Consequently, little emphasis was placed on developing a skin simulant which exhibited lacerative and tear strength properties similar to the human skin, as noted by Peterson et al. (9).

SKULL MATERIAL DEVELOPMENT

Observation of the early fracture tests of the polyester skull model from MPL indicated it was essential to drastically reduce the unrealistic brittleness of the original polyester, as seen under high rate of loading, and to increase its modulus of elasticity. A test program using small notched bar impact specimens, and subsequently hollow hemispherical forms simulating the radius of curvature of the cranium, led to a choice of skull material which consisted of a 90%-10% blend of the rigid and flexible polyester resins used by MPL mixed with 1/8-in. long milled glass fibers. The inclusion of glass fibers to constitute 10% by weight of the skull material was a practical maximum if the material was to remain fluid enough to be successfully cast. This proportion of polyester resins and glass fibers resulted in a stiffer material which was more resistant to crack propagation than the original skull model material.

RESULTS OF IMPACT TESTING

The skull shown in the right side of Fig. 2 with the skin-flesh material molded over it as described above was subjected to a series of localized impact tests similar to those reported in Refs. 1 and 2. During the early stages of these tests, the importance of skin-flesh thickness became apparent so that flesh thicknesses were adjusted to be in accordance with the values listed in Table 1. These thicknesses are within reported ranges (4), but the frontal thickness of 0.3 in. is larger than the 0.25 in. skin-flesh thickness which gave humanlike response in earlier development. The thicker skin-flesh was necessary for compliance of the frangible head model with the blunt impact fracture range of the human head.

TABLE 1

Soft Tissue Thickness Currently Used
on the GMR Frangible Dummy Head

Site	Thickness
Fontal	0.30 in.
Zygomatic	0.45 in.
Parietal	0.20 in.
Vertex	0.20 in.
Front of Maxilla	0.50 in.

The force levels which produced fractures in the frangible head under local impacts to the frontal and zygomatic areas with a 1-sq.-in. impactor are given in Table 2. Also included in Table 2 are the force levels which produced "clinically significant" fractures, together with suggested minimal tolerance values for these regions as reported by Gadd, Nahum et al. (1, 2). The frontal fracture forces for the frangible head were higher than the human responses. The zygomatic fracture levels for the frangible head, while within the human response, exhibited a large amount of scatter. A frangible head subjected to a zygomatic impact is shown in Fig. 4 with the skin-flesh material removed at the impact site to reveal the fractured structure.

TABLE 2

Skull Fracture Levels for Localized Impact of the
GMR Frangible Head and the Human Head

	Frontal	Zygomatic (left and right)
GMR Frangible Head	1400-1900 lb. (21 tests)	160-380 lb. (16 tests)
Clinically Signifcant Human Fractures (ref. 2)	1100 lb.	225 lb.
Minimum Fracture Tolerance Values (ref. 2)	900 lb.	200 lb.

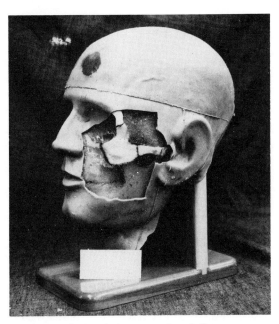

Fig. 4. Current skull with superfical soft tissues overlayed. Skin is cut away to reveal fracture of zygoma after impact.

Frangible heads were also dropped onto a steel bedplate to determine the tolerance of the frontal region to blunt impact. Fracture of the frontal region occurred at energy levels ranging from 405 to 450 in.-lb. These levels are in the low end of the 400 to 600 in.-lb. levels which were cited in the development of the Tramasaf headform (7). The low level of fracture energy for blunt impact to the frangible head and the high level of fracture force for localized impact to the frontal region indicate that further development is required if the frangible head is to become a useful tool to evaluate human skull fracture hazard.

CONCLUSION

This paper documents progress in the early development of a trauma-indicating dummy head. The primary objective of this program is to develop a useful tool for the assessment of human skull fracture hazard in impact environments. If these objectives are to be met, then a more extensive understanding of the structure and injury mechanics of the human head must be developed, and the technology of head model construction must be greatly advanced.

Work is currently under way to modify the geometry of the frangible head to comply with the work of Hubbard and McLeod (10). Included in these modifications is a head-neck attachment which is incorporated into the skull and does not rely on an aluminum occipital plate.

The present skull form is not as stiff as a human skull, and materials with higher moduli of elasticity must be employed in further development of the frangible head. In addition, an improved skin-flesh simulant is required to establish compatible fracture levels under both blunt and localized trauma, as well as providing more realistic skin laceration response.

REFERENCES

1. C. W. Gadd, J. P. Danforth, A. M. Nahum, and J. Gatts, "A Study of Head and Facial Bone Impact Tolerances," Proceedings of General Motors Automotive Safety Seminar, Milford, Michigan, 1968.

2. A. M. Nahum, J. D. Gatts, C. W. Gadd, and J. P. Danforth, "Impact Tolerance of the Skull and Face," Proceedings of 12th Stapp Car Crash Conference, Society of Automotive Engineers, 1968.

3. W. R. Hodgson, W. A. Lange, and R. K. Talwalker, "Injury to the Facial Bones," Proceedings of 9th Stapp Car Crash Conference, Society of Automotive Engineers, 1966.

4. C. W. Gadd, A. M. Nahum, D. C. Schneider, and R. G. Madeira, "Tolerance and Properties of Superficial Soft Tissues in Situ," Proceedings of 15th Stapp Car Crash Conference, Society of Automotive Engineers, 1970.

5. J. W. Melvin, J. C. McElhaney, and V. L. Roberts, "Development of a Mechanical Model of the Head – Determination of Tissue Properties and Synthetic Substitute Materials," Proceedings of 14th Stapp Car Crash Conference, Society of Automotive Engineers, 1970.

6. A. M. Nahum and D. C. Schneider, "Dynamic Impact Studies of the Facial Bones and Skull," Proceedings of 16th Stapp Car Crash Conference, Society of Automotive Engineers, 1972.

7. H. G. Holcombe and D. M. Herod, "Tramasaf – The Development of a Laboratory Instrument for Trauma Indication," Proceedings of a Laboratory Instrument for Trauma Indications," Proceedings of General Motors Automotive Safety Seminar, Milford, Michigan, 1968.

8. C. W. Gadd, W. A. Lange, and F. J. Peterson, "Strength of Skin and its Measurements, Biomechanics Monograph," American Society Mechanical Engineers, 1968.

9. F. J. Peterson, W. A. Lange, and C. W. Gadd, "Tear and Tensile Strength of Human Skin Under Static and Dynamic Loading," Winter Annual Meeting, American Society Mechanical Engineers, Nov. 1972.

10. R. P. Hubbard and D. G. McLeod, "A Basis for Crash Dummy Skull and Head Geometry," Proceedings of the Human Impact Response Symposium, General Motors Research Laboratories, Warren, Michigan, October 1972.

DISSCUSSION

G. W. Nyquist (GM Research Laboratories)

Are the points of impact well defined, beyond saying frontal and zygomatic? I'm sure that they are, but I guess I've never seen or heard any definitions for the impact points.

C. W. Gadd

We have had two papers (references 1 and 2) and there's a third one coming out in the Proceedings of the 16th (1972) Stapp Conference on the work in California, in cooperation with Dr. Nahum, using unembalmed subjects. I think most of those dimensions are covered in one or another of those papers. The latest paper is going to be in more detail. For example, in the zygoma we tried to get a square blow 30 degrees out from the sagittal plane. In the case of the forehead, again we tried to produce a square blow which has been defined in one of these papers. For the parietal region we have a horizontal blow, laterally. Actually I think the variations between people and the variations even in plastic models constitute a much greater variable.

L. M. Patrick (Wayne State University)

I'd like to make a comment. I think that the purpose of the dummy is to eliminate these variations. We realize that the variations exist in the human, but we certainly want to eliminate them in the dummy. Could the variations be due to the fact that you have such an intricate shape that it's impossible to hit it the same each time? This is probably what Dr. Nyquist was discussing.

C. W. Gadd

I guess that's a factor. I think that in humans there are dimensional variations from skull-to-skull whereas in our case we still get quite a spread or variation in the results which is not so much due to the gross dimensional variations which we try to minimize, but from the material variations. In the old days we ran tensile tests on steel and observed quite a variation, even though this was a very simple geometrical shape. Here we have a much more complicated shape and the polyester material itself has variations in properties. I don't know how we're going to eliminate variations in results even if we have real close geometrical accuracy.

S. H. Backaitis *(NHTSA, Department of Transportation)*

Are the fracture patterns similar between this artificial skull and a real skull?

C. W. Gadd

Well, in the zygomatic area the triple fracture appears to be quite similar. We haven't completely duplicated the fracture patterns of the cranial area, since an approximation of the proper force levels is what we have been shooting for so far. As I said, we started off with catastrophic fractures of the original polyester that were far from being natural looking, and we have made some progress in the direction of localizing the fractures to more nearly the degree that they are localized under hammer blows against the biological specimens.

L. M. Patrick

Hopefully, we're going to design vehicles in a manner to eliminate concentrated forces such as you're applying. Have you considered distributing the forces to see whether or not this reduces the spread, and perhaps gives a more repeatable result?

C. W. Gadd

Well, do you mean along the lines of the trend over the years of instrument panels to be made more free of sharp projections and ridges? There had been a trend in that direction. When we first started using the hemispherical headform years ago, we had to contend with a variety of relatively hard localized knobs and projections. A great deal has been done to try to minimize this type of thing over the last decade or so.

L. M. Patrick

Thinking about the vehicle, I can't visualize where you're going to get a one inch diameter rigid form to strike. Where is such an object in the head impact area?

C. W. Gadd

Well, that's true. We don't have them anymore. When we set out on this program we decided to test two extremes, so to speak. Therefore, a one inch diameter hammer blow and a flat impact were used.

D. G. McLeod

I have a comment with regard to one or two questions directed to Charlie concerning the type of failure we have seen with localized frontal impact. We have seen, not in general but in many cases, an interior stellar facture in the frontal zones. With regard to where we are going to find the knobs and projections; early in the program, and this program has been going on for some time, there were knobs and projections. The initial goal of this program was to develop a skull with the ability to indicate representative fractures from impacts associated with a one-square-inch knob or similar projection. Later on, as we achieved a fairly good localized impact response, blunt trauma evaluation was proposed. We are currently developing the model to respond properly to blunt trauma to the front of the head and we are considering blunt facial impacts; however, we have no data to report at this time.

J. H. McElhaney *(HSRI, University of Michigan)*

Did you observe the same things that Melvin did in his test on parietal sections where beyond 3/4 of an inch diameter, the loads to fracture didn't change significantly all the way up? Impacts using strikers with less than a 3/4 inch diameter caused local failures. For larger diameters the results were independent of striker size.

C. W. Gadd

We haven't actually gone as far as Dr. Melvin in the parietal region. The only thing we've done in this region so far is try to duplicate the one-inch diameter impactor blows which we have previously done out in California. It would be of interest, of course, to extend the program further to include larger and smaller impactors.

ANALYSIS OF A SLANTED-RIB MODEL
OF THE HUMAN THORAX

K. FOSTER

Consulting Engineer

ABSTRACT

This presentation is concerned with the analysis and applicability of a slanted-rib thorax for simulation of the human thorax. It is shown that the slanted-rib configuration, because of shearing induced in intercostal material, can produce the damping and nonlinear stiffness that are observed in the human thorax. A method of analysis is developed for evaluating the force-deflection characteristics. Rib rotation, bending, and torsion are taken into account in calculating deflections. Numerical results are examined to illustrate the effects of various parameters. Loading-rate sensitivity observed during testing is attributed to the viscoelastic properties of the intercostal material.

INTRODUCTION

One of the most difficult elements in the human simulator to design has proven to be the rib cage. Various configurations have been tried by various manufacturers, but none has exhibited the necessary characteristics. The shortcomings have been in two areas: nonlinearity and damping. Previously tried configurations using straight ribs were inherently linear, and only lightly damped. Attempts at providing damping by means of wrapping the ribs with plastic were not successful because the deformations produced in the plastic were flexural rather than shearing. Molded plastic rib cages also proved to be too linear and too lightly damped, as well as being susceptible to breakage. Several schemes have been generated to compensate for the shortcomings, including lead sheets and internal dampers. These have served to produce non-linearities and damping in the assembled thorax, but the adequacy of the results is open to question because the basic rib cage does not exhibit human-like characteristics.

In an effort to improve on the previous designs, the human skeleton was scrutinized closely. It was observed that a major factor which had not been taken into account previously was the general downward slope of the ribs. It was recognized that this slope would cause the ribs to rotate downward under the action of an anterior-posterior force. This rotation would produce a shearing deformation of the intercostal tissue, resulting in nonlinear force-deflection characteristics as well as high damping. Based on these observations, a dummy rib cage was designed with slanted ribs. Intercostal tissue was simulated by means of viscoelastic plastic. The posterior ends of the ribs were attached to the spine through articulated joints to allow vertical rotation of the ribs. The anterior ends of the ribs were attached to a leather sternum which would also allow rib rotation but would maintain spacing between rib ends.

The analysis presented here describes a method of calculating the nonlinear force-deflection characteristics of a slanted-rib cage. The evaluation of damping is not undertaken because it is recognized that the rib cage provides only part of the total damping observed in the human thorax, so that effort in this area did not seem justified. It is noted, however, that shearing deformation, such as experienced by the intercostal plastic, is the best way to produce damping in a viscoelastic material.

ANALYTICAL MODEL

For purposes of analysis, the rib cage is idealized as an assemblage of parallel semi-circular ribs connected to each other by means of shear-resistive material. The ribs are pinned at their ends to allow vertical rotations. The total external force is assumed to be horizontal, such as would be applied by a frictionless ram. Three deformation components are taken into account: rotation, torsion, and bending. As the external force is applied, each rib rotates downward, the rotation being resisted by the shear stiffness of the intercostal plastic. The component of the force which lies in the plane of the rib causes it to bend, while the component normal to the rib causes it to twist. The resulting displacements at the anterior end of the rib cause a net horizontal deflection of the applied force. The horizontal force and deflection are the quantities which define the stiffness characteristics of the rib cage.

It is recognized that actual rib cages, both human and simulated, include multiple ribs of various sizes. However, this analysis concerns the behavior of only a single rib. If a complete rib cage is to be evaluated, the results for individual ribs can easily be superimposed.

Fig. 1 shows the basic configuration. Each rib is a thin strip bent into semi-circular shape and pinned at each end. The intercostal plastic has an effective width, a. The initial slant angle is identified as θ_0.

STIFFNESS ANALYSIS

Rib Rotation — Consider the rib to have rotated through an angle $\Delta\theta$. This would require a tangential force, F_T, as shown in Fig. 2. The tangential force is equilibrated by shear forces in the plastic, F_s.

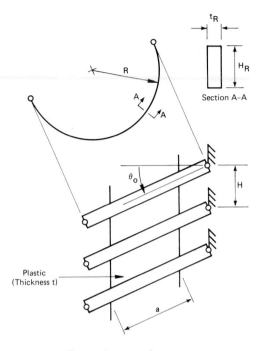

Fig. 1. Basic configuration.

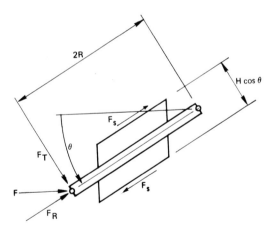

Fig. 2. Rotated rib.

Thus:

$$2RF_T = F_sH \cos\theta \qquad (1)$$

where

$$\theta = \theta_o + \Delta\theta \qquad (2)$$

The shear strain in the plastic is:

$$\gamma = \Delta\theta \left(1 + \frac{H_R}{H\cos\theta_o - H_R} \right) = \Delta\theta \left(\frac{H\cos\theta_o}{H\cos\theta_o - H_R} \right) \tag{3}$$

so that the shear force can be written as:

$$F_s = ta\tau = taG\gamma = taG \, \Delta\theta \left(\frac{H\cos\theta_o}{H\cos\theta_o - H_R} \right) \tag{4}$$

Thus, the tangential force can be written as:

$$F_T = \frac{F_s H \cos\theta}{2R} = \frac{taGH}{2R} \left(\frac{H\cos\theta_o}{H\cos\theta_o - H_R} \right) \Delta\theta \, \cos(\theta_o + \Delta\theta) \tag{5}$$

where G = shear modulus of the plastic material.

Since this is a transcendental equation, the rotation angle $\Delta\theta$ cannot be expressed as a closed-form function of F_T. However, the reverse approach can be used. That is, for any rotation angle $\Delta\theta$, the force F_T can be calculated.

It should be noted that the relationships among the forces are:

$$F_R = F \cos\theta \tag{6}$$

$$F_T = F \sin\theta \tag{7}$$

$$F_R = \frac{F_T}{\tan\theta} \tag{8}$$

$$F = \sqrt{F_R^2 + F_T^2} = \frac{F_T}{\sin\theta} \tag{9}$$

Thus, for each value of $\Delta\theta$, it is possible to determine F_T from Eq. (5), F_R from Eq. (8), and F from Eq. (9).

Rib Bending — The in-plane force F_R causes the rib to bend, as shown in Fig. 3. The deflection is calculated by equating the work done by the force to the strain energy in the ring:

$$1/2 \; F_R \; \Delta_R = \int \frac{M^2 \, dx}{2EI} = \int_o^\pi \frac{(RF_R \sin\beta)^2 \; Rd\beta}{2EI} = \frac{\pi R^3 F_R^2}{4EI} \tag{10}$$

$$\Delta_R = \frac{\pi R^3 \; F_R}{2EI} \tag{11}$$

where E = modulus of elasticity of rib material

and $\quad I = \dfrac{1}{12}(H_R)(t_R)^3$

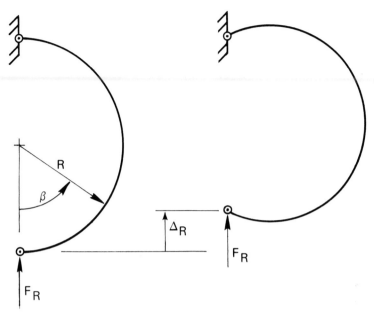

Fig. 3. Bending deformation.

Ring Torsion — The out-of-plane force F_T causes torques and bending moments at all points in the ring. However, since the torsional flexibility is much greater than the bending flexibility, only the torsional deflection, as shown in Fig. 4, is considered here. Again, the work done by the force F_T is equated to the strain energy n the ring:

$$1/2 \, F_T \Delta_T = \int \frac{T^2 \, dx}{2G_R K} = 2 \int_0^{\frac{\pi}{2}} \frac{[RF_T (1-\cos\beta)]^2 \, Rd\beta}{2G_R K} = (\frac{3\pi}{4} - 2)\frac{R^3 F_T^2}{G_R K} \qquad (12)$$

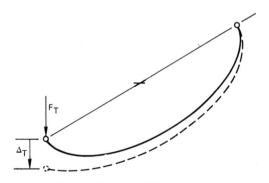

Fig. 4. Torsional deformation.

$$\Delta_T = 2(\frac{3\pi}{4} - 2) \frac{R^3}{G_R K} F_T = \frac{1.42\, R^3}{2\, G_R K} F_T \tag{13}$$

where G_R = modulus of rigidity of rib material

and K = torsion constant $= \frac{1}{3}(H_R)(t_R)^3$

Total Deflection — Referring to Fig. 5, it is seen that the total horizontal deflection is:

$$D = \Delta_1 + \Delta_2 + \Delta_3$$

$$= 2R(\cos\theta_o - \cos\theta) + \Delta_R \cos\theta + \Delta_T \sin\theta$$

$$= 2R(\cos\theta_o - \cos\theta) + \frac{\pi R^3 \cos\theta}{2EI} F_R + \frac{1.42\, R^3 \sin\theta}{2G_R K} F_T \tag{14}$$

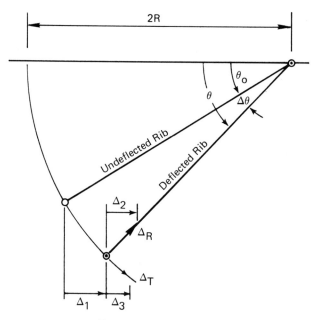

Fig. 5. Total deflection.

From Eqs. (5) and (8) it is seen that:

$$F_T = \frac{taGH}{2R} \left(\frac{H\cos\theta_o}{H\cos\theta_o - H_R} \right) \Delta\theta \cos\theta \tag{15}$$

$$F_R = \frac{F_T}{\tan\theta} = \frac{taGH}{2R} \left(\frac{H\cos\theta_o}{H\cos\theta_o - H_R} \right) \Delta\theta \frac{\cos^2\theta}{\sin\theta} \tag{16}$$

so that Eq. (14) becomes:

$$D = 2R(\cos\theta_o - \cos\theta) + \frac{\pi R^2 \, taGH}{4EI} \left(\frac{H\cos\theta_o}{H\cos\theta_o - H_R} \right) \Delta\theta \frac{\cos^3\theta}{\sin\theta}$$

$$+ \frac{1.42 \, R^2 \, taGH}{4G_R K} \left(\frac{H\cos\theta_o}{H\cos\theta_o - H_R} \right) \Delta\theta \, \sin\theta \, \cos\theta \qquad (17)$$

This equation is written entirely in terms of θ and $\Delta\theta$, where $\theta = \theta_o + \Delta\theta$. The companion equation for determining applied force is botained from Eq. (9):

$$F = \frac{F_T}{\sin\theta} = \frac{taGH}{2R} \left(\frac{H\cos\theta_o}{H\cos\theta_o - H_R} \right) \Delta\theta \frac{\cos\theta}{\sin\theta} \qquad (18)$$

The calculation procedure is to select values of $\Delta\theta$, and determine the values of D and F for each value of $\Delta\theta$. Then corresponding values of D and F can be plotted to obtain a force-deflection curve.

NUMERICAL RESULTS

The foregoing equations were evaluated numerically by means of a small computer program. The results are shown in Fig. 6 for several different sets of variables. The force values have been multiplied by a factor of 8 to matach an actual configuration designed by Sierra Engineering Company.

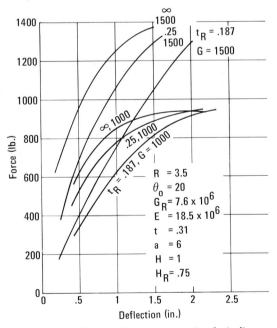

Fig. 6. Force-deflection curves (analytical).

Several interesting aspects of the results may be observed. The curves, as expected, are nonlinear. The plastic modulus G is seen to have a strong influence on stiffness. Since this is also the source of damping, it is to be expected that loading rate would have a marked effect on apparent stiffness. The effect of rib thickness is less pronounced. Unlike the effects of plastic modulus, variations of rib thickness produce variations in stiffness but not in maximum force. Thus, variation of rib thickness provides a means for "fine tuning" of rib cage characteristics.

TEST RESULTS

Fig. 7 shows a summary of test results for a particular rib cage built and tested by Sierra Engineering Company. The curves for 11 and 16 mph are averages of many tests. The static curve represents a single static test. It is seen that, because of the viscoelastic property of the intercostal plastic, stiffness varies greatly with impact velocity. This is desirable, as indicated by the GMR/Mertz Corridors. The visco-elasticity also produces the great amount of damping indicated by the return segments of the curves. The nonlinearity indicated by analysis and static test is accentuated by impact loading, the resulting curves conforming well to the indicated corridors.

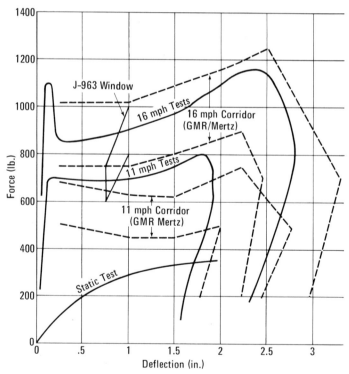

Fig. 7. Summary of test results.

CONCLUSIONS

Based on the foregoing discussions, several conclusions can be drawn. The slanted-rib concept can produce a rib cage which exhibits the desired human-like mechanical properties. The viscoelastic properties of the intercostal plastic are brought into full effectiveness because of the shearing deformation produced by rib rotation. The method of analysis developed so far, while applicable only to static loading, can be used in conjunction with test results to determine what design changes are required to produce particular changes in characteristics.

DISCUSSION

P. G. Fouts *(Chrysler Proving Ground)*

You indicated on your last slide an average of test results. Can you give us some idea of the boundaries of those test data; in other words, the scatter rather than the average?

K. Foster

The set of data that I saw, and I'm not sure I saw all the data available, is a summary of tests by Sierra. The data are somewhat biased on the high side and there were some spikes near the beginning. The load at 1 in. deflection varied from 860 to 920 pounds for the 16 mph tests and from 650 to 750 pounds for the 11 mph tests. The test data envelopes are indicated in the accompanying figure by the cross hatched areas for the 11 and 16 mph impacts.

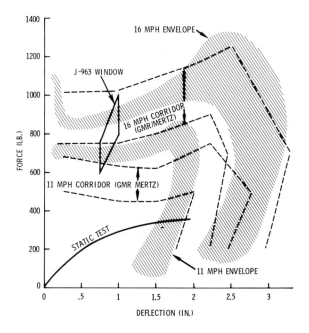

J. H. McElhaney *(HSRI, University of Michigan)*

Are you concerned, as I am, in this slant rib design, about what happens as f(t) changes or as the angle of the impulse changes? For an impulse in line with the angle of the rib, the stiffness is quite different if there's no f(t) to tilt the rib. Also, it seems in some automotive environment, one might have an f(t) acting in the opposite direction than you show.

K. Foster

This analysis did not attempt to cover all possible directions of loading. A basic assumption was that the applied force is horizontal, as would be produced by a frictionless ram, for example. No attempt was made to determine what the effect of the direction of impact force would be. You're right. In the event that a force is applied in the plane of the rib, there would be no rotation and the chest would appear much stiffer than if the force were perpendicular.

L. M. Patrick *(Wayne State University)*

Isn't this stiffening likely occurring in the human also?

K. Foster

Very possibly, but what we have attempted to do is design a dummy which will exhibit the required performance characteristics in a ram test. The dummy then will hopefully respond properly and any variations would correspond with the variation in the human thorax.

J. W. Justusson *(GM Research Laboratories)*

I have two questions. Concerning the 16 mph test, can you explain the sharp drop-off in load after what appears to be a fairly elastic response? The second question concerns your validation tests. How many rib elements did you have in the validation model? I may have missed that point.

K. Foster

Let me answer the second question first. The model that was fabricated had 10 ribs; that is, 5 pairs of semicircles. The calculations here were done for 8 ribs because of the fact that the area of impact of the ram excluded the uppermost rib in the particular test configuration used. The analytical results correspond to a particular set of ribs which consisted of four pairs. I can't answer the first question, regarding the rate at which the curve drops off except to conclude that it's related to the amount of damping in the system. I don't know of a kinematic explanation for the drop-off.

L. M. Patrick

Wouldn't you expect some interaction between the ribs due to the shear elements between the ribs, even though the impact surface did not cover all of them?

K. Foster

Are you referring to the one set that we left out?

L. M. Patrick

Yes. You said that only four pairs were affected. As I understand it, this was because the impactor did not cover the entire set of ribs. Isn't there intereaction between the ribs even though the impactor does not contact all of them?

K. Foster

There is interaction and that top rib will rotate with the others. Its presence produces the shear force of the plastic; but with no force on the front of that rib, the total force is only divided-up among four pairs. It serves to establish a boundary condition in effect.

J. N. Silver (*GM Proving Ground*)

With respect to your validation tests, and this may be getting slightly off your particular topic, under what test conditions, aside from 16 and 11 miles per hour, were those conducted? What about impactor mass and size and so forth?

K. Foster

I don't know. I can't answer that question. The tests were done by Cornell and Sierra, and I'm not familiar with all the details on the tests.

W. Goldsmith (*University of California*)

I wonder if you might tell us how the response would be affected if, in place of the berillium copper ribs, which you appear to have employed in view of your moduli, a lower modulus material would have been used which might be a little bit more representative of the bones in the rib cage? Secondly, you expressed some reservations about being able to make a dynamic analysis because you don't know the properties of the material under dynamic conditions. There is sufficient information available today over a very wide range of loading rates for similar materials to make a pretty good educated guess of its behavior, so perhaps the dynamic analysis is not so far removed. Thirdly, if you're not willing to make a dynamic analysis at this point, perhaps you could make a static analysis since you do have a static test.

K. Foster

Let's cover the third point first. We did do a static analysis and that's what the set of curves in Fig. 6 of the paper represent. The static test, therefore, corresponds to a calculable configuration. I don't know how they agree, because the particular static test configuration was not analyzed. We analyzed a set of configurations in order to identify the types of behavior to expect. Then a configuration was built-up. The final configuration was not analyzed, so we don't have a comparison for that curve. It would indeed be possible to conduct a dynamic analysis without great difficulty. Regarding the comment about the rib material: in the event that this analysis was done for materials more representative of ribs, the result would be, I believe, a more flexible rib cage. That would correspond to the lower set of curves presented in Fig. 6, where the maximum force did not change, but rather the slope of the curve decreased so that the maximum force occurred at a larger deflection. In other words, it would appear to be softer, but the maximum force wouldn't change much.

G. W. Nyquist *(GM Research Laboratories)*

It seems to me that the shear strains you're encountering in this chest are getting pretty large. You're probably pushing the limit of the small strain analysis and getting to the point where one should start looking at things such as stretch tensors. Am I correct in saying that?

K. Foster

Yes. You're correct in saying that.

L. M. Patrick

What about the reproducibility? Have you impacted the chest under identical conditions several times to evaluate the repeatability?

K. Foster

Again, I can't answer this question. Let me refer this question to John Roshala of Sierra.

John Roshala *(Sierra Engineering)*

Yes. We've reproduced the thorax in limited quantity, approximately 15. Several of these went to customers for evaluation. All of them were prototypes for testing. Reproducibility is primarily due to manufacturing problems. The viscoelastic material has properties heavily dependent on the mix ratio. This is one area where we have had problems. We're currently solving that problem. The other problem area is the heat treating and forming of the ribs. The ribs have a very peculiar configuration. They are formed in three dimensions. This three-dimensional forming leads to problems. Again, the problems are being resolved. It's a matter of quality control and tooling. The prototype rib cages were hand built one at a time by engineering technicians. With regard to repeatability, when the manufacturing variables and their limitations are effectively established and controlled, then repeatability will be demonstrated.

SESSION III

IMPACT RESPONSE AND APPRAISAL CRITERIA THORAX II AND NECK

Session Chairman
V. L. ROBERTS

*University of Michigan
Ann Arbor, Michigan*

HUMAN TORSO RESPONSE TO BLUNT TRAUMA

R. L. STALNAKER, J. H. McELHANEY, and V. L. ROBERTS

University of Michigan, Ann Arbor, Michigan

M. L. TROLLOPE

University of Michigan Medical Center, Ann Arbor and the Wayne County General Hospital, Eloise, Michigan

ABSTRACT

The most frequent causes of blunt abdominal injury are the steering wheel and the lap belt. The organs most often injured are the liver, pancreas, spleen and intestines. Based on this information, a series of animal abdominal impacts were designed to study the relationship between shape and type of impactor, velocity and direction of impact, body region impacted and injury level. The results of this study are given in the form of an experimental scaling factor which relates the sensitivity to impacts of the various body regions. This scaling factor was found to be dependent of body weight, making it applicable for evaluating human tolerances to abdominal impacts.

A cadaver program was initiated as part of the human response to impact study. Cadavers were received approximately one to three days after death, and were unembalmed. These cadavers were impacted on the front of the chest over the fourth rib. The velocity of the impact and the mass of the impactor were held constant. Static compression tests were also conducted on the chest in the anterior-posterior direction using both cadavers and human volunteers. The results are given as load-deflection curves with comparison to previous published data.

INTRODUCTION

Blunt torso trauma is a major cause of death in the United States. However, little experimental work has been done to clarify the mechanism of blunt torso injury, and that which has been done was not always well controlled. (1, 11) A large series of abdominal impacts were designed to study the forces involved in blunt abdominal trauma in general, and to quantitate the forces involved in liver injuries in particular. A second series of tests were carried out to study the response of the human thorax to both static and dynamic loading under controlled conditions.

References pp. 197–198

METHODS AND MATERIALS

Experiment 1 — In order to provide a range of primate masses and body proportions, squirrel monkeys, Rhesus monkeys, and baboons were used. In addition, fifteen pigs (Suis Scrofa) weighing approximately 100 pounds each were used because their abdominal visceral weights are known to approximate those of man. Each of these animals were exposed to a highly controlled and monitored abdominal impact. The animal to be tested was kept without food or water for twelve hours prior to the test and was anesthesized with intravenous Phenobarbital or Ketamine. An attempt was made to provide analgesia and yet preserve skeletal muscle tone.

The anesthesized animal was shaved and targeted for high speed photographic analysis. A complete set of anthropometric measurements was obtained.

The subject was seated and held in an erect position with silk threads through the ears. The upper limbs were secured out of the field with masking tape. This method of support makes the animal essentially a free body, provides reproducible results, and eliminates the complicated boundary conditions of a seat or sling.

Various sized impactors were used to simulate common automotive injuries (Fig. 1). Both rigid and padded bars were used to simulate steering wheels, door handles, and arm rests. A large surface area impactor was used to simulate passenger impacts with the dashboard and auto-pedestrian injuries. In other tests, scaled lap belts were used either to slap against the abdomen of the suspended animal or to contain the animal in a sled run.

Fig. 1. Various impactors used to produce injury.

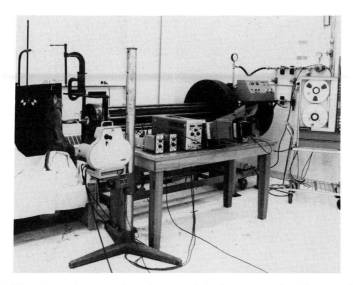

Fig. 2. The pneumatic cannon is used to propel the impactor against the test subject.

All impacts were carried out by a pneumatically-operated cannon specially constructed for impact studies (Fig. 2). The machine consists of a ground and honed cylinder, an air reservoir, and two carefully fitted pistons. The transfer piston (22 lbs) is propelled by compressed air through the cylinder and transfers its momentum to the impact piston (22 lbs.). A striker plate, attached to the impact piston, travels a distance of about ten inches where an inversion tube absorbs the energy of the impact piston and halts its movement. The stroke of the impactor is controlled by its initial positioning, and its velocity is controlled by the reservoir pressure. The impactor is instrumented with an accelerometer and an inertia-compensated force transducer. High speed motion pictures at 3000 to 5000 frames per second were taken for photographic analysis. Contact forces and pulse duration were photographically recorded from an oscilloscope. A sample data record is shown in Fig. 3. Impactor velocity was determined from the high-speed motion picture analysis. The animals were positioned to limit the depth of penetration to approximately 50 percent of body width, and a one-foot-thick soft foam pad was arranged to absorb the momentum of the monkey after impact.

Anterior impact areas were varied in location from the mid-epigastric to the supra-pubic region. These regions were defined as Region 1 — upper one-third of abdomen, Region II — mid one-third of abdomen, and Region III — lower one-third of abdomen. In order to produce liver injuries, some impacts were made in the right anterior oblique position one-half way between the xyphoid and the 12th rib.

All animals were autopsied following the impact. Injuries were then classified and rated on an Estimated Severity of Injury scale 1-5 as shown in Table 1. The relative importance of various injuries was subjectively determined. When several organs were

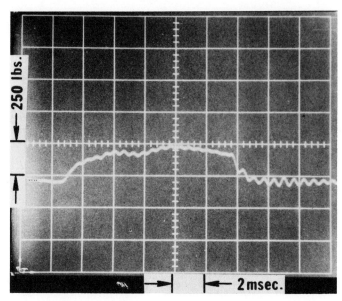

Fig. 3. A sample oscilloscope display recording the force and pulse duration of an impact.

injured, the Estimated Severity of Injury scale was made to reflect the severity of the total injury. Some injuries were judged to be at a level between two whole numbers and were given ratings accordingly. All injuries were photographed so that direct comparisons between animals could be made and the indexing could be double checked for consistency throughout the experiment. A sample of a 3+ injury is shown in Fig. 4, a 4+ injury in Fig. 5, and a 5+ injury in Fig. 6.

TABLE 1

Definition of Estimated Severity of Injury (ESI)

1+ — Minor Trauma	— Retroperitoneal hematoma, mesenteric abrasion, subcapsular hematoma of liver.
2+ — Mild Trauma	— Splenic hematoma, intestinal hematoma, small non-bleeding liver laceration or capsular hematoma.
3+ — Moderate Trauma	— Spenectomy or liver injury requiring suture repair.
4+ — Major Trauma	— Hepatic resection, pancreatic fracture — survival only with maximum surgical care.
5+ — Massive Trauma	— Complete maceralion of liver, spleen, or pancreas. Potentially lethal at accident scene.

Experiment II — A cadaver program was initiated to consolidate the merits of both live animal and unembalmed cadaver testing into a single theory of human response to impact.

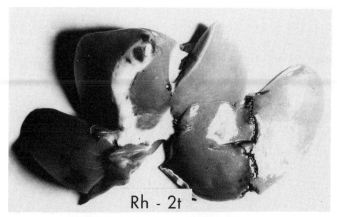

Fig. 4. A 3+ liver injury with a capsular tear.

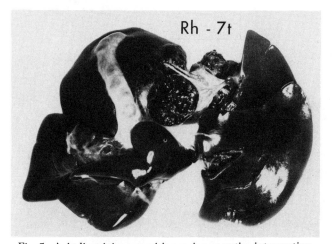

Fig. 5. A 4+ liver injury requiring major operative intervention.

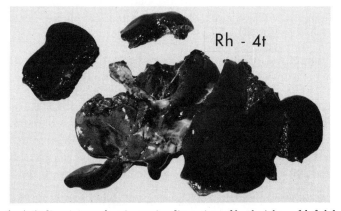

Fig. 6. A 5+ liver injury showing major disruption of both right and left lobes.

The cadavers were received on the average four days after death, and were unbalmed. The bodies were kept at 37 °F until two hours prior to testing, when it was allowed to reach room temperature. The cadaver was then thoroughly manipulated so that the flexibility of its neck, arms, and torso approached that of a relaxed living human.

For the dynamic test, the cadaver was positioned in a heavily padded chair with no back support and supported in an upright sitting posture by positioning its arms on a pair of specially designed arm rests (Fig. 7).

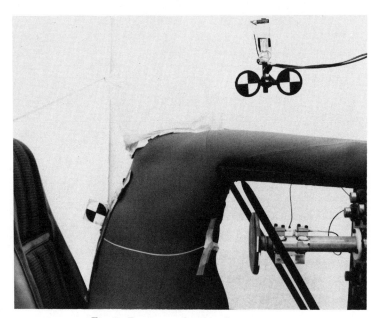

Fig. 7. Test set-up for chest impact test.

The cadaver's head was held out of the path of the impactor by an adjustable elastic band. Care was taken not to put undue tension on the neck, which would distort the cadaver's torso from its relaxed position.

A 2" x 2" target was fixed to the cadaver's back at the point directly opposite the point of impact. The post-impact travel of the piston was four inches.

The impactor was a six-inch diameter rigid cylinder with a 1/2-inch edge radius weighing 22 lbs. It was centered medially over the fourth interspace. Impact velocities were 13 mph ± 1 mph. Impact forces were measured with an inertia compensated load cell. Chest displacements were measured photometrically by comparing targets on the impactor and a cadaver back target aligned with the axis of the impactor.

The static tests were conducted by placing the cadaver between the crosshead and the load frame of an Instron Low Speed Testing machine (Fig. 8). The loading was applied through a 6-inch diameter metal disc, and the rate of loading was a constant 2.0 inches per minute. The maximum deflection was set at approximately 25 percent thorax width and measured by a linear variable differential transformer. Both the load and deflection were recorded photographically from an oscilloscope.

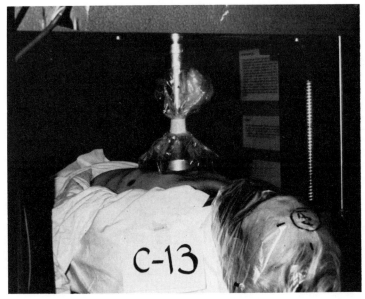

Fig. 8. Test set-up for static chest load-deflection test.

A series of tests were conducted on human volunteers to establish the load-deflection characteristics of the chest in the relaxed and tense state. The volunteer was placed against the wall and a 6-inch diameter metal disc with a load cell and a displacement measuring potentiometer was centered over the fourth interspace. The loads were then applied by an adjustable lever arm pinned to one side of the test subject. The maximum displacement that the subject was willing to undergo was set by the subject by placing a stop against the lever arm. The loading was always done at the end of expiration. The first run was made on each subject in the relaxed state. The volunteer was then asked to exhale and resist the load on his chest as hard as he could. The subject was then allowed to move around for five to ten minutes and the test was repeated.

RESULTS

Experiment I — The data from Experiment I is recorded in Table 2 and is arranged by impactor type in order of increasing impactor velocity. For purposes of analysis, this data was grouped by species and further subdivided by region of abdominal impact within each species. The data from each subgroup was submitted to a computer-assisted statistical analysis to obtain correlations between the various impact parameters and estimated severity of injury. It was found that the peak force, pulse duration and contact pressure had a high level of correlation with the ESI for each subgroup studied. Then, linear regression analysis was conducted to obtain relationships between the above parameters and ESI. The relationship of peak forces and pulse durations to ESI are shown in Figs. 9 and 10 respectively for Region I impacts on pigs and Rhesus monkeys. Similar relationships were obtained for impacts in other regions for all the other species studied. ESI is plotted vs. contact pressure in Fig. 11 for Region I impacts on Rhesus monkeys. Again, impacts in other regions for other species also showed similar results.

TABLE 2

Summary of Abdominal Impact Data

Number	Impactor Type	Animal	Mass lbs.	Impact Region	Contact Area Inches2	Piston Velocity Ft/Sec	Peak Contact Force-Pounds	Pulse Duration msec	Peak Contact Pressure PSI	ESI
1	6" x 1/2" Flexible	R	6.8	API	1.07	29.6	376	5.0	352	3
2	6" x 1/2" Flexible	R	6.0	API	1.04	36.6	LOD	LOD	LOD	1
3	6" x 1/2" Flexible	R	6.4	API	1.06	37.8	408	6.6	387	2
4	6" x 1/2" Flexible	R	5.1	APII	1.02	25.5	204	4.8	201	2
5	6" x 1/2" Flexible	R	7.5	APII	1.09	38.4	380	4.4	351	2
6	6" x 1/2" Flexible	R	6.4	APIII	1.06	34.3	266	6.0	252	0
7	6" x 1/2" Flexible	R	7.1	APIII	1.07	50.1	430	5.6	402	2
8	6" x 1" Flexible	R	8.4	API	2.22	30.3	260	10.4	117	1
9	6" x 1" Flexible	R	7.1	API	2.14	33.6	304	8.4	142	2.5
10	6" x 1" Flexible	R	4.2	APII	1.98	35.2	254	10.0	128	1
11	6" x 1" Flexible	R	5.1	APIII	2.03	34.9	364	8.6	179	0
12	6" x 2" Flexible	R	8.2	API	4.42	26.2	186	7.4	42.2	0
13	6" x 2" Flexible	R	7.5	API	4.34	37.5	354	9.2	81.6	3.3
14	6" x 2" Flexible	R	7.7	APII	4.36	29.3	166	9.4	38.1	0
15	6" x 2" Flexible	R	4.2	APII	3.96	35.4	220	5.6	55.6	4.3
16	6" x 2" Flexible	R	7.5	APIII	4.28	39.2	354	9.4	82.7	1
17	8" x 1/2" Rigid	R	6.8	API	1.07	40.3	450	9.4	422	2.5
18	8" x 1/2" Rigid	R	8.4	APII	1.11	37.2	320	10.6	288	1
19	8" x 1/2" Rigid	R	7.9	APIII	1.10	47.6	282	6.4	256	1
20	8" x 1/2" Rigid	R	7.3	APIII	1.08	51.0	525	7.6	486	2
21	8" x 1/2" Rigid	R	11.0	API	1.86	31.6	325	11.6	175	5
22	8" x 1/2" Rigid	R	10.0	API	1.81	36.4	275	11.4	152	4
23	8" x 1/2" Rigid	R	14.5	API	1.99	46.4	825	13.0	414	4.3
24	8" x 1/2" Rigid	R	12.2	API	1.91	47.4	700	7.6	266	5
25	8" x 1/2" Rigid	R	14.2	APII	1.97	41.8	525	10.5	266	3
26	8" x 1/2" Rigid	R	11.0	R	1.86	39.6	400	11.2	215	LOD
27	8" x 1/2" Rigid	R	11.6	R	1.87	40.5	400	7.6	214	4

(Continued)

TABLE 2 (Continued)

Summary of Abdominal Impact Data

Number	Impactor Type	Animal	Mass lbs.	Impact Region	Contact Area Inches^2	Piston Velocity Ft/Sec	Peak Contact Force-Pounds	Pulse Duration msec	Peak Contact Pressure PSI	ESI
28	8" x 1/2" Rigid	R	12.2	R	1.92	41.1	475	7.4	247	3
29	8" x 1/2" Rigid	R	11.0	R	1.85	41.1	400	7.5	216	2.5
30	8" x 1" Rigid	R	12.2	API	3.83	29.5	LOD	LOD	LOD	2
31	8" x 1" Rigid	R	12.2	APII	3.83	32.0	375	7.8	97.8	3
32	8" x 1" Rigid	R	6.8	API	2.17	30.0	170	11.6	78.4	3
33	8" x 1" Rigid	R	6.0	API	2.08	32.0	133	7.2	64.0	1.3
34	8" x 1" Rigid	R	9.5	API	2.28	32.2	300	11.4	131.5	3
35	8" x 1" Rigid	R	9.5	APII	2.28	27.6	116	7.6	50.9	1
36	8" x 1" Rigid	R	7.9	APII	2.20	41.4	431	11.2	196	4
37	8" x 1" Rigid	R	8.4	APIII	2.22	39.6	560	11.2	228	1
38	8" x 1" Rigid	R	12.2	R	3.80	33.7	LOD	11.6	LOD	1
39	8" x 1" Rigid	R	11.3	R	3.72	39.5	LOD	LOD	LOD	3
40	8" x 1" Rigid	R	10.3	R	3.66	40.9	775	15.5	212	4
41	8" x 1" Rigid	R	13.2	R	3.90	42.0	LOD	LOD	LOD	4
42	8" x 2" Rigid	R	5.3	API	4.08	31.1	190	13.0	46.5	2.5
43	8" x 2" Rigid	R	11.4	API	4.80	36.6	225	7.2	47.0	4
44	8" x 2" Rigid	R	7.9	APII	4.40	29.7	280	12.6	63.6	0
45	8" x 2" Rigid	R	11.0	APII	7.40	35.6	400	8.8	54.1	1
46	8" x 2" Rigid	R	10.1	APII	4.64	41.5	413	10.0	89.0	3
47	8" x 2" Rigid	R	7.3	APIII	4.32	40.2	489	9.4	115.4	1
48	8" x 2" Rigid	R	7.9	AIII	4.40	51.5	450	8.0	102.0	2
49	3" Diam. Circ. Rigid	P	143	RAOI	7.2	34.9	650	11.2	92.0	0
50	3" Diam. Circ. Rigid	P	117	RAOI	7.2	42.8	600	LOD	85.0	1
51	3" Diam. Circ. Rigid	P	104	RAOII	7.2	30.1	600	16.0	85.0	1
52	3" Diam. Circ. Rigid	P	106	RAOII	7.2	34.3	LOD	LOD	LOD	0
53	3" Diam. Circ. Rigid	P	128	RAOII	7.2	40.5	700	17.5	99.2	4.5
54	3" Diam. Circ. Rigid	P	110	RAOII	7.2	44.0	550	9.5	77.8	2
55	3" Diam. Circ. Rigid	P	97	RAOIII	7.2	45.5	850	14.5	120.0	3.7
56	3" Diam. Circ. Rigid	P	108	RAOIII	7.2	47.0	700	11.0	99.3	2
57	3" Diam. Circ. Rigid	P	106	LAOI	7.2	35.6	600	11.5	85.0	2
58	3" Diam. Circ. Rigid	P	112	LAOI	7.2	55.5	650	10.5	92.0	0
59	3" Diam. Circ. Rigid	P	104	LAOII	7.2	34.8	600	11.0	85.0	4.3
60	3" Diam. Circ. Rigid	P	99	LAOII	7.2	45.2	850	10.0	120.0	4.7
61	3" Diam. Circ. Rigid	P	110	LAOIII	7.2	42.0	LOD	LOD	LOD	1
62	3" Diam. Circ. Rigid	P	95	LAOIII	7.2	45.8	425	10.0	60.2	1
63	3" Diam. Circ. Rigid	P	112	LAOIII	7.2	47.5	950	12.5	134.0	3
64	3" Diam. c 1-1/2" Foam	R	7.7	API	5.71	41.9	250	8.0	43.8	4
65	3" Diam. c 1-1/2" Foam	R	5.7	APII	5.60	35.8	150	10.0	26.8	2

(Continued)

References pp. 197–198

TABLE 2 (Continued)

Summary of Abdominal Impact Data

Number	Impactor Type	Animal	Mass lbs.	Impact Region	Contact Area Inches^2	Piston Velocity Ft/Sec	Peak Contact Force-Pounds	Pulse Duration msec	Peak Contact Pressure PSI	ESI
66	6" Diam. Circ. Rigid	R	11.6	RAOI	21.16	37.1	1000	13.5	47.5	4
67	6" Diam. Circ. Rigid	R	13.2	RAOI	21.62	38.8	1050	14.4	48.6	1
68	6" Diam. Circ. Rigid	R	14.2	RAOI	20.38	41.6	1200	14.0	59.0	5
69	6" Diam. Circ. Rigid	R	8.4	RAOI	20.42	45.6	1050	8.2	51.4	4.3
70	6" Diam. Circ. Rigid	R	11.6	RAOI	20.90	49.3	1950	8.0	93.4	4.5
71	6" Diam. Circ. Rigid	R	10.3	RAOI	20.44	49.3	1250	8.0	61.2	4.3
72	6" Diam. Circ. Rigid	R	11.0	RAOI	20.40	50.6	1400	14.4	68.7	4.7
73	3" x 2-1/4" Wedge Pad	S	1.32	RI	6.3	32.0	80	8.0	12.6	1
74	3" x 2-1/4" Wedge Pad	S	1.16	RI	6.3	35.5	140	5.8	22.3	0
75	3" x 2-1/4" Wedge Pad	S	1.30	LI	6.3	32.6	60	5.2	9.5	0
76	3" x 3" Wedge Pad	R	9.3	RI	9.0	27.8	LOD	2.6	11.1	0.3
77	3" x 3" Wedge Pad	R	11.3	RI	9.0	34.9	260	10.6	28.9	3.5
78	3" x 3" Wedge Pad	R	13.4	RI	9.0	36.5	280	7.8	31.1	4
79	3" x 3" Wedge Pad	R	8.8	LI	9.0	41.6	360	9.8	40.0	0.3
80	3" x 3" Wedge Pad	R	8.9	LI	9.0	43.8	260	9.2	28.9	3.7
81	3" x 3" Wedge Pad	R	9.0	LI	9.0	45.0	280	8.0	31.1	0.2
82	3" x 3" Wedge Pad	R	8.4	LI	9.0	55.0	180	9.8	20.0	2.0
83	10" x 4" Wedge Pad	B	31.0	RI	22.0	28.2	505	10.0	23.0	2
84	10" x 4" Wedge Pad	B	34.8	RI	23.3	41.0	756	12.6	32.4	1.0
85	10" x 4" Wedge Pad	B	32.0	RI	22.3	44.4	816	7.8	36.6	2.7
86	10" x 4" Wedge Pad	B	47.5	RI	26.8	47.0	1140	8.0	42.6	4.3
87	10" x 4" Wedge Pad	B	31.0	RI	21.9	53.0	1020	3.6	46.6	2.0
88	10" x 4" Wedge Pad	B	25.7	LI	20.0	45.5	755	12.0	37.7	2
89	10" x 4" Wedge Pad	B	33.5	LI	22.9	47.5	750	11.6	32.8	2.3
90	10" x 4" Wedge Pad	B	32.6	LI	22.6	56.0	1220	3.1	54.3	4.7
91	10" x 4" Wedge Pad	B	30.8	LI	21.8	56.0	1020	8.4	46.8	1.7
92	10" x 4" Wedge Pad	B	42.7	LI	26.2	51.8	890	8.0	34.0	4
93	Lap Belt Sled	R	8.8	APII		65.8	340	120		2
94	Lap Belt Sled	R	9.5	APIII		54.5	480	100		1
95	Lap Belt Sled	R	10.1	APIII		65.4	490	104		0
96	Lap Belt Sled	R	9.5	APIII		72.4	472	106		0

ANIMAL		POSITION		REGION	
R	– Rhesus Monkey	AP	– Anterior	I	– High Abdomen
S	– Squirrel Monkey	R	– Right Side	II	– Mid Abdomen
B	– Baboon	L	– Left Side	III	– Low Abdomen
P	– Pig	RAO– Right Anterior Oblique			
		LAO– Left Anterior Oblique			

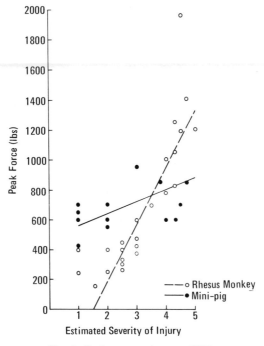

Fig. 9. Peak contact force vs. EST.

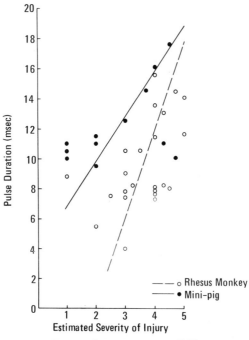

Fig. 10. Pulse duration vs. EST.

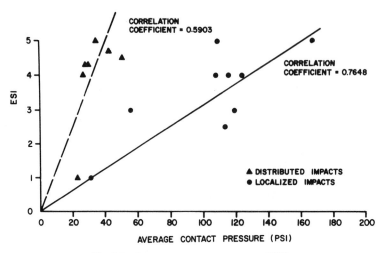

Fig. 11. Average contact pressure vs. EST.

The animals used in these experiments varied in size and weight. In order to integrate these data, dimensional analysis techniques were used with the aid of a computer. This resulted in a formula which compensates for weight and contact area:

$$\text{ESI} \; \alpha \; \text{Log} \; \frac{f_\tau{}^2}{m\,a}$$

f = force of impact
τ = duration of impact
m = mass of the animal
a = contact area

The data for all four species were found to correlate with this formula as shown in Fig. 12. (Region I displays upper abdominal impacts; Region II, mid-abdominal impacts; and Region III, lower abdominal impacts.)

Experiment II — Ten cadavers were used in this study. The average age of the subjects used in this study was 66.8 years, the youngest being 49 years and the oldest 88 years old. A description of the cadavers is shown in Table 3. The results of the chest impacts are shown in Fig. 13. None of these chest impacts were found to cause rib fractures. The maximum force of impact was 1400 pounds while the maximum deflection was approximately 2.1 inches. A variation of approximately 400 pounds was found in the average chest loads recorded for these ten cadavers. The variation in the penetration was found to be 0.3 inches.

The range of values recorded for the static load deflection tests on cadavers are shown in Fig. 14. The average stiffness for the unembalmed cadavers tested was found to be 70 pounds per inch. Rib fractures were recorded at loads of approximately 500 pounds and higher, and deflections of approximately three inches.

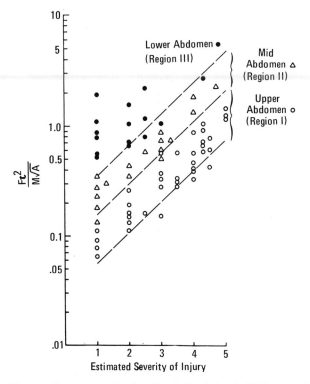

Fig. 12. Experimental scaling factor for abdominal injury sensitivity.

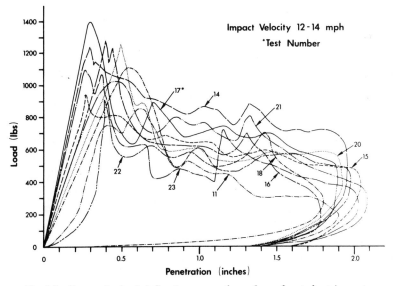

Fig. 13. Composite load-deflection curves for cadaver front chest impacts.

References pp. 197–198

TABLE 3

Description of Cadavers

No.	Sex	Age	Height	Body Wt. lbs.	Head Wt. lbs.	W in.	L in.	H in.	Circ. in.	Brain Wt. lbs.	Chest in.	Days Dead	Cause of Death	Comments
C11	M	70	5'6"	123.8	10.7	5.9	6.8	5.4	21.0	3.09	34.1	2	Acute Coronary	
C14	M	73	—	—	10.8	6.0	7.8	4.9	22.0	3.00	31.3	3	Unknown	Double Amputee
C15	M	65	5'2"	77	8.3	4.5	6.3	4.4	20.0	2.94	30.1	3	Bilateral Extensive Broncho-pneumonia, Fibrosis, Emphysema	
C16	M	88	5'8"	149.4	10.9	6.3	8.4	5.5	23.6	3.11	37.1	4	Unknown	
C17	M	49	5'11"	155.0	11.3	6.2	7.3	5.9	22.3	2.81	38.2	2	Acute Myocardial Infarction	
C18	F	65	5'3.5"	100.0	9.6	5.4	6.8	4.8	20.3	2.97	30.8	3	Generalized Carcinomatosis	
C19	M	52	5'10"	203.0	11.7	6.3	7.9	5.5	22.5	3.12	42.0	3	Coronary Heart Disease	
C20	F	75	4'8"	88.0	9.3	6.2	8.1	5.6	20.8	3.11	28.5	5	Coronary Thrombosis	
C21	M	62	6'0"	113.0	10.5	6.1	7.7	4.7	21.5	2.89	31.7	2	Coronary Heart Disease	
C22	M	63	5'7"	128.0	11.0	6.1	7.4	5.3	22.0	2.84	33.3	1	Cardiosenic Shock	
C23	M	58	5'10"	155.0	12.1	6.1	8.0	6.3	23.0	3.11	39.5	3	Aspiration of Pneumonia	

The results of two human volunteer tests are also shown in Fig. 14. Subject A was 32 years old, 6 feet tall and weight approximately 240 pounds, while Subject B was 27 years old, 5 feet 6 inches tall and weighed 136 pounds. The average stiffness for both subjects in the relaxed state was found to be approximately 230 pounds per inch while the stiffness for the tense state was found to be approximately 650 pounds per inch indicating a factor of three between the relaxed and the tensed state. Subject A was able to sustain a load of about 175 pounds and a chest deflection of about 0.8 inches before experiencing any discomfort in the relax state. In the tense state Subject A was able to undergo deflections of approximately 0.6 inches while sustaining loads of approximately 340 pounds before feeling discomfort.

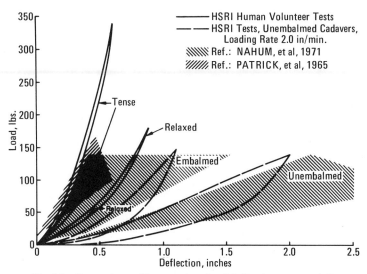

Fig. 14. Comparison of human chest load-deflection curves, A-P.

Subject B was able to sustain a load of about 50 pounds and a chest deflection of about 0.4 inches before experiencing any discomfort in the relax state. In the tense state Subject B was able to tolerate a load of 150 pounds at the same deflection level as for the relaxed state before feeling discomfort.

DISCUSSION

Experiment I — When the location of impact and mass of the subject was taken into account, the composite function (Estimated Severity Index $\alpha \, \mathrm{Log} \, \dfrac{f_r^2}{m \, a}$) related well to the degree of injury produced in an abdominal impact. This formula might be expected to apply to man for the following reasons: 1) the visceral weights of the pigs approximate those of man, 2) the other species studied were primates, and 3) the mass of the subject was taken into account in the formula.

The injury produced by a given force was found to be a function of the surface area, duration of impact, and mass of the animal. Narrow impactors applied to the body surface overlying the liver resulted in tearing and transection fractures of the liver. This was due to the localized and concentrated forces produced in the area. However, the liver tended to burst when the impact was made with a large diameter impactor. If the same force and velocity of impact were applied to a narrow bar impactor and a large surface impactor, the small bar tended to produce greater liver injury. When the impact pressure of the large diameter impactor and the bar impactor were the same, the greater injury was produced by the large impactor. Equal injuries (ESI-3) were produced when a pressure of 97 psi was applied with a bar impactor and 28 psi applied with a large surface impactor. These semmingly inconsistent results can

References pp. 197–198

be explained by the fact that liver tissue is viscoelastic. The viscoelastic properties of the liver cause high surface stresses resulting in failure over a large area. This property also results in injury being dependent on the velocity of impact. The injury produced by a force applied over a long period of time is much more severe than an injury produced by a briefly applied force, at the same velocity. Large diameter impactors produce a widely distributed load, because the structures adjacent to the liver such as ribs and muscle help distribute the load.

The location of the impact also influences the injury produced. Relatively small forces were required to produce severe injuries of the solid abdominal viscera when the impact was made in the upper abdomen (Region I, Fig. 9). Much greater forces were required to produce comparable injuries when force was applied to the lower abdomen (Region III).

The low incidence of injury of solid viscera associated with the use of lap belts confirms their value as a restraint system in the automobile. Few injuries other than hematomas of the mesentery and retroperitoneum were noted in these experiments. Intestinal injuries were rarely produced in this series and only occurred when an accidental maximum penetration was obtained. This agrees with Williams who postulated that lap belt injuries were only associated with a crushing of the intestine between the abdominal wall and the spine. (1, 10).

Experiment II — Results of the HSRI chest impact study are compared to impacts conducted by Kroell and Nahum (12, 13, 14) in Fig. 15. In their study, impactor weights of 42.5, 52, 4.1, 3.6, and 12.1 pounds were used to impact both embalmed and unembalmed cadavers. The impact velocities studied by Kroell and Nahum range from 11.1 miles per hour to 30.9 miles per hour. In the reduction of their data, peak forces, contributing to load cell ringing, were rounded off. The peak forces indicated in Fig. 13 were not due to ringing of the load cell because the resident frequency of

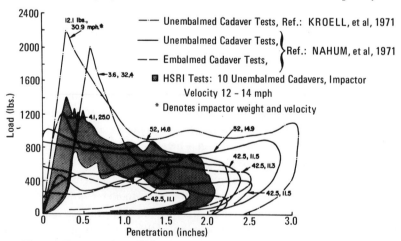

Fig. 15. Dynamic force-deflection curves for human chest.

the impactor was well above 3000 hertz. These high spikes at the onset of impact are contributed to the inertial loads required to accelerate the sternum up to speed.

The results of the static chest compression tests for embalmed cadavers, unembalmed cadavers and human volunteers indicates that fresh unembalmed cadavers have the lowest stiffness, while the embalmed cadavers were found to be significantly stiffer. The human volunteers in the relaxed state were found to have a stiffness comparable to that of the embalmed cadavers tested by Patrick (15, 16). The tensed volunteer load deflection curves were found to be considerably stiffer than any of the embalmed cadavers reported in the literature.

SUMMARY

There is a positive correlation between the degree of abdominal injury produced in blunt trauma and the force surface area, animal mass, and duration of impact in pigs and three species of primates. When the location of impact and mass of the sufject are taken into account, a composite function, ESI $\alpha \, \mathrm{Log} \, \dfrac{f_r{}^2}{m \, a}$, relates well to the degree of injury produced in an impact.

In the static compression test of the chest the embalmed cadaver was found to be stiffer than the chest of an unembalmed cadaver. The chest of a human volunteer in the tense state was found to be three times as stiff as when relaxed. The chest stiffness for the relaxed volunteer was found to be about the same as the embalmed cadavers.

Although the forces in the dynamic test varied widely, the rib fractures were found to be mostly dependent on the amount of deflection of the rib cages. .

REFERENCES

1. C. F. Baxter, and R. D. Williams, "Blunt Abdominal Trauma," J. Trauma 1: pp. 241-247, 1961.
2. D. L. Beckman, J. H. McElhaney, V. L. Roberts, and R. L. Stalnaker, "Impact Tolerance — Abdominal Injury," NTIS No. PB204171, 1971.
3. F. Glenn, Z. Mujahed, and W. R. Grafe, "Graded Trauma in Liver Injury," J. Trauma 3: p. 288, 1963.
4. L. Grimelius and G. Hellstrom, "Patho-Anatomical Changes After Closed Liver Injury: An Experimental Study in Dogs," Acta. Chir. Scand., 131: pp. 485-494, 1966.
5. G. Hellstrom, "Intra-Vascular Pressure Response to Closed Liver Injury: An Experimental Study in Dogs," Acta. Soc. Med. Upsal. 70: pp. 167-190, 1965.
6. G. Hellstrom, "Electrocardiographic, Blood Pressure, Respiratory and Electroencephalographic Responses to Closed Liver Injury: An Experimental Study in Dogs," Acta. Soc. Med. Upsal., 70: pp. 152-166, 1965.
7. E. T. Mays, "Bursting Injuries of the Liver." Arch. Surg. 93: p. 92, 1966.
8. H. Mertz, and C. Kroell, "Tolerance of Thorax and Abdomen," In Gurdjian, Impact and Crash Protection, C. Thomas, Springfield, Ill., pp. 372-401, 1970.

9. R. L. Stalnaker, J. H. McElhaney, R. G. Snyder and V. L. Roberts, "Door Crashworthiness Criteria," NTIS No. PB203721, 1971.

10. R. C. Williams and F. T. Sargent, "The Mechanism of Intestinal Injury in Trauma," J. Trauma 3: pp. 288-294, 1963.

11. P. G. Windquist, P. W. Stumm and R. Hansen, "Crash Injury Experiments with the Monorail Decelerator," AF Technical Report No. AFFTC 53-7, April 27, 1953.

12. C. K. Kroell, D. C. Schneider and A. M. Nahum, "Impact Tolerance and Response of the Human Thorax," Proceedings of the Fifteeth Stapp Car Crash Conference, p. 84, Paper No. 710851, Society of Automotive Engineers, Inc., New York, 1971.

13. A. M. Nahum, C. W. Gadd, D. C. Schnwider and C. K. Kroell, "The Biomechanical Basis for Chest Impact Protection: I. Force-Deflection Characteristics of the Thorax," J. Trauma 11: p. 874, 1971.

14. A. M. Nahum, C. W. Gadd, D. C. Schneider and C. K. Kroell, "Deflection of the Human Thorax under Sternal Impact," 1971 International Automobile Safety Conference Compendium, p 30, Paper No. 700400, Society of Automotive Engineers, Inc., New York, 1970.

15. L. M. Patrick, H. J. Mertz and C. K. Kroell, "Cadaver Knee, Chest and Head Impact Loads," Proceedingss of the Eleventh Stapp Car Crash Conference, p. 106, Paper No. 670913. Society of Automotive Engineers, Inc., New York, 1967.

16. L. M. Patrick, C. K. Kroell and H. J. Mertz, "Forces on the Human Body in Simulated Crashes," Proceeding of the Ninth Stapp Car Crash Conferences. p. 237, University of Minnesota, Minneapolis, 1965.

DISCUSSION

K. R. Trosien (Wayne State University)

I'm interested in your rating scale for liver impact injury. We're working with baboons and chimps where a series of tests are run on a given animal and therefore, it is not always practical to sacrifice immediately after a specific test. Also, will you comment, if possible, on the feasibility of extending this animal data to cover either children or adults subjected to a similar injury hazard?

R. L. Stalnaker

The evaluation of all injuries in this study was made within two hours of impact. The injury level was based on the mechanical damage caused by the impact. All injury ratings were made by an abdominal surgeon.

The scaling factor might be expected to apply to adult humans for three reasons: 1. The visceral weights of the pigs approximate those of man, 2. the other species studied were primates, and 3. the mass of the subject is taken into account in the scaling relationships.

T. L. Black (GM Design Staff)

I have a question about your thorax tests. Can you explain the difference between your load-deflection curves and the curves which were just presented by Larry Patrick? Also can you explain what the cracking sounds were that you mentioned?

R. L. Stalnaker

The cracking sounds in the static tests were the ribs failing. There were no rib fractures in the dynamic test.

I think the difference between our load-deflection curves and Patrick's are due to a couple of things. First, the angle at which the cadaver hit the impactor was different in the sled test than in our tests. Second, the amount of available energy in the two test set-ups were considerably different; our impactor weighs only 22 lbs. These two differences in test set-ups could make a large difference in the load-deflection curves.

H. E. VonGierke *(Aerospace Medical Research Laboratory)*

In your animal tests, the back of the animal was not supported by a rigid plate. In your human tests, however, the cadaver apparently was supported by a rigid plate. Would you explain why you used different techniques?

R. L. Stalnaker

The cadaver test as well as all animal tests were free; no back support was used in any test.

IMPACT RESPONSE OF THE HUMAN THORAX

**T. E. LOBDELL, C. K. KROELL, D. C. SCHNEIDER
and W. E. HERING**

General Motors Research Laboratories, Warren, Michigan

A. M. NAHUM, M.D.

University of California, San Diego, California

ABSTRACT

Part I — Biomechanics Response Data — Thoracic impact response data for unembalmed human cadavers previously published by three of the authors are reviewed. These data are then "averaged," adjusted to reflect an estimate for muscle tensing, and used as the basis for recommended force-deflection corridors to serve as dummy design guidelines. A volunteer study of muscle tensing, as related to thoracic stiffness at low force and deflection levels, is discussed, and comments are made concerning additional response data recently acquired by other investigators. Finally, consideration is given to possible "second order" refinements for future generations of a high fidelity dummy thorax.

Part II — Response of Current Dummy Chests — The chest structures of five currently available dummies were evaluated for blunt impact force-deflection response. Testing was conducted in essentially the same manner as was used to acquire the cadaver data of Part I. The resulting force-deflection characteristics were then compared with the GMR recommended performance corridors. In all cases, the existing structures were found to develop excessive resisting forces at deflection levels beyond 3/4 inch, clearly indicating the need for an improved design.

Part III — Mathematical Model for Thoracic Impact — A mathematical model has been developed which simulates the dynamic force-deflection response of the human thorax under blunt impact. The model consists of four differential equations derived from a mechanical thoracic analog formed from springs, masses, and dashpots. Equation parameters were adjusted to give desired responses. It was found that when parameters were set to give responses correlating closely with cadaver data previously published by three of the authors, the model response also correlated well with The

References pp. 242 - 243

University of Michigan HSRI cadaver data when the proper impact conditions were used. The model was used to show the relationship between various types of blunt thoracic impact and for dummy thorax design.

PART I – BIOMECHANICS RESPONSE DATA

INTRODUCTION

The General Motors Research Laboratories has maintained a collaborative biomechanics research program with the University of California since 1968. One of several in depth research studies which continues to constitute a major part of this program is addressing the question of human thoracic response to impact. Response in terms of both injury tolerance and mechanical properties is of interest, the latter being pertinent to this Symposium.

Thoracic mechanical response data for unembalmed cadavers were presented at the 1970 International Automobile Safety Conference (1) and at last year's 15th Stapp Conference (2) by three of the authors. Since that time, these data have been "averaged," adjusted to estimate the effect of muscle tensing, and used as the basis of load-deflection corridors which have been recommended as state-of-the-art impact response criteria for dummy chest design.

During this same period, several currently available dummy chest structures were tested for evaluation and comparison with the guidelines, and none was found adequate. The details and results of this study will be summarized in the second part of this paper.

In addition, a mathematical model for thoracic impact has been developed which exhibits response characteristics correlating well with both recommended force-deflection corridors, the latter having been established for two different combinations of impacting mass and velocity. It is encouraging also that this same model agrees reasonably well with the HSRI data just previously presented, which represents yet another combination of mass and velocity and, in addition, is presented in terms of *total* rather than *skeletal* deflection. This mathematical modeling work is documented in Part III of this paper.

Part I of the paper concerns biomechanics response data and will include:

1. A review of the General Motors Research/University of California published data,

2. A discussion of the recommended response corridors developed from these data,

3. Comments concerning a limited study of muscle tensing using volunteers,

4. A discussion of other biomechanics data concerning thoracic compliance, and finally,

5. A comment on possible future refinements to be considered for a truly high fidelity anthropomorphic thorax.

REVIEW OF AUTHORS' PUBLISHED DATA

Methodology — The published work covers two phases, chronologically documented in references (1) and (2). Reference (1) includes results from static as well as lower velocity impact tests performed on four embalmed and six unembalmed cadavers. Reference (2) includes data for higher velocity impact tests performed on fourteen additional specimens, all unembalmed. The test procedures and measurement techniques employed in the two studies were generally similar. One major exception concerns the measurement of thoracic deflection and is discussed below.

Figs. 1 and 2 illustrate the technique used for applying blunt sternal loading to the cadaver specimens. The subjects were seated in front of the impactor with arms outstretched and lightly taped to horizontal supports for maintaining the torso posture. During impact, this minimal restraint was broken with negligible effect upon the resulting whole body kinematics.

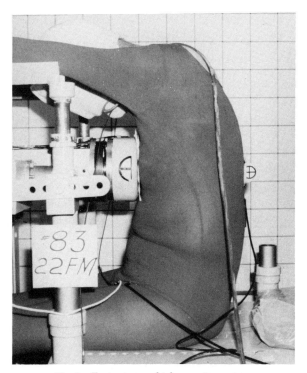

Fig. 1. Test set-up — high mass impactor.

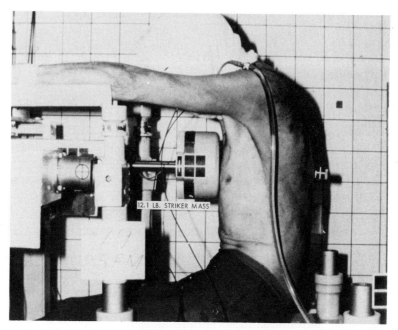

Fig. 2. Test set-up — low mass impactor.

The impactor mass was guided by horizontal rails and accelerated to test velocity by elastic shock cords. Two basic impactor configurations were employed. For the heavier striker weights (43-52 lb) a single mass was accelerated to speed, released and allowed to impinge directly against the chest (Fig. 1). For the lighter weights (3.6-12.1 lb) a carrier/striker combination (Figs. 2 and 3) was used. With this arrangement both masses were simultaneously accelerated to test velocity. The carrier was then stopped, and the lighter weight striker continued forward to provide the impact. The two masses did not separate completely, the striker being constrained to fore and aft horizontal motion by a pair of stainless steel shafts riding in bushings affixed to the sides of the carrier (Fig. 3). This impactor configuration was used to provide higher velocity, shorter duration pulses; and to provide for a minimum striker weight, measured acceleration and striker mass were used to compute the interface force. The heavier, single mass configuration was used more frequently and was equipped with a load cell for measuring the impact force directly.

In all cases, the impact was centered over the fourth interspace and delivered through a 6-inch diameter unpadded, flat wooden form with 1/2-inch edge radius.

When the load cell was used, triaxial components of the impact force were measured. Also recorded were precision timing and velocity measuring pulses, striker displacement, and signals to effect synchronization of the oscillograph record and high-speed film. In addition, high-speed photographic documentation was

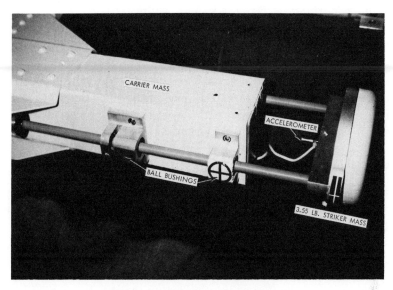

Fig. 3. Details of low mass impactor.

accomplished at a nominal framing rate of 1000 pictures per second for the lower velocity tests and at 2000 pps for the higher velocities. Analysis of these films was considered the best technique for determining the chest deflection versus time relationships desired. As noted above, different film analysis procedures were employed in the two phases of this study. In the earlier work, chest deflection was most frequently measured by photographically monitoring the displacement of a rod attached anteriorly adjacent to the sternum, passing through the thoracic cavity and emerging through a bushing sutured to the back. By measuring the displacement of the free end of this rod relative to the back of the subject, the *skeletal* deflection (sternum relative to spine) was closely approximated.

In the second phase work, use of this technique was precluded by the desire to autopsy for visceral damage subsequent to the impact. Thus, *total* chest deflection was determined from film analysis by measuring the differential displacement of the striker and the X-component of the subject's back motion. This measurement at any time was taken as the instantaneous *total* thoracic deflection. This *total* deflection includes interface effects and will always be somewhat greater than the more meaningful *skeletal* deflection (i.e., more meaningful as an injury criterion and dummy design parameter). By "interface effects" is meant the compressibility of superficial tissues overlying the bony thorax and the deviation from a square impact, which is determined by the anterior thoracic anatomy and posture of the subject.

The differential displacement analysis used to determine *total* chest deflection was begun with the film frame corresponding to load-time-zero on the oscillograph record. This instant was almost always clearly defined on the force-time curve by an abrupt rise from the baseline. Correspondence with a specific high-speed film frame could be achieved within 1/2 ms by use of either a timing pulse frequency switch or an event signal — redundant methods for film/record synchronization. In both cases, a conspicuous mark is impressed simultaneously upon record and film which, together with the precision timing marks, is used to establish the film frame corresponding most closely to load-time-zero. From this frame then, which frequently did not show the striker face in full area contact with the chest, measurement of the differential displacement was begun. It is considered that in most cases a value of from 1/2-3/4 inch should be subtracted from the *total* deflection thus measured to approximate the true *skeletal* deflection desired.

In Phase 1, in addition to the dynamic tests, several cadavers were loaded statically to low-force levels, and corresponding force-deflection relationships were determined. This was accomplished by advancing the impactor against the chest of a subject rigidly supported by a backstop. The interface was the same as used dynamically, and a jack screw was used to develop the loading. A six-pound preload was set and used as the zero reference point for measuring force and deflection. The latter was measured as the displacement of the impactor relative to the rigid backstop, and as such represents *total* chest deflection. It was found that if a given deflection level were applied and held constant, the associated load would fall off with time. Thus, the force-deflection curves were determined by loading to a given deflection level, immediately reading the force, unloading back to zero and then repeating the sequence to the next deflection level.

The force-deflection data from both phases of the study are presented in the RESULTS section in the original format. In addition, the *estimated skeletal* deflection for the Phase 1 static curves and for the averaged Phase 2 15-16-1/2 mph dynamic curves are given. For additional details and qualifications relating to the deflection measuring techniques and to the methodology in general, refer to the original publications, (1) and (2).

Table 1 provides data on the cadavers used. Testing was carried out from two to ten days after death, most often within four days. The specimens were kept refrigerated except for the periods of preparation and testing. Specimen identification terminology consists of a number (used in chronological sequence) followed by two letters, the first referring to state of preservation (E-embalmed and F-unembalmed) and the second to sex (M-male, F-female).

Results — The Phase 1 results of interest include both static and lower velocity dynamic (9-14 mph/43-lb striker) force-deflection characteristics. These are presented in Figs. 4 and 5, respectively.

As noted, the Phase 1 static data of Fig. 4a, like all the Phase 2 dynamic data, were obtained and plotted in terms of *total* thoracic deflection and thus include the interface effects previously described. Unfortunately, it is simply not possible to assign a well established correction factor to these data to express them in terms of *skeletal* deflection. As previously explained under METHODOLOGY, in the case of the Phase 2 dynamic data, it is recommended that 1/2-3/4 inch be subtracted from all

TABLE 1

Specimen Data

Cadaver No.	Sex	Age	Height	Wgt	Chest Girth	Chest A-P Diameter	Date of Death	Dates of Testing	Cause of Death
1 EM	M	64	5'7-1/2"	120	35-1/2	10	1/12/69	1,2/69	Myocardial infarction
2 EM	M	80	5'7"	175	40-3/4	9-1/2	6/18/68	2,3/69	Cardiac standstill arteriosclerosis
3 EM	M	70	5'4"	125	39	9-3/4	2/23/69	4,5,6/69	Heart block due to pulmonary emphysema
4 EM	M	73	5'9"	150	38-1/2	9-1/4	3/16/69	5,6/69	Carcinomatosis
5 FM	M	60	6'1"	190	44	10-1/8	7/2/69	7/69	Myocardial infarction
6 FM	M	83	6'0"	170	42	10	9/6/69	9/69	Coronary and cerebral arteriosclerosis
7 FF	F	86	5'6"	83	30-1/2	7-7/8	9/15/69	9/69	Cerebral thrombosis resulting from arteriosclerosis
9 FM	M	73	6'1"	168	40	9-3/8	10/29/69	10/69	Myocardial infarction Coronary arteriosclerosis
10 FF	F	82	5'3"	95	29	6-5/8	11/18/69	11/69	Cerebral anoxia Coronary insufficiency
11 FF	F	60	5'3"	130	–	–	12/13/69	12/69	Malnutrition Liver failure
12 FF	F	67	5'4"	138	33	7-3/8	2/28/70	3/10/70	Heart disease
13 FM	M	81	5'6"	168	38	9-11/16	3/9/70	3/13/70	Renal failure
14 FF	F	76	5'1-1/2"	127	35	8-1/2	4/6/70	4/10/70	Myocardial infarction
15 FM	M	80	5'5"	117	34	7-7/8	4/17/70	4/24/70	Bronchopneumonia
18 FM	M	78	5'9-1/2"	145	36	8-5/8	6/16/70	6/18/70	Strangulation
19 FM	M	19		145	35-1/2	8	7/10/70	7/14/70	Leukemia
20 FM	M	29	5'11"	125	32	8	7/26/70	7/28/70	Metastatic pulmonary carcinoma
21 FF	F	45	5'8-1/2"	151	39	8-3/8	8/7/70	8/10/70	Cerebral edema
22 FM	M	72	6'0"	165	40-1/2	8-7/8	9/2/70	9/4/70	Myocardial infarction
23 FF	F	58	5'4"	135	34-1/2	8-7/8	10/18/70	10/22/70	Ovarian carcinoma
24 FM	M	65	6'0"	180	40	9-7/8	10/28/70	11/2/70	Strangulation
25 FM	M	65	5'6"	120	34	8-1/8	1/27/71	1/29/71	Coronary thrombosis
26 FM	M	75	5'8"	140	37	9-3/4	2/28/71	3/3/71	Cerebral occlusion
28 FM	M	54	6'0"	150	34-1/2	9-3/8	5/5/71	5/7/71	Large bowel carcinoma

References pp. 242 - 243

deflection values (i.e., the curves be shifted to the left or the ordinate axis to the right by this amount). This, of course, is a simplified estimate also, but reflects a best judgment based upon photographic description of the interface contact geometry and some measurements of superficial tissue thickness. These latter data do not exist for

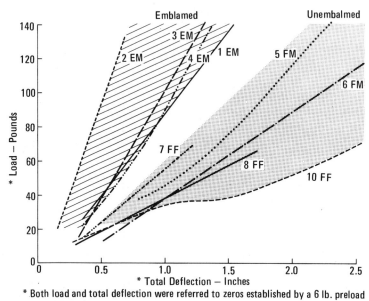

Fig. 4a. Static force-deflection characteristics for embalmed and unembalmed cadavers — Reference (1).

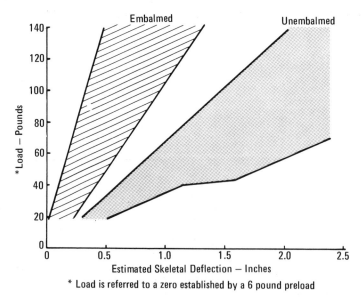

Fig. 4b. Static curves of Fig. 4a adjusted to estimate skeletal deflection.

the Phase 1 static curves. Neither is it quite clear how best to make an approximate correction. No other factors considered, the lower load levels of these static curves would logically require a smaller correction than applicable for the higher level

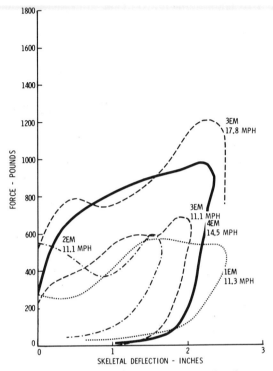

Fig. 5a. Dynamic force-deflection characteristics for embalmed cadavers using 43-lb. striker — Reference (1).

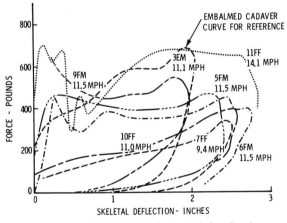

Fig. 5b. Dynamic force-deflection characteristics for unembalmed cadavers using 43-lb. striker — Reference (1).

dynamic tests. On the other hand, because of the low rate of loading, without the presence of viscous and inertial effects, the static curves do not follow the characteristic shape of the dynamic curves (rapid, early rise in load to a plateau level, which is maintained to large deflections). They are essentially linear, and in such a case the interface correction would be expected to increase with increasing load. Also the correction required would be reduced by the six-pound pre-load used as the zero reference level.

In the light of the above and in view of the substantial scatter demonstrated among specimens, a *suggested* correction is to subtract 1/8 inch from the deflection level at, say, 20 lb and 1/4 inch from the level at $\sim$ 140 lb. If this process is applied to the curves defining the scatter bands (embalmed and unembalmed) of the static data, the result is not a very significant change in the static corridors (Fig. 4b).

In Figs. 5a and 5b are plotted the Phase 1 dynamic force-deflection results — for embalmed specimens in 5a and for unembalmed in 5b. Here, as the result of using the probe deflectometer technique, the data are plotted directly in terms of *skeletal* deflection.

The Phase 2 tests, at higher impact velocities, have been very thoroughly documented in reference (2). Fig. 6 illustrates the format used to summarize each test in that reference, and detailed qualifications of the data as presented are also included therein. For the purposes of the review presented here, only the composites of force-time and force-deflection characteristics are given. Fig. 7 displays copies of the original force-time curves, and Figs. 8a and 8b are composite plots of the corresponding *total* force-deflection relationships. All of the curves in Fig. 8a

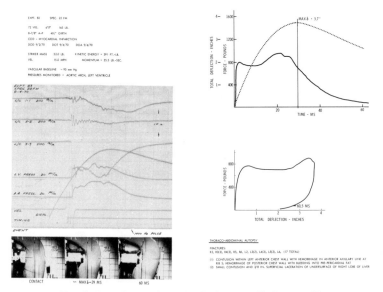

Fig. 6. Data format for individual tests in Reference (2).

correspond to impact velocities in the range of 15-16-1/2 mph and to a striker weight of 50-52 lb. In Fig. 8b are plotted other curves representing a variety of impact conditions, as noted.

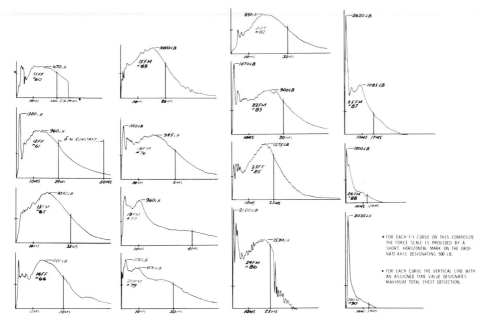

Fig. 7. Force-time curves from Reference (2).

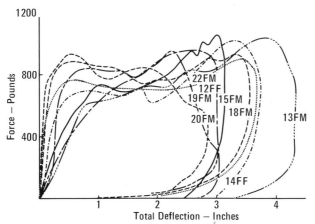

Fig. 8a. Dynamic force-deflection characteristics for unembalmed cadavers using 50-52 lb. striker at ∿16 MPH – Reference (2).

RECOMMENDED RESPONSE CORRIDORS

The dynamic force-deflection characteristics of Figs. 5b and 8a have been used as the basis for a pair of corridors recommended as performance guidelines for the

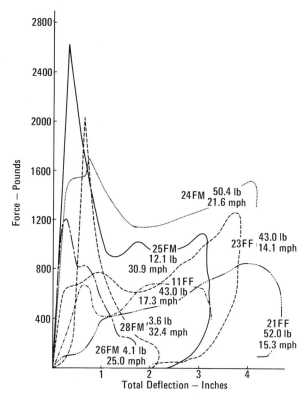

Fig. 8b. Dynamic force-deflection characteristics for unembalmed cadavers using various combinations of striker mass and velocity — Reference (2).

design of improved dummy chest structures. "Eyeball" averaging of the curves for subjects 5 FM, 6 FM and 9 FM of Fig. 5b (43-lb striker @ ∿ 11 mph) and of all the curves in Fig. 8a (∿ 51-lb striker @ ∿ 16 mph) was first performed. The resulting 16-mph "average" curve for *total* deflection was displaced to the left by one-half inch to represent *skeletal* deflection (as discussed under METHODOLOGY). Then, to approximate muscular stiffening for a maximally tensed condition (assumed as probable for most occupants at the onset of a collision), a constant force augment of 150 lb was applied out to maximum load for both the 11- and 16-mph levels. This is definitely a current suggestion based upon very limited quantitative data. There exist numerous unanswered questions relative to the thoracic stiffening augment from muscle tensing, and these are discussed in greater detail in the following section, MUSCLE TENSING STUDIES. It is anticipated that further clarification of these questions will emerge from future studies. However, it is also believed that some adjustment of the existing unembalmed cadaver data is warranted to provide the most reasonable guidelines for design and performance evaluation purposes.

The unloading portions of the curves thus adjusted were simply faired to the same general shapes as those of the original averaged curves. It is conceivable that the effect of muscle tensing would be to cause a more rapid unloading and a concomitant reduction in hysteresis; however, there simply are no pertinent quantitative data available at the present time.

Because of uncertainties as to the fidelity of the leading edges of the force-time histories involved (see reference 2, page 101 for detailed discussion), the recommended performance corridors do not extend below 1/4 inch of *skeletal* deflection. Thus the waveshape during the first 1/4 inch of *skeletal* deflection is not restricted so long as the remainder of the curve falls within the corridor boundaries. Beyond 1/4 inch, the corridors were constructed by straddling the averaged, adjusted design goal curve by approximately ± 15% of the load out to peak load and, during unloading, ± 15% of the deflection, from maximum deflection down to the 200-lb level. The general wave form of the original force-deflection curves is essentially established by the time the load has fallen to this level and below which the curves become less well defined.

The ± 15% value does not apply point for point over the entire curve but was used to establish the basic corridor configuration in the regions of 1) the extended load plateau, and 2) maximum deflection. These segments were then joined to complete the design/performance corridor. To simplify specification, the corridor boundaries have been reduced to a series of line segments with coordinate values assigned to their points of intersection. Fig. 9 displays both the averaged, adjusted curves for both velocity levels and the associated corridors as described above.

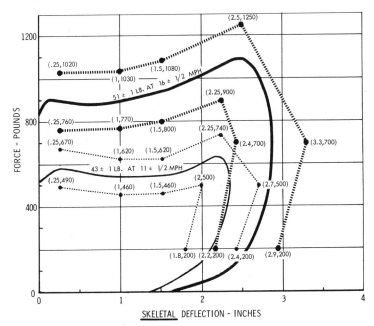

Fig. 9. Averaged, adjusted force-skeletal deflection curves and recommended response corridors.

References pp. 242 – 243

It is recommended that the thoracic impact response characteristics of any dummy designed in accordance with the above proposed guidelines be ascertained experimentally using a test method as nearly as possible identical to that employed in acquiring the original cadaver data. This has been summarized above under METHODOLOGY and covered in detail in references (1) and (2).

In the cadaver "mid-sternum" was defined as the level of the fourth interspace (space between the fourth and fifth ribs). For the fiftieth percentile male dummy, it is currently specified by SAE J963 as lying 18 ± 0.5 inches below the top of the head.

Unfortunately, because of necessary midstream equipment modifications, the striker masses associated with the Phase 1 data (11 mph, 43 lb) and the Phase 2 data (16 mph, 50.4 and 52 lb) differ by as much as 9 lb. Although it is believed, and also indicated theoretically, that mass differences of this amount would not appreciably extend the scatter beyond that presently seen to result from specimen variations, this cannot be established positively until additional data are obtained. Therefore, it is recommended that, for the present, dummy testing be performed at $11 \pm 1/2$ mph using a 43 ± 1 lb mass and at $16 \pm 1/2$ mph using a 51 ± 1 lb mass.

As discussed under METHODOLOGY, the primary measurements required are 1) the component of the impact force perpendicular to the striker contact surface as a function of time, and 2) the mid-sternal *skeletal* A-P deflection as a function of time (frequency response should be channel class 1000 as per SAE J211). Finally the crossplot of these functions, determined either directly on-line or through data reduction, is necessary to provide the desired force-deflection characteristics.

MUSCLE TENSING STUDIES

A limited investigation has been performed concerning the effects upon thoracic stiffness of deliberate, generalized muscle tensing. Male volunteers were loaded mid-sternally in a quasi-dynamic fashion ($\sim$ 100 ms to peak load and deflection) using the same interface configuration (hard, flat, 6-in. diameter surface with 1/2 in. edge radius) as had been used with the cadavers. As shown in Fig. 10, the subject was seated with his back bearing against a rigid, adjustable support. An abrupt deflection input of fixed amplitude was applied manually using the lever arrangement shown. Normal force against the chest and absolute AP displacement of a point on the loading bar directly behind the center of the interface were measured using the standard load cell and a string potentiometer, respectively. This displacement was taken to be the *total* AP chest deflection. A complete data record consisted of the time histories of both force and *total* deflection as well as the force-deflection crossplot. These were read out simultaneously on a dual beam oscilloscope with one beam input supplied through a multi-channel chopper amplifier. A typical record is illustrated in Fig. 11.

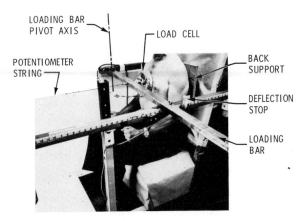

Fig. 10. Test set-up for volunteer muscle tensing tests.

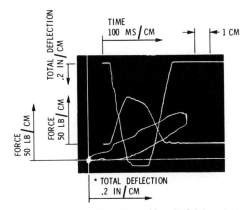

* Deflection measured from the condition of a 5-6 lb. preload;
  force measured from true zero.

Fig. 11. Typical complete data record for volunteer muscle tensing test.

Each volunteer was asked to assume an erect torso posture with his back bearing against the plywood support. The loading bar pivot axis was then adjusted fore and aft so that the loading bar would be approximately parallel to the back support at mid-deflection; and the pitch angle of the load cell was set to compensate for rake of the anterior rib cage (of the order of 5-10°). The loading bar was then displaced until a preload of approximately five pounds was developed, and a positive stop was set to allow a maximum additional AP displacement of 1 to 1-1/4 inches. For this position of the loading bar, the potentiometer was then adjusted either for zero output or, in some cases, with enough pre-deflection to ensure that the force-deflection curves generated would include no preload at indicated zero deflection.

Measurements were made first with the subject in a fully relaxed state and taken unaware. Then he was asked to maximally tense the generalized musculature of his upper body (thorax, shoulders, arms, back, and neck) and, when ready, signal for the application of load. This was accomplished in the absence of any deep respiratory inspiration. However, in all cases there was an increase in AP chest diameter associated with the tensing maneuver alone. As a result, two techniques were used for applying the deflection input. First, subsequent to testing in the relaxed state, all equipment adjustments were maintained, and an experiment was run in the tensed state. Then, the loading arm pivot and stop locations and the zero deflection reference were readjusted for the tensed state as had been done for the relaxed, and a second test was run under these conditions.

Figs. 12a, 12b, and 12c show typical force-deflection curves for the relaxed and the two tensed loading conditions, respectively. Figure 12b corresponds to a tensed state loading using the equipment adjustments established for the relaxed state of Fig. 12a. Fig. 12c is for a tensed state loading with tensed state adjustments (0-5 lb preload at zero deflection). The loading curve slopes (stiffnesses) for both tensed conditions are essentially the same and are substantially greater than that for the relaxed state (Fig. 12d). The main difference between the curves of Figs. 12b and 12c is the significant preload ($\sim$ 70 lb) existing at zero indicated deflection. This is the direct result of using a zero deflection reference established in the relaxed state when loading a chest expanded as the result of muscle tensing. In other words, chest contact and load buildup have already begun by the time deflection monitoring begins.

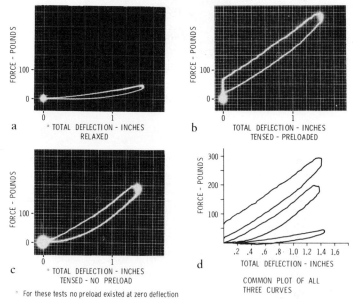

Comparison of Thoracic Stiffness in Relaxed and Tensed Modes

Fig. 12a. Relaxed.   Fig. 12b. Tensed – Preloaded.

Fig. 12c. Tensed – No preload.   Fig. 12d. Common plot of all three curves.

The mechanism of chest expansion due to generalized muscle tensing without deep inspiration is not thoroughly understood. A combination of rib motion and bulging musculature of both chest and back appears to be involved. It is plausible that any contribution due to rib motion — i.e., true skeletal expansion — could represent added protective potential derived from tensing beyond that due to a direct increase in stiffness. It seems that the internal distance between sternum and spine should be an important dimension from the standpoint of thoracic trauma, and to deform the chest sufficiently to produce a given separation distance (sternum to spine) would require higher forces and greater energy as the result of both increased stiffness and initial skeletal diameter.

Figs. 13a and 13b are corridor type plots for the volunteer loading curves in both the relaxed and tensed states. In Fig. 13a, the tensed state corridor is for the readjusted 0-5 lb preload condition, and in Fig. 13b, it corresponds to tests for which relaxed state adjustments were used. For any given *total* deflection level, the comparative distances between sternum and spine for the relaxed and tensed states are not known. However, it seems probable that in Fig. 13a this dimension for the tensed state should exceed that for the relaxed, whereas in Fig. 13b it should be slightly less. In other words, a selected *total* deflection reference level corresponds to a certain spine to sternum separation distance and associated force level in the relaxed state. The same separation distance for the tensed state requires a different *total* deflection level, and the associated force should fall between the two tensed state force levels corresponding to the selected deflection reference level (i.e., between the tensed state force levels from Figs. 13a and 13b as measured at the selected *total* deflection reference level).

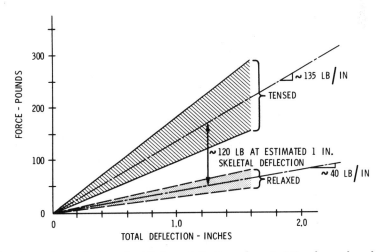

Fig. 13a. Comparison of thoracic AP force-deflection characteristics for male volunteers in relaxed and tensed conditions — zero deflection set at nominal 5 lb. preload in both cases.

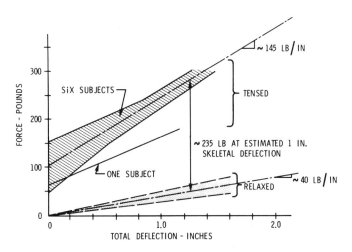

Fig. 13b. Comparison of thoracic AP force-deflection characteristics for male volunteers in relaxed and tensed conditions — zero deflection set at nominal 5 lb. preload for relaxed state and not altered for tensed state.

For the loading conditions used in this study, there is clearly a significant effect upon thoracic stiffness imposed by tensing of the upper body musculature. For example, at a *total* deflection reference level of 1-1/4 inches (assumed to approximate a *skeletal* deflection of ∿ 1 inch), from Figs. 13a and 13b the tensed state force level should be somewhere between 120 - 235 lb greater than that for the relaxed state. Table 2 provides certain anatomical data for the participating volunteers.

There remain numerous unanswered questions concerning the effects of muscle tensing. No data are known of which indicate the frequency, nature, and degree of muscle tensing which typifies vehicle occupants at the onset of an accident. It is considered reasonable, however, that the average forewarned occupant would be in a tensed condition. It is not known how the stiffening effect of muscle tensing might vary with impact velocity, level of chest deflection, degree of skeletal damage, or respiratory state. The effect of tensing upon the unloading curve (and thus hysteresis) at higher impact levels is not known. Also, there is the method of loading question — free body impact with a transthoracic inertial gradient in the field versus compressive loading between two surfaces in the volunteer experiments described. However, as noted in the preceeding section, it is believed that some load level adjustment should be applied to the unembalmed cadaver curves, and as a first approximation, the constant 150-lb augment has been suggested.

OTHER NEW BIOMECHANICS DATA

Recently The University of Michigan Highway Safety Research Institute has presented dynamic force-deflection characteristics for a series of ten unembalmed human cadaver subjects. As described, the test method used was very similar to that

TABLE 2

Volun-teer	Age Wt-lb Ht-in.	Chest Girth in.				† Chest AP Diameter in.				† Chest Lateral Diameter in.				Sternal Depression* in.			
		E**	R	C	T	E	R	C	T	E	R	C	T	E	R	C	T
GH	33 177 69.1	41.0	39.5	37.8	40.5	10.6	9.9	9.5	11.2	14.1	13.3	13.3	14.9	.88	.75	.63	.56
JDH	39 147 65.9	37.5	36.5	35.8	37.8	9.4	8.9	8.6	9.5	12.3	11.7	11.6	13.1	.50	.38	.38	.56
WEH	35 145 67.5	36	34.5	33.5	37.4	8.9	8.4	7.8	9.1	12.1	11.8	11.8	12.8	.31	.19	.19	.44
CKK	39 174 70.5	40.1	39.0	37.4	39.1	10.4	9.3	9.0	10.0	13.8	12.8	12.4	14	.75	.56	.56	.63
WCL	29 198 71.6	40.4	39.0	38.3	39.9	9.9	9.4	9.2	9.8	12.9	13.3	13.2	13.2	.19	.25	.25	.38
JDM	29 207 75.4	43.0	41.5	39.5	42.8	11.3	10.4	10.5	11.3	14.1	13.5	13.3	14.3	.50	.38	.38	.50
JER	26 182 73.3	40.6	39.0	38.0	39.4	10.3	9.0	8.8	9.8	13.6	13.7	12.8	13.8	.25	.31	.28	.25

* *Sternal depression is defined here as the distance from the surface of the skin covering the sternum to a straight edge laid across the pectoral protuberances. This is sensitive to variations in thoracic posture.*

** *E – expanded maximally through inspiration; no generalized muscle tensing.*

R – relaxed

C – contracted maximally

T – generalized upper body muscle tensing; no deep inspiration.

In E, R, and C states, measurements were made with subject standing and hands on hips; in T state, arms were bent ⌄ 90° at the elbows and held in the AP direction.

† *Measured by parallel straight edges at nipple level.*

of the GMR/U of C study, the major difference being the selection of impact parameters. In the HSRI tests an impactor mass of 22-pounds and an impact velocity of 12-14 mph were used. The cadaver specimens involved appear to have been generally similar from the standpoint of age and anatomy to those used in the GMR/U of C work. However, despite the advanced age, a complete absence of skeletal damage was reported, whereas the data of Figs. 5 and 8 correspond, except in two cases of cadavers under thirty years of age, to significant degrees of fracture damage.

References pp. 242 – 243

Because of the lower kinetic energy of the HSRI impactor, the maximum total deflection levels were only of the order of 2-inches. The lack of skeletal fracture damage *at this deflection level* is consistent with the findings reported in references (1) and (2).

Generally, the characteristic curve shapes of the HSRI data are similar to those presented in Figs. 5 and 8, the best comparison being with the curves for 19 FM and 20 FM of Fig. 8a. These are for specimens of 19 and 29 years, respectively, and for which there was no skeletal damage despite maximum deflections of 3.0 and 2.8 inches.

The major obvious difference is the more pronounced force peak defining the early portion of the curves. Subsequently, the force decreases with increasing deflection, with no upward hooking near maximum deflection, such as is typical in Figs. 5 and 8. A plausible explanation for the latter is that, at *total* deflection levels of up to 2-inches, "bottoming out" of the thoracic viscera has not yet begun to occur; and the decline in load with increasing deflection is a manifestation of the highly viscous character of the thoracic system. The curves for 26 FM and 28 FM, Fig. 8b, with maximum *total* deflection levels of the order of 2-inches clearly demonstrate this same behavior.

The question of the early load peak is interesting and pertinent. The peaks in the HSRI data are fully developed and on the decline within a *total* deflection level of 1/2 inch. Time wise this corresponds to only 2-3 ms at the velocity levels in question. It will be seen that several of the load cell generated curves of Fig. 7 exhibit early load spikes of this same general nature. However, as is explained in detail in reference (2), because of uncertainty as to whether these spikes were real or spurious (load cell response), it was decided to delete them when crossplotting to produce the force-deflection curves. On the other hand, the curves for 25 FM, 26 FM and 28 FM in Fig. 7, generated by an accelerometer on a lightweight, very rigid impactor mass at higher velocity levels (25-32 mph), display very early load peaks the validity of which is not questioned. Thus, it is possible that the other spikes seen in the curves of Fig. 7 are valid also. As a proportional part of the entire curve, such spikes are expanded when plotted against displacement (deflection) as compared to time. Thus, including them in the crossplots of Fig. 8 would render the initial portions of the curves so affected a better match to the HSRI data. Knowledge concerning the performance specifications of the load cell used to acquire the latter data would be pertinent and helpful.

In any event it will be noted that only about half of the Fig. 7 curves do show the pronounced spikes. Further, even if these do represent true interface forces, they occur at low values of *total* chest deflection and have essentially vanished by the onset of perceptible skeletal deflection.

The new data are regarded as a useful contribution by extending the definition of human blunt thoracic impact response to yet another set of striker mass and velocity conditions. The wider the spectrum of knowledge in this respect the better. It is felt that for most effective utilization as dummy design guidelines all such data should be adjusted in the best manner possible to reflect *skeletal* deflection and include an augmentation for muscle tensing.

Some investigators have expressed concern over the use of data for dummy design purposes obtained from tests producing significant skeletal damage in the cadavers. It is submitted here that such tests represent accident situations sufficiently severe to cause serious thoracic trauma and, as such, are of definite interest in collision simulation work. In such cases, it is highly desirable that the fidelity of dummy chest response extend into the deflection domain corresponding to skeletal damage. Actually, even during these more severe exposures, the response exhibited is that of a structurally sound thorax out to the large deflection levels at which skeletal failure begins. A truly high fidelity artificial chest structure will exhibit proper response not only at these more severe exposure levels but at lower levels of severity and for other combinations of impactor mass and velocity as well.

CONSIDERATIONS FOR THE FUTURE

Although for the present the task of designing and constructing a dummy chest which will demonstrate good blunt impact performance correlation with the existing biomechanics data is amply challenging, it is not too early to consider other design/performance criteria which, in the future, may well contribute to the refinement of a truly high fidelity chest structure.

Thus far, efforts have been mainly concentrated on defining the response characteristics (and injury tolerance picture) for blunt, anteroposterior, mid-sternal impact. Definition of these characteristics over the impact velocity range from static to 30+ mph is still being carried out. It would be desirable, ultimately, to vary the size and configuration of the loading interface, and beyond that its location on the body and the direction of load application. Particularly, the question of lateral impact has not been addressed, nor that of the "shoveling" effect of Voight (3). A limited study of highly localized loading ($\sim$ 1 square inch, applied over individual ribs) was reported in reference (1), but additional data are needed.

Finally, of paramount importance is the interfacing of the thoracic structure with equally high fidelity shoulder and abdominal sections. For, the end product must function not merely as a collection of isolated parts but as a highly complex mechanical amalgam of closely coupled, intricate components.

ACKNOWLEDGMENTS

The authors gratefully acknowledge the contributions of Messrs Robert Barton and George Nakamura whose assistance with the experimental work was invaluable.

References pp. 242 - 243

PART II — RESPONSE CHARACTERISTICS OF CURRENT DUMMY CHESTS
INTRODUCTION

The occupant protection requirements set forth in MVSS 208 are based on human tolerance data. In order for the measurements obtained from dummies to have any meaning relative to these tolerance levels, the dynamic response characteristics of the dummy must be similar to the human.

Thoracic response requirements are given in the SAE Recommended Practice J963. This specification places a restriction on chest resisting force at a given deflection level. The current commercial chest structures of Alderson and Sierra dummies meet this requirement.

Recent cadaver response data (1,2), upon which the GMR recommended response corridors (4) are based, indicate a somewhat different force-compression response. For a dummy chest to be acceptable on the basis of these data, its force-compression response must be within the respective corridor for each impact condition. These performance corridors are considered as necessary but not sufficient requirements to define a humanlike chest structure.

Dummy chest structures evaluated for this paper are compared with these performance corridors.

Test Procedure — The test procedure and facility for evaluating the thoracic structures are similar to those used to obtain the human response data. This consists

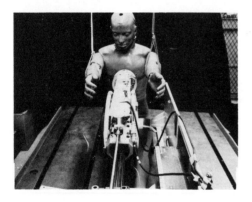

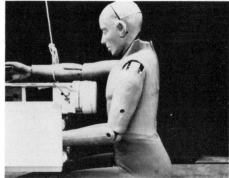

Fig. 14. Typical test set-up for blunt chest impacts.

of impacting the completely assembled dummy with a mass guided horizontally, essentially in a free flight condition. A typical test setup is shown in Fig. 14. The dummy is seated erect, sternum perpendicular to the impacting surface, and arms extended forward and held by lightweight cords hung from the ceiling to simulate the cadaver tests. The arm has been lowered in the side view to show the impacting face. Impact location was within the SAE J963 specification of 18 ± 0.5 in. from the top of the head to the center of the impacting face. This vertical impact point was controlled by placement of foam pads under the buttocks. The torso was free to rotate rearward upon impact.

An interface of 0.5 in. thick Rubatex Corporation's R-310V closed-cell foam was used (see Fig. 14) to reduce load cell ringing. Effect of the padding is illustrated in the shape of the force-compression curves shown in Fig. 15. The foam pad was attached to a 6-in. diameter wood block with an edge radius of 0.5 in. Impact force was measured with a load cell located between the wood-foam interface and the body of the impactor. Axial force only was measured in this series of tests. An accelerometer, mounted directly behind the load cell, was used for verification of the load cell response.

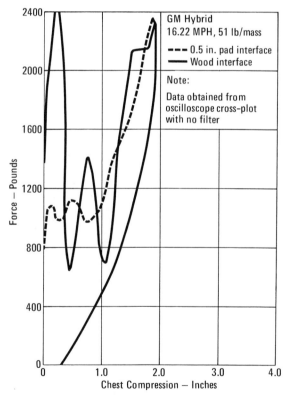

Fig. 15. Effect of padded interface on response characteristics.

The impacting mass was accelerated up to speed with two parallel bungee cords; a 10-inch elongation of the cords produced approximately 16 mph with a 51-lb mass. Velocity was obtained from a magnetic pickup signal which indicated 0.5 in. increments of impactor travel over a three-inch range.

Response characteristics of the chest structures in terms of force-compression were obtained by cross plotting force-time and compression-time traces. Indicated loads were corrected for the mass located in front of the load cell to give force applied to the chest. Chest compression was measured essentially collinear with centerline of the impacting target and was obtained from either a linear potentiometer, if space allowed, or a rotary potentiometer mounted on the spine and turned by a link which was connected to the sternum by a slider mechanism. Both force and compression were recorded with 0-1000 Hz flat response.

Chest Structures Evaluated — Five chest structures were evaluated: an Alderson VIP-50A, a Sierra 1050, a Sierra 850, a GM Hybrid, and an Ogle-MIRA. All chest structures in this investigation are commercially available "as tested" with the exception of GM Hybrid. However, similar chest structures can now be purchased from the Alderson Research Laboratories with the same modifications. It is doubtful that there is any significant difference between a GM Hybrid and Alderson modified chest structure.

The GM Hybrid is essentially an Alderson VIP-50A dummy modified as recommended by the GM Proving Ground Safety R&D Lab. The modification consists of attaching 1/4-inch thick Lord, LD-400, damping material to the inside of each rib, replacing the standard sternum with a 1/4-inch thick leather sternum, and removing the Ensolite pads and pad supports from behind the anterior rib structures. Fig. 16 shows a GM Hybrid rib structure along with an Alderson VIP-50A. The GM

Fig. 16. GM hybrid chest (left) compared to the Alderson VIP-50A chest with Ensolite pads.

Hybrid ribs were provided by the GM Proving Ground from their supply purchased from the Inland Division.

The Alderson Model VIP-50A dummy chest structure developed by Alderson Research Laboratories is compatible with SAE J963 specifications. The Ensolite pads and pad supports, used for damping in this structure, limit chest compression to approximately two inches. Therefore, it is impossible to achieve proper chest compression of almost three inches for a 16-mph impact with this design.

The Sierra Model 292-1050 chest structure consists of a polyethylene shell of relatively uniform thickness with a somewhat humanlike shape, as shown in Fig. 17. An earlier Sierra design, Model 292-850, was selected for evaluation even though outdated by the Model 292-1050 because this was the first chest structure designed on the basis of cadaver response data (5), and there are several such models in existence which are being used. The Sierra 850 chest is shown in Fig. 18. The structure was modified from the standard production model only by the addition of vertical shoulder supports, as shown in Fig. 18b. These were added to prevent contact between the shoulder structure and the top ribs.

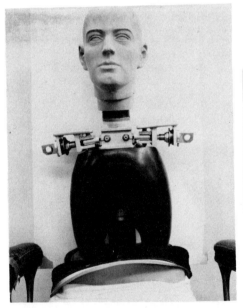
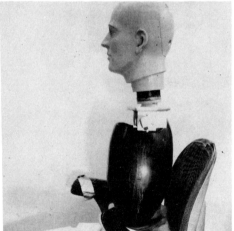

Fig. 17. Chest structure of the Sierra Model 292-1050 dummy.

The Ogle-MIRA M50/71 chest structure was developed in England by The Motor Industry Research Association (MIRA) (6) and is being manufactured by David Ogle, Ltd. The chest was designed to simulate force-deflection characteristics of the

References pp. 242 - 243

embalmed cadaver data reported by Patrick et al. (5). This structure closely resembles a human skeletal structure in terms of shape, number of individual ribs, ribs present between the shoulders, and the attachment of the shoulders as shown in Fig. 19. The top of the sternum is joined to each shoulder by means of a steel clavicle. The opposite end of the clavicle is connected to a scapula.

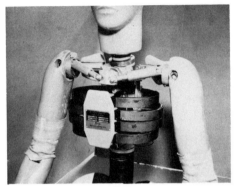

FRONT VIEW   BACK VIEW

Fig. 18. Chest structure of the Sierra Model 292-850 dummy.

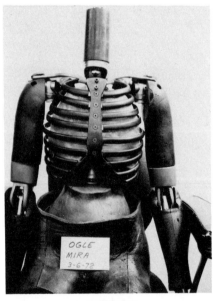
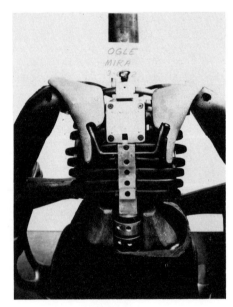

FRONT VIEW   BACK VIEW

Fig. 19. Chest structure of the Ogle-MIRA M 50/71 dummy.

Inside the front of the rib cage are three lead sheets separated with polyurethane foam. The front lead sheet represents the mass associated with the chest wall, while the rear two represent the mass of the internal organs.

Comparison of Responses to Corridors — Response characteristics of each dummy chest structure were obtained for nominal impact velocities of 11 and 16 mph. Chest compression (deflection) will refer to motion of the sternum relative to the back structure in all cases.

The Alderson VIP-50A responses fall outside the corridors at both impact speeds, as shown in Fig. 20. Peak force is approximately twice as great as it should be at both velocities. The response of the initial portion of each curve is relatively good up to 0.75 in. at 11 mph and 1.25 in. at 16 mph. Although not meaningful for total compliance, it does indicate the loading rate sensitivity required for proper compliance.

A similar response of the GM Hybrid is shown in Fig. 21. There is slightly more desirable hysteresis with the Hybrid, but the higher load for the 16-mph impact is

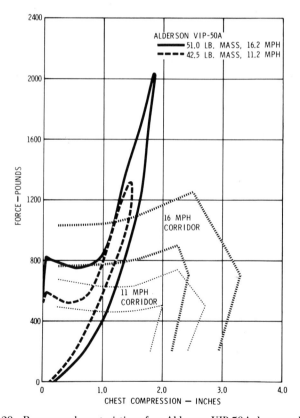

Fig. 20. Response characteristics of an Alderson VIP-50A dummy chest.

References pp. 242 - 243

undesirable. It is not possible to distinguish significant response differences between the GM Hybrid and the Alderson VIP-50A chest structures based on these response curves.

The Sierra 850 chest response also falls outside the corridors, as shown in Fig. 22. Maximum deflection at both impact velocities are less than those obtained from all other chest structures discussed in this paper.

The Sierra 1050 chest response also falls outside the recommended corridors, as shown in Fig. 23. Lack of allowable chest compression is illustrated by the sharp rise in load during the 16-mph impact. This problem could be solved by modifying the simulated thoracic spine to eliminate contact between the spine and the upper edge of the chest structure.

The Ogle-MIRA chest response is shown in Fig. 24. It is evident that this chest does not respond within the human response corridors at either impact velocity. Due to the presence of a chest accelerometer box, rib deflection is limited as indicated by the sharp rise in force at 1-1/4 in. deflection.

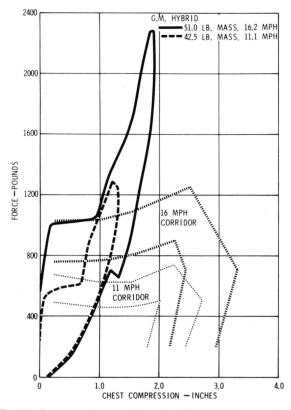

Fig. 21. Response characteristics of a GM Hybrid dummy chest.

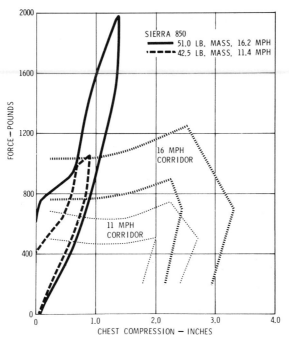

Fig. 22. Response characteristics of a Sierra 850 dummy chest.

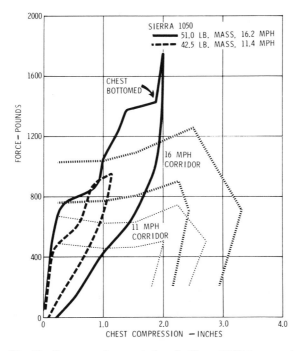

Fig. 23. Response characteristics of a Sierra 1050 dummy chest.

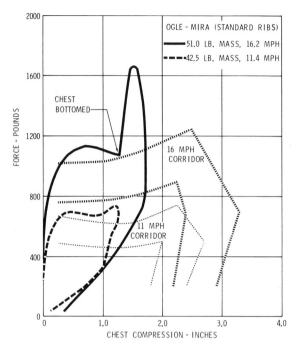

Fig. 24. Response characteristics of the chest of an Ogle-MIRA M50/71 dummy.

DISCUSSION

Chest Structures — All the commercial chest structures have deficiencies which must be corrected before they can be considered humanlike, both chest compliance and physical design must be considered. All but the Sierra 1050 and Ogle-MIRA have a poor shape in that the flat sternum, generally parallel to the simulated spine, produces a box shape profile. Difficulty was encountered in installing adequate chest deflection measuring devices in all but the Sierra 850 because of space limitations. Although installation of such a device is not required at present, a more careful design of the thoracic structure to allow for deflection measurement and three inches of chest compression is necessary for determining dummy chest performance.

Use of any of these stiff chest structures could result in excessively high chest loads and accelerations. Thus, a perfectly safe impact environment could easily be judged unsafe with these chest structures. All chests, except the Ogle-MIRA, lack rib structures above the arm attachment level, which in the human is represented by ribs Number 1, 2, and 3. This region is particularly important for steering assembly evaluations and tests in which the chest-chin interaction occurs. Correction of this problem will probably require a completely new shoulder design in most cases.

Alderson VIP-50A: This particular design relies on Ensolite foam pads inside the chest cavity for compliance with SAE J963. This in itself is not necessarily a bad design because this rate sensitive foam does contribute to the rate sensitivity needed for proper chest compliance. However, the braces supporting the Ensolite act as a deflection limiting device. Consequently, allowable deflection required for compliance with the response corridors cannot be attained. Bottoming of this chest structure could occur in an impact environment with resulting high unrealistic chest accelerations.

GM Hybrid: Absence of the Alderson braces in this design is a definite improvement with regard to maximum available deflection of 2-7/8 inches. However, additional deflections were not demonstrated in the response curves because of the additional stiffness afforded the simulated ribs by the damping material. The deflection measuring device would have prevented more than 2-3/8 inches of deflection for a 16-mph impact even with a compliant chest because of space limitations. Static force-deflection measurements indicate both the Alderson and the GM Hybrid structure to be similar up to 1.75 in. of deflection.

Sierra 850: No advantage was found to support the use of the Sierra 850 chest. In addition to poor compliance, the large flat aluminum sternum is not humanlike and should be improved upon. The eight-rib design (four on each side) provides an insufficient vertical height for correct physical shape and size.

Significant differences in force-time responses were observed with the Sierra 850 chest structure from two consecutive 16-mph impacts. The time for maximum force varied from approximately one millisecond to a five millisecond plateau as illustrated in Fig. 25. Similar responses were intentionally produced using the GM Hybrid

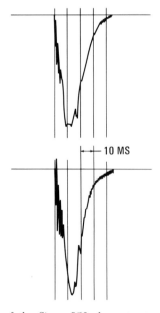

Fig. 25. Force-time responses of the Sierra 850 chest structure from two consecutive 16-mph impacts.

References pp. 242-243

dummy with foam, as little as 1/4-in. thickness, placed directly on the sternum and no additional covering on the upper torso. Varying the type of foam resulted in changing the force-time trace from a single peak of short duration to flat and even double peaks. Therefore, it is felt the foam placed around the rib cage in the Sierra 850 dummy contributed to variations in the force-time and force-deflection responses.

Sierra 1050: Physical shape of this chest structure is its most desirable feature, in that it somewhat resembles a human. This can be important for correct evaluations of automobile components, such as steering systems and shoulder harness restraints.

One disadvantage of this design is the lack of individual ribs. In particular, small impact areas of an automobile interior can be misinterpreted since the spring rate of the entire chest rather than an individual rib would be applied to a small area.

A bad feature of this polyethylene chest is the long time required for returning to its initial shape, if at all. Repeated tests with this chest may require several hours between tests depending on the severity of the test. A permanent set of 1/16 in. was recorded 24 hours after the last of four tests in this investigation; two tests were run at each velocity. If the amount of set is cumulative, the test life of the structure could be very short depending on the severity of use.

Ogle-MIRA: The standard rib design for this structure does not comply with the J963 specifications. However, rib sets have been made available by the manufacturer to meet this requirement. Since the J963 ribs provide a chest with a higher spring rate, the impact response characteristics will obviously fall outside the recommended response corridors.

The provision of a box in which to mount chest accelerometers results in an unrealistic allowable chest deflection. Removal of this box would be a definite improvement. Location of this box and the lead-foam mass behind the sternum resulted in difficulty in mounting a chest deflection measuring device.

SUMMARY

Five current state-of-the-art dummy chests have been impacted at velocities of 11 and 16 mph with impact masses of 42.5 lbs and 51 lbs, respectively. Force-deflection characteristics have been compared with human response corridors.

None of the chest structures exhibit impact response characteristics within the human response corridors during the loading and unloading cycle.

ACKNOWLEDGMENTS

The author gratefully acknowledges the assistance given by Messrs G. Horn, J. McCleary, and W. McPhail in obtaining the data necessary for this project.

Acknowledgment is also given to the GM Proving Ground for supplying dummy components not available at the Research Laboratories.

PART III — MATHEMATICAL MODEL FOR THORACIC IMPACT

INTRODUCTION

A mathematical model for blunt thoracic impact has been developed as an aid in the understanding of the impact response of the human thorax. The model can also be used as a design tool in the development of dummy thoracic structures. A mechanical thoracic analog was postulated and differential equations were derived from the analog. The model was evolved from the simplest possible combination of springs, masses and dashpots through an iterative procedure. Elements were added only when required to give prescribed impact response. Linear elements were used wherever possible. The impact configuration simulated is limited to types similar to that described in the earlier parts of this paper.

DISCUSSION

Model Derivation — The thoracic impact model was derived by postulating a mechanical analog system and then comparing the system impact response with the desired response. Mechanical elements were added and their characteristics were adjusted until the system response correlated with the desired response.

The final mechanical analog system is shown in Fig. 26. The system is free to move on a frictionless horizontal plane. Mass one (m_1) represents the mass of the impactor and is given an initial velocity equal to that of the impactor. Mass two (m_2) represents the effective mass of the sternum and a portion of the rib structure and thoracic contents. Mass three (m_3) represents the effective mass of the remaining portion of the thorax and the part of the total body mass that is coupled to the thorax by the

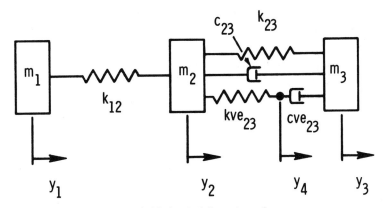

Fig. 26. Mechanical thoracic analog.

vertebral column. Masses two and three are initially at rest. The spring (k_{12}) coupling masses one and two represents the skin and other compliant material between the impacting mass and the sternum. Mass one is considered to be attached to the spring.

The elements coupling masses two and three represent the elasticity and viscous damping of the rib cage and thoracic viscera. Masses two and three are connected directly by a spring (k_{23}) which represents the effective elasticity of the rib cage and directly coupled viscera such as the heart. The dashpot (c_{23}) connecting the two masses represents thoracic damping derived, for example, from air in the lungs and blood in thoracic vessels, both of which leave the thorax during impact. Viscoelastic elements, such as might be present in thoracic muscular tissue, for example, are represented by the series combination of spring (kve_{23}) and dashpot (cve_{23}).

The system has four degrees of freedom. These are noted by the four "y" variables in the figure. Using the notation given above, the equations of motion are as follows:

$$m_1 \ddot{y}_1 = -k_{12} (y_1 - y_2) \tag{1}$$

$$m_2 \ddot{y}_2 = k_{12} (y_1 - y_2) - k_{23} (y_2 - y_3)$$
$$- kve_{23} (y_2 - y_4) - c_{23} (\dot{y}_2 - \dot{y}_3) \tag{2}$$

$$m_3 \ddot{y}_3 = k_{23} (y_2 - y_3) + cve_{23} (\dot{y}_4 - \dot{y}_3)$$
$$+ c_{23} (\dot{y}_2 - \dot{y}_3) \tag{3}$$

$$cve_{23} (\dot{y}_4 - \dot{y}_3) = kve_{23} (y_2 - y_4) \tag{4}$$

where

$$\dot{y}_i = \frac{d}{dt} (y_i)$$

and

$$\ddot{y}_i = \frac{d^2}{dt^2} (y_i)$$

The equations are solved in a digital computer by means of Hamming's modified predictor-corrector method. This method is a fourth order procedure and has the capability of choosing and changing the calculation step size as required for a predetermined accuracy. The maximum step size is set equal to 1 millisecond. The calculation starts with given initial conditions and proceeds until the force in spring k_{12} becomes tensile. In the actual impact event, such a tensile force is not possible and at that point the impactor separates from the thoracic structure. The computer output is displayed directly in graphical form and may be easily compared with cadaver data as well as with specified dummy thorax responses.

Dummy Thorax Simulation — Model parameters were adjusted until the simulated impact response fell within the force-deflection corridors recommended for dummy thorax performance. Preliminary dummy thorax design concepts were then derived from the model configuration.

The recommended force-deflection corridors are specified for impactor force and skeletal deflection. Impactor force is simulated in the model by the force on mass one which is, in turn, the force in spring k_{12}. Skeletal deflection is simulated in the model by the differential displacement between y_2 and y_3. Therefore, graphical computer output was set to give these quantities.

Impactor mass and initial velocity are fixed by the impact conditions corresponding to the recommended performance corridors. The combined mass of masses two and three must be within reasonable limits. An initial value of 45 pounds was selected in light of results from cadaver impact experiments.

The stiffness of spring k_{12} was set to give one-half inch deflection at a force level of 800 pounds. For the case of 16-mph cadaver impacts, it has been estimated that after the impactor first strikes the skin it moves about one-half inch before the sternum begins to move. A combination of a deviation from square impact and a compression of superficial tissues accounts for the one-half inch. A force level of 800 pounds is about midway between the two force plateaus in the recommended response. The value for k_{12} was then 1600 pounds per inch.

The spring constant for spring k_{23} was estimated by fitting a load-deflection line through the origin and the points of maximum allowable deflection for the recommended response. A line having a slope of 200 pounds per inch was used.

As model parameters were varied over ranges of values, it was noted that each parameter affected a specific portion of the force deflection curve. The amount of damping provided by the directly coupled dashpot (c_{23}) primarily affects the mid-range force level, maximum deflection and shape of the curve during force decay. The viscoelastic components (cve_{23} and kve_{23}) influence the mid-range force level and the shape and location of the "corner" at maximum force and deflection. Mass two affects the shape of the initial portion of the curve, while mass three affects the maximum deflection and the force at that deflection. Spring k_{12} influences the response at low deflections and spring k_{23} influences the remainder of the response.

It was necessary to use a nonlinear spring rate for spring k_{23} so that model response at large deflection would pass through the recommended corridors. The spring rate consists of two linear portions. Up to a deflection of 1.5 inches, a value of 150 pounds per inch is used. Beyond 1.5 inches, a value of 450 pounds per inch is used. Equations 2 and 3 were then modified such that the term representing the force in spring k_{23} was calculated according to this nonlinear function.

It was also necessary to increase mass three to 60.0 pounds so that there would be sufficient deflection. Although 45.0 pounds seems to be the effective mass of the thorax alone, it is possible that an additional 15.0 pounds could be coupled to the thorax from the rest of the body by means of the vertebral column.

Dashpot c_{23} was set to have more damping in extension than in compression so that the force decay portion of the curve would pass through the recommended corridors. In compression, the value is 3.0 pounds per inch per second and in extension the value is 7.0 pounds per inch per second.

Values for the remaining elements were found using the same iterative procedure of trial and error. Mass two was set equal to 1.0 pound. Spring kve_{23} was found to have a spring rate of 75.0 pounds per inch and dashpot cve_{23} was set at 1.0 pound per inch per second. It was determined that the spring rate of 1600 pounds per inch for spring k_{12} was suitable. The parameter values are summarized under dummy heading in Table 3.

TABLE 3

Parameter Values

Thorax Simulated	k_{12} lb/in	m_2 lb	k_{23} lb/in	c_{23} lb/in/sec	kve_{23} lb/in	cve_{23} lb/in/sec	m_3 lb
Dummy	1600	1.0	150 or 450 Change at 1.5 inch	comp. 3.0 exten. 7.0	75.	1.0	60.
Cadaver	1600	0.7	60 or 410 Change at 1.3 inch	comp. 2.3 exten. 12.5	75.	1.0	40.

The simulated dummy thorax response is shown in Fig. 27. This figure is derived directly from the graphical computer output. The recommended response corridors are plotted on the output. Each successive point in the plot represents the state of the system at times 1 millisecond apart. It can be seen that the response at 11 miles per hour has just sufficient maximum deflection, while at 16 miles per hour the maximum deflection is near the upper limit. The model is a system whose parameters do not change as a function of sternal deflection (other than the k_{23} parameter). But during cadaver impact tests, the internal thoracic structures were altered destructively as sternal deflection increased. This may be the reason for the lack of a good fit between the simulated response and the recommended corridors.

It is possible to derive a preliminary dummy thorax design from the model configuration. A dummy thorax could consist of a rib cage whose elastic response matches the spring rate of spring k_{23}. The secondary spring rate could be implemented by use of a second spring system which is picked up at a sternal deflection of 1.5 inches. A sternal mass of one pound, less one-half the weight of the ribs, could be used to complete the rib cage. The dummy skin and padding could have a spring rate of 1600 pounds per inch. The damping and viscoelastic elements could be represented by a rate-sensitive foam with directional control of the escape and in flow of air into the foam. The foam would be placed inside the rib structure. Thoracic weight could be set to an effective 60 pounds by adjustment of the dummy lumbar spine. Such a dummy thorax should give impact response that is close to that simulated by the mathematical model.

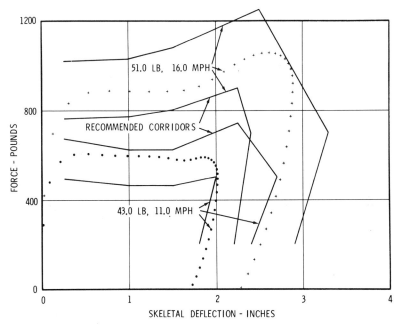

Fig. 27. Simulated dummy thorax response.

Cadaver Data Simulation — The thoracic impact model was also used to simulate force-deflection data obtained from unembalmed cadaver impact tests. These data were unmodified and did not include adjustment for muscle tensing. It was possible to set model parameters such that model response correlated well with the data. The model was then used to examine other cadaver impact data for similar and alternate impact configurations.

An average force-deflection curve was fitted to the data contained in Reference (2) for 16 miles per hour impacts. These data give total deflection, that is, the deflection of material between the sternum and the impactor is included. Another average force-deflection curve was fitted to the data for subjects 5FM, 6FM and 9FM reported in Reference (1) for 11 miles per hour impacts. The latter data give skeletal deflection. Therefore, the average curve was displaced to the right by approximately one-quarter inch such that a line extrapolated to the origin coincided with the 16 miles per hour average up to 250 pounds. The resulting curves are shown in Fig. 28, together with the response of the simulation for corresponding impact conditions.

References pp. 242–243

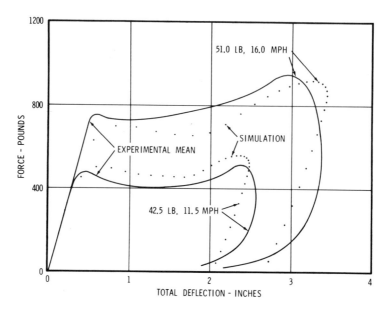

Fig. 28. Simulated cadaver thorax response.

The model parameters were adjusted in an iterative procedure until a reasonable fit to the average curves for the 11 mph and 16 mph impacts were obtained. Values for k_{12}, kve_{23}, and cve_{23} were unchanged. Effective thoracic weight was reduced such that mass two equalled 0.7 pounds and mass three equalled 40.0 pounds. The spring rate for spring k_{23} was altered to give 60.0 pounds per inch up to a sternal deflection of 1.3 inches and 410 pounds per inch beyond that point. Damping provided by the dashpot c_{23} was changed such that it gave 2.3 pounds per inch per second in compression and 12.5 pounds per inch per second in extension. These values are tabulated under the cadaver heading in Table 3.

Spring rates for springs k_{12} and k_{23} were derived directly from cadaver data. The initial slope of the average cadaver force-deflection curves is approximately 1600 pounds per inch, so that value was used for k_{12}. Static sternal deflection in cadavers, as reported in Reference (1), was found to show a mean value of 60 pounds per inch. Therefore, this value was used for k_{23}. During parameter variation, it was necessary to introduce a secondary spring rate after 1.3 inches of sternal deflection. This seems consistent with the idea that under dynamic conditions the sternum moves and thoracic viscera are progressively compressed. Under static loading, it is possible that the viscera may have time to move out of the way. Thus dynamic loading could show a stiffening spring rate.

The differential damping used for dashpot c_{23} is in agreement with the idea that fluids leave the thorax during the loading phase of the impact much easier than they return during unloading. A secondary explanation is that rib fracture occurred for nearly all cadaver tests in the two cases used and that the sternum could not return to its original position because of this damage.

It was not possible to fit both curves closely with the same set of parameters so the final set is a compromise. The correlation is still fairly good except for the portion of the curves where the force level is decaying. Total deflection includes deflection of spring k_{12}. This spring is compressed by approximately one-third to one-half of one inch at the time of maximum total deflection, depending on the velocity of the impact. The response stops when the force in spring k_{12} falls to zero, that is, when the spring returns to its unloaded length. Therefore, the total deflection must decrease by at least the amount the spring is compressed and cannot remain fixed at the maximum value.

Although model parameters were set to give specific force-deflection responses, the model also gave time domain responses that correlated well with actual cadaver data. For example, the 16-mile-per-hour simulation gave a maximum force at 17 milliseconds and a maximum deflection at 22 milliseconds, while the corresponding times for a typical cadaver test were 20 and 30 milliseconds respectively.

Other impactor masses and velocities were used in cadaver impact tests. These masses and velocities were used as input to the model while using parameter values for the cadaver shown in Table 3. The results are shown in Figs. 29 through 33. The first four curves are from Reference (2) and the fifth curve is a composite derived from unpublished data from The University of Michigan HSRI. Correlation is generally

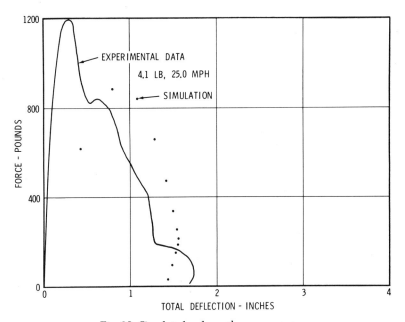

Fig. 29. Simulated cadaver thorax response.

good in deflection, but force levels in Figs. 32 and 33 are too low in the model. Differences in Fig. 33 may be due to differences in experimental procedure, but the cause of the discrepancy in Fig. 32 is not known. In addition, initial force peaks present in the cadaver data were not present in the simulated response. These peaks may be caused either by damping in the tissues between impactor and sternum or by an abrupt change in the loading geometry at the beginning of the impact event. In the model, spring k_{12} is undamped and can give only a gradual rise in force level with increasing total deflection.

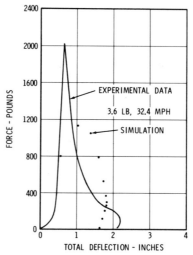

Fig. 30. Simulated cadaver thorax response.

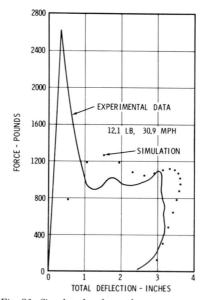

Fig. 31. Simulated cadaver thorax response.

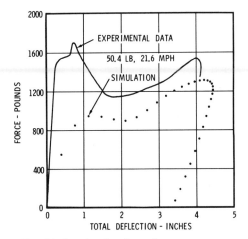

Fig. 32. Simulated cadaver thorax response.

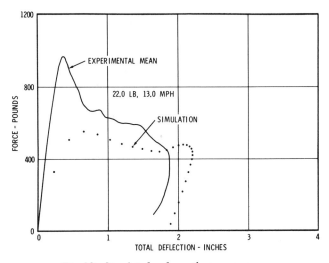

Fig. 33. Simulated cadaver thorax response.

The model was used in a preliminary study to relate the type of thoracic impact described in this paper to the direct type of impact which occurs when a freely moving cadaver thorax strikes a thoracic target. Mass one was given a zero initial velocity and a very large (10^8 pounds) mass. Masses two and three were given an initial velocity of 11 miles per hour toward mass one. Model parameters were set at the cadaver values listed in Table 3. The simulated response is shown in Fig. 34 together with the response for 11 miles per hour shown in Fig. 28. It can be seen that the responses are similar up to a deflection of approximately 2.0 inches. Beyond that point, force and deflection values are much greater for the simulated direct impact than those present in the other, reversed type of impact. This difference is consistent

References pp. 242-243

with the fact that all of the kinetic energy of the moving cadaver thorax is absorbed by the thorax, while in the reverse impact situation about one-half of the impactor energy is absorbed by the thorax and the other one-half is transferred as thoracic kinetic energy. The rising trend of the force level beyond 2.0 inches of deflection agrees with the trend exhibited by embalmed cadaver thoraxes under direct impact as reported in Reference (7). No exact comparison between simulation and experimental data was possible due to lack of the latter.

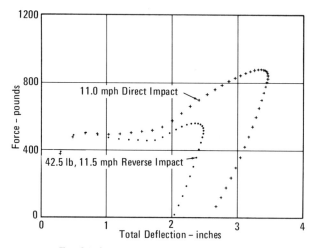

Fig. 34. Simulated cadaver thorax response.

CONCLUSION

A mathematical thoracic impact model was developed and was shown to be capable of simulating the impact response of the cadaver thorax. It was also shown that the model could be used as a tool in the design of a dummy thorax. In addition, the model was used to correlate thoracic response data obtained using a variety of impact configurations. It is planned to expand the model to include nonlinear elements such that correlation with experimental data will be improved.

REFERENCES

1. A. M. Nahum, C. W. Gadd, D. C. Schneider, C. K. Kroell, "Deflection of the Human Thorax Under Sternal Impact," 1970 International Automobile Safety Conference Compendium, SAE Paper No. 700400, 1970.

2. C. K. Kroell, D. C. Schneider, A. M. Nahum, "Impact Tolerance and Response of the Human Thorax," Proceedings of Fifteenth Stapp Car Crash Conference, 1971.

3. K. Wilfert, G. Voigt, "Mechanisms of Injuries to Unrestrained Front Seat Passengers and Their Prevention by Progressive Instrument Panel Design," Proceedings of Fifteenth Stapp Car Crash Conference, 1971.

4. C. K. Kroell, "GMR Proposed Blunt Impact Response Characteristics and Associated Test Procedure for the Dummy Thorax," Presented to the SAE Safety Advisory Committee, Better Dummy Panel, Detroit, Michigan, March 1972.

5. L. M. Patrick, C. K. Kroell, and H. J. Mertz, Jr., "Forces on the Human Body in Simulated Crashes," Proceedings of the Ninth Stapp Car Crash Conference, 1965.
6. J. A. Searle, and C. M. Haslegrave, "Improvements in the Design of Anthropometric/Anthropomorphic Dummies," MIRA Bulletin No. 5, Sept./Oct. 1970.
7. L. M. Patrick, H. J. Mertz, Jr., C. K. Kroell, "Cadaver Knee, Chest and Head Impact Loads," Proceedings of Eleventh Stapp Car Crash Conference, 1967.

DISCUSSION

S. W. Alderson *(Alderson Biotechnology Corporation)*

I think I've made the same comment at each of the last four or five meetings. It should be made clear that chest characteristics of such dummies as the VIP 50a match J963 data to some extent. Their characteristics are not a case of bad design but a case of new additional data. Now the problem changes to that discussed by the speakers — making chests obey the new data.

D. H. Robbins *(HSRI, University of Michigan)*

I would like to address the first question to Charlie Kroell. If I understand it correctly, your forces are measured from a transducer located in the impactor and the deflections are measured from targets on the front and back of the body. Is this correct?

C. K. Kroell

When the total deflection was measured, that is correct. Skeletal deflections were measured by a probe through the thorax.

D. H. Robbins

On that basis, in your force-deformation curves have you tried to separate out the dynamic effects due to the inertial properties of the chest and those due to the visoelastic properties of the structure?

C. K. Kroell

No. Actually I think this is well explained in the Stapp paper. About half of our force-time histories from which the force-deflection cross-plots were developed did have, to some degree, high early load spikes not at all dissimilar to those reported in the HSRI data. These lasted for as long as 1.5 milliseconds, which I think is equivalent to about half an inch of total deflection. We did observe that the natural half-period of the load cell arrangement which we were using was in the general range of this time duration; and therefore, we simply aren't absolutely sure whether these were valid inertial type spikes or artifacts, and we elected at that time to delete them. We did include all the original data, hoping that this can be resolved eventually. We haven't really tried to separate the contribution due to inertial reactions and viscous effects.

D. H. Robbins

I suppose separation of inertial effects from structural effects would be useful in trying to decide whether you need a dynamic test or a static test to determine the deflection properties, if they could be uncoupled somehow.

C. K. Kroell

It certainly would be helpful, yes.

D. H. Robbins

Question for Dr. Lobdell. How were the masses chosen for input to the model? I noticed, I believe, a 60 lb mass for a dummy and a 40 lb mass for a cadaver. What was the effect of variance of this particular quantity on the other input parameters such as the k's and c's.

T. E. Lobdell

In previous cadaver impact experiments it was estimated that the effective mass of the thorax was around 30 to 50 lbs, and so I started the simulation with a value of 45 lbs. In the case of the dummy corridors it was not possible to get enough deflection with a mass that low. I felt that the 60 lbs that I ended up with was not unrealistic because the rest of the body mass is transmitted to a certain degree by the lumbar-spine. So perhaps it's going to turn out that the specification of the thorax is going to have to include the lumbar spine. I think the 40 lbs that was used for the cadaver simulations is fairly reasonable.

D. H. Robbins

I was thinking that by using the model it might be possible somehow to separate-out the inertial effects.

T. E. Lobdell

Yes, I've put some viscoelastic material in a preliminary step beyond this model, but it's not been thoroughly developed yet.

Y. K. Liu *(Tulane University)*

I would like to raise a fundamental objection to this type of parameter variation. No one has said anything about what criteria are used in making these variations. It seems to me that you run a forward analysis problem, parameterize the data, eyeball a fit, and say that this is what my best fit is. There is no scheme in terms of what the performance index is, how do you vary it, and what changes take place as you vary the parameters, for instance. What appears to be fundamentally a system identification problem turns out to be a forward analysis one in which you do the optimization by eyeballing. I don't understand.

T. E. Lobdell

The purpose of the analysis was to identify constants needed to design a mechanical chest structure. I really didn't have a chance to go into the problem in depth as much as I would have liked. I think that you have a valid point.

C. K. Kroell

I think that's the whole point. The idea was to get a tool to help in identifying the parameters that could be altered in terms of the mechanical analogue that was desired. There was a need for a tool to help select parameters at this stage of the game.

W. Goldsmith *(University of California)*

It's always possible to object to any modeling, and the degree of objection depends entirely upon the results that are desired. As far as sweeping parameters is concerned, this is also currently a very active fad and again the objections to that depend, similarly, upon ones own wishes. But there are some limits on what should and should not be incorporated in a particular model. Specifically, if one thinks that the effects of certain quantities are going to cause a rather significant change of the behavior of the system, then they should at least be considered. On this particular model, what worries me very much is that it is isolated from the rest of the human body. It would seem to me that consideration should have to be given to the attachments. If you want to keep it very simple, springs and masses can represent the head-neck junction and the rest of the torso can be represented by the means of another spring and mass. This should have been done to see whether or not the characteristic dynamic behavior would have been drastically altered by the addition of such elements.

There are other objections one can raise (anyone can raise objections). I personally might have liked to have seen a different kind of element representative of a rheological fluid. As was mentioned, the damping is very largely due to the emission of the fluid from the arteries and veins during impact. I doubt that a viscoelastic element characteristic of a linear solid would describe that very well; but I think that's a detail. I think the addition of the masses, however, is something that at least from the dynamic point of view would grossly change the behavioral characteristic of the model, and I would like to have seen an investigation of that.

EVALUATION OF DUMMY NECK PERFORMANCE

J. W. MELVIN, J. H. McELHANEY and V. L. ROBERTS

University of Michigan, Ann Arbor, Michigan

ABSTRACT

As the structural link between the head and chest, the neck plays a vital role in determining the dynamics of the head during indirect impact. This paper discusses the factors involved in specifying and quantifying the dynamic performance of dummy neck simulations.

The literature on cervical spine mechanics, including recent x-ray studies on cervical spine mobility in human volunteers and human volunteer sled tests is reviewed and summarized with respect to human head-neck response to dynamic loading. The mechanics of the head-neck system is discussed and a rationale is developed for utilizing human volunteer response data in assessing neck simulation performance. Additional factors influencing dummy neck design are indicated and the results of a program to develop an improved neck simulation utilizing these concepts is presented.

INTRODUCTION

As the structural linkage between the head and chest of an anthropometric dummy, the neck simulation plays a vital role in determining the dynamics of the head during indirect impacts. The anatomy of the human neck is complex both in the structure of the cervical spine and in the musculature attached to the spine. A mechanical simulation of the human neck must satisfy the need for producing realistic motions of the head and, at the same time, possess highly reproducible characteristics. Although a complex mechanical structure similar to the actual human neck could produce realistic head motions, the necessity of reliable, reproducible performance under sometimes severe acceleration environments dictates that neck

References p. 258

simulations be of a rugged and simple nature. Thus, the design of a neck simulation should, in general, be the simplest structure necessary to produce the desired dynamic head motions.

The specification of basic dummy neck performance requires consideration of three aspects of human head motion:

1. The range of angular motion of the head relative to the torso.

2. The trajectory of the center of gravity of the head relative to the torso.

3. The resistance of the neck to motion of the head during loading and rebound.

Each of these three requirements is discussed in the following sections with primary emphasis on motion in the mid-sagittal plane. The results of a program to develop an improved neck simulation at HSRI (1) are used as an illustrative example in discussing the performance requirements.

HEAD ANGULAR RANGE OF MOTION

The angular motion of the head relative to the torso in the mid-sagittal plane has received a great deal of attention in the literature on the motion and mobility of the cervical spine. The data presented in most motion and mobility studies represents human voluntary motion and is primarily of use in establishing the mean levels of total range of angular motion for various populations grouped according to sex, age and physical condition. In addition to studies on total range of motion, some research has been directed at determining the angular range of motion of each of the vertebral segments of the cervical spine through the use of radiographic techniques. The results of this type of study can be used to give an indication of the distribution of angular motion along the cervical spine. A summary of the more complete studies (2-10) on total motion and segmented motion is given in Table 1.

For the case of dynamic involuntary loading, it can be expected that the limits of voluntary motion will be exceeded somewhat during severe loading due to additional

TABLE 1

Total Head/Torso Motion Range

Reference	Sex of Subjects	Flexion Angle (degrees)	Extension Angle (degrees)
Delahaye (1)	male	72.2	39.4
Buck (2)	male	66	73
Glanville (3)	male	59.8	61.2
Defibaugh (4)	male	58	79
Bhalla (5)	female	58	34
Buck (2)	female	69	81
Bennett (6)	female	54.4	93.2

TABLE 1 (Continued)

Segmented Motion Range

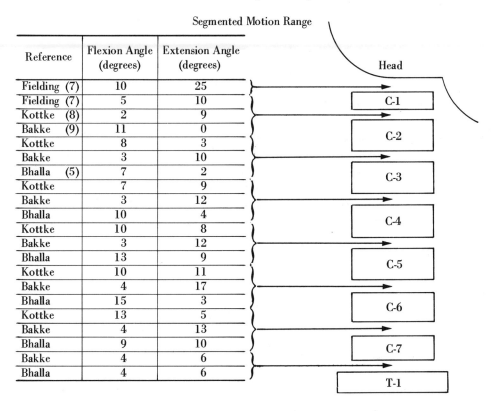

Reference	Flexion Angle (degrees)	Extension Angle (degrees)
Fielding (7)	10	25
Fielding (7)	5	10
Kottke (8)	2	9
Bakke (9)	11	0
Kottke	8	3
Bakke	3	10
Bhalla (5)	7	2
Kottke	7	9
Bakke	3	12
Bhalla	10	4
Kottke	10	8
Bakke	3	12
Bhalla	13	9
Kottke	10	11
Bakke	4	17
Bhalla	15	3
Kottke	13	5
Bakke	4	13
Bhalla	9	10
Bakke	4	6
Bhalla	4	6

motion at each vertebral level. This would be evident particularly in the case of hyperextension where there is no mechanism to arrest head motion equivalent to the situation of chin-chest contact in hyperflexion. The recently reported work of Ewing and Thomas (11) represents the most comprehensive study to date on the head-neck dynamics of human volunteers in flexion. Analysis of the head angular motion relative to the first thoracic vertebra (T-1) for 11 male volunteers at the highest acceleration levels (8-10 G) in the study indicates an average maximum forced flexion angle of 70° with a range of 60° - 85°. Similar data on human volunterrs in forced extension does not exist, however, data is available on human cadavers. Lange (12) reports an average maximum head angle relative to the torso of 105° in hyperextension for nine fresh human cadavers at very severe rear-end collision simulations of 19-29 G. The range of hyperextension angles was 75° - 140°. Mertz and Patrick (13) used two embalmed cadavers to obtain a maximum hyperextension angle of 85° under moderately severe loading conditions. In view of the absence of muscle tone in the fresh cadavers and hardening of soft tissues in the embalmed cadavers, it would appear that forced maximum hyperextension angles could be estimated to lie in the range of 80° - 100° for the living human.

References p. 258

HEAD TRAJECTORY

The path that the head takes in the mid-sagittal plane as it moves through its range of angular motion depends on the linkage system of the cervical spine and upon the forces acting on the head and neck. The complexity of the cervical spine and the variability of the different combinations of loads, muscle action and resistance to motion do not allow a unique determination of the trajectory of the head during its motion. However, it is useful to consider the nature of existing data on head center of gravity trajectories as a guide to assessing the general effectiveness of a neck simulation in producing realistic trajectories. The work of Ewing and Thomas (11) includes graphical data on the paths of the center of gravity of the heads of human volunteers under both dynamic forced flexion and voluntary head nodding. The paths were plotted relative to an orthogonal coordinate system fixed on the T-1 vertebra. A comparison of forced and voluntary trajectories for one subject is shown in Fig. 1. Note that the dynamic trajectory lies within the range of the voluntary trajectories,

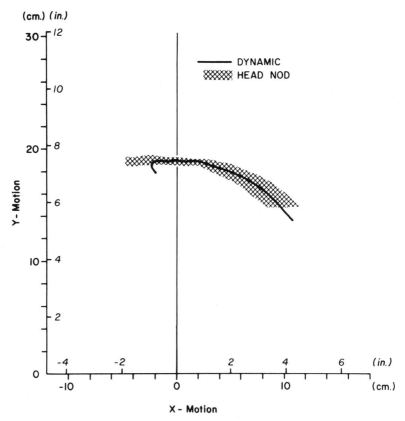

Fig. 1. Comparison of voluntary and dynamic head C.G. trajectories, Ewing and Thomas (11) Subject No. 3, 9.5 G Run.

indicating that, at the acceleration levels studied, the dynamic loading did not produce significant geometrical changes in the spinal linkage. The trajectories reported by Ewing and Thomas appear to be quite similar for most of the volunteers in terms of the shape of the path while the different neck lengths shifted the absolute values of the coordinates of the trajectories. Fig. 2 shows the trajectories for the three

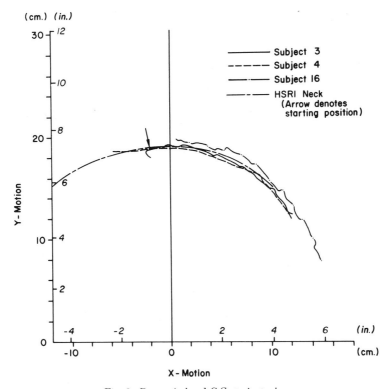

Fig. 2. Dynamic head C.G. trajectories.

volunteers closest to the 50th percentile range anthropometrically for similar acceleration inputs (9-10 G peak). Also shown for comparison is the trajectory produced by the HSRI neck simulation under similar acceleration conditions. The trajectory for the HSRI neck has been rotated clockwise about the origin by 20° to obtain a coordinate system equivalent to the system used by Ewing and Thomas. This rotation is the result of the tendency for dummy neck simulations to be straight in the unloaded state while the cervical spine has lordosis or curvature in its normal state. Thus, the dummy neck simulation should be considered a secant approximation of the spine system between T-1 and the occipital condyles as shown for the case of the upright seating position used by the subjects in the work by Ewing and Thomas in Fig. 3. The secant approximation of the human neck must also be considered for the typical automotive test configuration of a seat back inclined 20° - 25° rearward from

References p. 258

vertical. If the head is to remain approximately level while the torso is inclined along the seat, the angle at which the neck simulation is inclined should reflect the equivalent configuration of the human cervical spine. Flexion of the lower spine in the human can account for much of the angular adjustment necessary to provide a level head attitude while reclining, however, some angular adjustment of the head and cervical spine is also necessary. Analysis of limited radiographic data indicates that angles of 5° - 10° forward of vertical for the reclined occupant are necessary to approximate the configuration of the cervical spine.

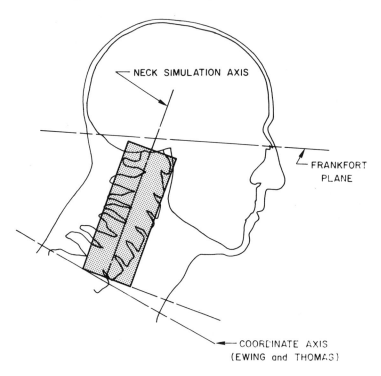

Fig. 3. Relative configurations of the human cervical spine and the equivalent dummy neck simulation.

RESISTANCE OF THE NECK TO HEAD MOTION

The objective of a neck simulation is to produce human-like head kinematics and dynamics in realistic acceleration environments. When subjected to indirect impact, the head may be considered as a rigid body with the forces and moments that act upon it transmitted to the head by the neck structure. In the case of hyperflexion, the chin may also transmit force to the head by contacting the chest. These conditions are depicted in the free body diagrams for both extension and flexion shown in Fig. 4. For motion in the mid-sagittal plane, the motion of the head is completely determined by the equations

$$\Sigma F_x = m_h a_x ; \quad \Sigma F_y = m_h a_y ; \quad \text{and } \Sigma M_{cg} = I_{cg} \alpha \qquad (1)$$

where ΣF_x and ΣF_y are the sums of the x and y projections of all the external forces acting on the head, m_h is the mass of the head, a_x and a_y are the x and y projections of the absolute accelerations of the center of gravity of the head, I_{cg} is the mass moment of inertia of the head about its center of gravity, and α is the absolute angular acceleration of the head. From the free body diagrams shown in Fig. 4, the equations can be written as follows:

For extension

$$\Sigma F = m_h a_x = F_s - W_{hx} ; \quad \Sigma F_y = m_h a_y = -F_a - W_{hy} \qquad (2)$$

and

$$\Sigma M_{cg} = I_{cg} \alpha = F_a d_a + F_s d_s + T_o$$

For flexion

$$\Sigma F_x = m_h a_x = F_s - F_{cx} + W_{hx}$$

$$\Sigma F_y = m_h a_y = -F_a - W_{hy} + F_{cy} \qquad (3)$$

$$\Sigma M_{cg} = I_{cg} \alpha = F_a d_a + F_s d_s + T_o + F_c d_c$$

where W_{hx}, W_{hy}, F_{cx} and F_{cy} are the x and y components of the head weight and chin contact force respectively.

Note that in both cases the moment equations contain terms that represent all of the forces and moments transmitted to the head by the neck and the chin (that is, neck shear force F_s, neck axial force F_a, chin contact force F_c, and neck resisting

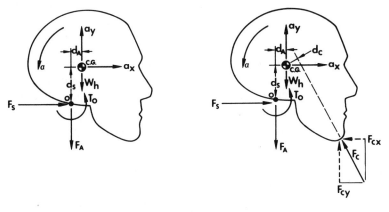

Extension   Flexion

Fig. 4. Free body diagram of the head.

torque T_o). This suggests that the term $I_{cg}\alpha$, which is indicative of the total moment acting on the head, may be the most effective single measure of force and moment input into the head by the neck. The angular acceleration α can be directly calculated from linear accelerometer outputs and thus requires no estimation of anatomical landmarks interior to the body of a human volunteer or estimation of the inertial characteristics of the head of the volunteer. Specification of the angular acceleration and linear accelerations of the head as it moves through its range of motion during loading and rebound due to a specific acceleration input to the T-1 vertebra will describe the effect of the resistance of the neck on the motion of the head through the use of the above equations.

Another approach to specifying neck resistance to motion has been suggested by Mertz and Patrick (14). They specify the neck resisting torque T_o at the occipital condyles as the indicator of neck performance under dynamic loading conditions. The point defined as the occipital condyles for the 50th percentile male is 0.75 in. posterior and 2.45 inferior of the c.g. of the head in the mid-sagittal plane (15). In addition to specifying the neck resisting torque by means of performance loading envelopes of T_o versus head angle relative to the torso, Mertz and Patrick specify the hysteresis associated with the unloading of the neck during rebound.

EVALUATION OF NECK SIMULATION PERFORMANCE BASED ON HUMAN VOLUNTEER TEST DATA

The use of human volunteer test data on head-neck response under dynamic loading in evaluating neck simulation performance is the most effective way to insure realistic simulations. In addition to the previously described study of Ewing and Thomas (11), Run 79, a flexion test, is well detailed by Mertz and Patrick (14), and a series of 12 volunteer shoulder harness tests are reported by Clarke, et. al (16) in terms of mean peak data. In the case of Run 79, the sled pulse had an average trapezoidal plateau deceleration of 9.6 G with a velocity of 21.8 ft/sec (6.6 m/sec) while the sled pulse used by Clarke was approximately half sine with a mean peak of 9.6 G and a velocity on the order of 25 ft/sec (7.6 m/sec).

These pulses and the Ewing and Thomas sled pulse are shown in Fig. 5 along with a mini-sled pulse used in the development of the HSRI neck simulation and a pulse suggested by the Department of Transportation for neck performance evaluation. The general similarity of the mini-sled pulse and the human volunteer pulses allows an evaluation to be made based on the performance criteria discussed in the previous sections.

Plotting head angular accelerations versus head/torso angle as shown in Fig. 6 allows both head-neck dynamics and range of angular motion to be compared for the various volunteer studies and the HSRI neck simulation. Also shown on Fig. 6 are the results of a photometric analysis of Run No. 4578 (H-2) of a Holloman test series conducted by the National Bureau of Standards (17). The subject was forward facing

and restrained by a Type 2 harness system. The sled pulse was 12.6 peak G with a
duration of 120 msec and a velocity of 21.8 ft/sec (6.6 m/sec). The remainder of the
angular acceleration data in Fig. 6 was obtained directly with accelerometers or in the
case of Subject 3 (Ewing and Thomas), differentiation of rate gyro data.

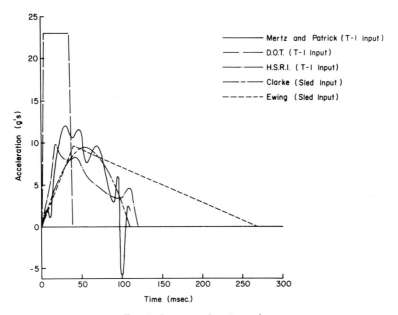

Fig. 5. Input acceleration pulses.

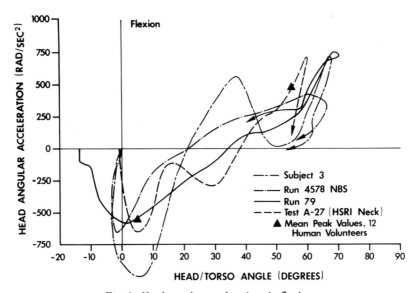

Fig. 6. Head angular accelerations in flexion.

References p. 258

A comparison of resultant head c.g. linear accelerations for two of the volunteer tests and the neck simulation are shown in Fig. 7. Wide variations exist in the accelerations associated with the initiation of head motion and may be due to different levels of muscle condition. However, the accelerations associated with the limit of motion are quite comparable and are in the same range as the 12-20 G reported by Clarke, et. al. A plot of the calculated moment at the occipital condyles versus head/torso angle is shown in Fig. 8 with both test data and the suggested performance envelope of Mertz and Patrick indicated.

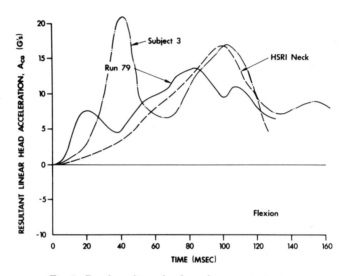

Fig. 7. Resultant linear head accelerations in flexion.

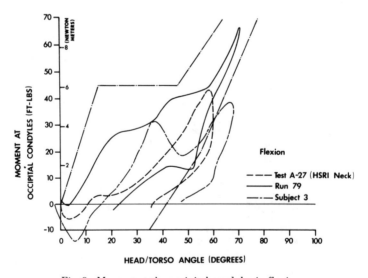

Fig. 8. Moment at the occipital condyles in flexion.

In contrast to the variety of data available on dynamic flexion of human volunteers, data on dynamic extension of volunteers is severely lacking. The data used for comparison with simulation response was obtained from the work of Mertz and Patrick (13) and consists of two cadaver tests and one volunteer at low acceleration levels of 5.2 g and 3.2 g respectively. The test pulses for the neck simulation were 5 g (A-33) and 9.5 g (A-30). The head angular acceleration responses for Test A-33 and the cadaver tests can be directly compared due to the similar acceleration pulses as shown in Fig. 9. The response of the HSRI neck show it to be stiffer than the cadaver necks and also to have less angular motion at the 5 g level. The response at the more severe 9.5 g pulse of Test A-30 shows that the range of motion of the neck reaches the range of the cadaver motions while Fig. 10 indicates that the moment at the occipital condyles produced by the HSRI neck matches the suggested response envelope of Mertz and Patrick.

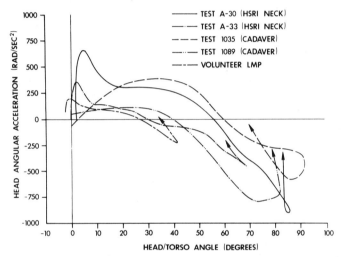

Fig. 9. Head angular accelerations in extension.

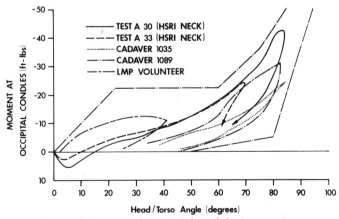

Fig. 10. Moment at the occipital condyles in extension.

References p. 258

CONCLUSIONS

Dummy neck simulation performance can be evaluated by comparison of existing human volunteer test data with neck simulation test data obtained under similar test conditions. The criteria used to evaluate the performance are:

1. The range of angular motion of the head relative to the torso.
2. The trajectory of the center of gravity of the head relative to the torso.
3. The resistance of the neck to motion of the head during loading and rebound.

REFERENCES

1. J. W. Melvin, J. H. McElhaney, and V. L. Roberts, "Improved Neck Simulation for Anthropometric Dummies," 16th Stapp Car Crash and Field Demonstration Conference Proceedings, SAE Paper No. 720958, 1972.
2. R. P. Delahaye, et. al., "Dynamic Radiology of the Cervical Spine of Flying Military Personnel: The Special Case of Jet Pilots," Rev. Corps Santa Armes, 9: pp. 593-614, 1968.
3. C. A. Buck, F. B. Dameron, M. J. Dow and H. V. Skowlund, "Study of Normal Range of Motion in the Neck Utilizing a Bubble Goniometer," Arch. Phys. Med., 40: pp. 390-392, 1959.
4. A. D. Glanville and G. Kreezer, "The Maximum Amplitude and Velocity of Joint Movement in Normal Male Human Adults," Hum. Biol., 9: pp. 197-211, 1937.
5. J. J. Defibaugh, "Measurement of Head Motion. Part I: A Review of Methods of Measuring Joint Motion. Part II: An Experimental Study of Head Motion in Adult Males." Physical Therapy, 44: pp. 157-168, 1964.
6. S. K. Bhalla and E. H. Simmons, "Normal Ranges of Intervertebral-Joint Motion of Cervical Spine." Canad. J. Surg., 12: pp. 181-187, 1969.
7. J. G. Bennett, L. E. Bergmanis, J. K. Carpenter and H. V. Skowlund, "Range of Motion of the Neck." J. Amer. Phys. Ther. Assn., 43: pp. 45-47, 1963.
8. J. W. Fielding, "Cineroentgenography of the Normal Cervical Spine," J. Bone Joint Surg., 39 A: pp. 1280-1288, 1957.
9. F. J. Kottke and M. O. Mundale, "Range of Mobility of the Cervical Spine," Arch. Phys. Med., 40: pp. 379-382, 1959.
10. S. N. Bakke, "Roentgenologische Observations on the Movement of the Spinal Column," Acta Radiol., Suppl. 13, 1931 (German).
11. C. L. Ewing and D. J. Thomas, "Human Head and Neck Response to Impact Acceleration," NAMRL Monograph 21, USAARL 73-1, August 1972.
12. W. Lange, "Mechanical and Physiological Response of the Human Cervical Vertebral Column to Severe Impacts Applied to the Torso," Symposium on Biodynamic Models and their Applications, AMRL-TR-71-29, December 1971.
13. H. J. 21, Mertz and L. M. Patrick, "Investigation of the Kinematics and Kinetics of Whiplash," SAE Paper No. 670919, 1967.
14. H. J. Mertz and L. M. Patrick, "Strength and Response of the Human Neck," 15th Stapp Car Crash and Field Demonstration Conference Proceedings, SAE Paper No. 710855, 1971.
15. C. C. Culver, R. F. Neathery and H. J. Mertz, "Mechanical Necks with Humanlike Responses," 16th Stapp Car Crash and Field Demonstration Conference Proceedings, SAE Paper No. 720959, 1972.
16. T. D. Clarke, et. al., "Human Head Linear and Angular Accelerations During Impact," 15th Stapp Car Crash and Field Demonstration Conference Proceedings, SAE Paper No. 710857, 1971.
17. R. F. Chandler and R. A. Christian, "Crash Testing of Humans in Automobile Seats," SAE Paper No. 700361, 1970.

DISCUSSION

J. D. States *(University of Rochester)*

Does your neck stretch at all?

J. W. Melvin

No, it does not stretch axially. There has been a lot of discussion on neck stretch. I think if you look at the configuration of the human neck in the normal position it has a Lordosis. In flexion, getting rid of that curvature is interpreted many times as stretch. I think its the change in length, but the neck isn't really stretching very much. In a dynamic situation, not allowing stretch, but having the right trajectory, would accomplish the same end.

J. D. States

Is it the mechanical elementes in your neck that would prevent stretch? Is there a rubber interposition?

J. W. Melvin

Stretch could be designed in. It isn't now. The mechanical linkages do not allow for axial motion.

W. Goldsmith *(University of California)*

For purposes of comparison, have you been able to obtain the data that Case Western Reserve is supposed to have collected on human volunteers with regard to the free vibration of the head-neck system? I believe the volunteers were strapped to a vertically suspended table, and the head was then displaced to produce a free vibration. The angularity of the various joints was then measured. I understand this work was supposed to have been completed about three years ago, but there is apparently no report. I wonder if you have had access to any of these data through personal communication?

J. W. Melvin

I think the fellows from the Safety Systems Lab may have some information on that. I seem to recall that they did look at the data. Perhaps they have some comments on this.

L. M. Patrick *(Wayne State University)*

In our human volunteer experiments there was considerable curvilinear translation before rotation started. Have you noticed this? It's a substantial period in the human. I think that this should be incorporated into the neck reaction if possible.

J. W. Melvin

Yes. There is a short period of lag, so to speak. We see this in the simulations.

J. N. Silver *(GM Proving Ground)*

Have you made any measurements on the hysteresis of your neck and compared it to human data?

J. W. Melvin

Well, defining hysteresis is a little bit difficult under those systems. About the best way probably is to look at the moment-angle curve. What we found, however, is that many times the shear force and axial force time histories can have an influence, and can give us what looks like a figure eight hysteresis loop, which is a little hard to interpret. Under those test conditions that don't give us figure eights, this neck gives us hysteresis in the range suggested by Mertz and Patrick.

R. J. Vargovick *(Ford Motor Company)*

Was there any configuration difference between your run A-30 and A-33?

J. W. Melvin

There was no configuration difference. Only the acceleration level was different. One was half the other. One is a 5g run and one is a 10g run. We expect the system to drive further, have more energy and have a higher deceleration peak here on the higher acceleration run.

R. J. Vargovick

Then you meant it to be different.

J. W. Melvin

I meant it to be different, yes. There's no comparison there. It's just showing you characteristics. There's very little data like this available for volunteers. That's why we don't have a good comparison.

S. H. Backaitis *(NHTSA, Department of Transportation)*

Professor Patrick pointed out in his movie yesterday a number of times that the head undergoes considerable translation before rotation takes place. Since the HSRI neck is built on a fixed pivot basis, this phenomena may not be in existence. Have I missed something?

J. W. Melvin

With three joints this can happen. It can obtain configurations that allow the head to translate relative to the base, since there can be relative rotations between the three segments.

S. W. Alderson *(Alderson Biotechnology Corporation)*

Have you done any work on the repeatability of the neck?

J. W. Melvin

Yes. We've subjected this system to a series of similar acceleration inputs. The peak acceleration values were within 3 percent variation on three successive runs.

PERFORMANCE REQUIREMENTS AND CHARACTERISTICS OF MECHANICAL NECKS

H. J. MERTZ, R. F. NEATHERY and C. C. CULVER

General Motors Research Laboratories, Warren, Michigan

ABSTRACT

A short history of the development of mechanical necks for anthropomorphic dummies is given. The response envelopes recommended by Mertz and Patrick are reviewed. A modified performance requirement for mechanical necks based on this data, but emphasizing loading corridors, is set forth. Autogenous and dynamic neck trajectories for the volunteer LMP are presented, and the difficulties of establishing a trajectory performance requirement are discussed.

Commerical necks were tested and found to be incompatible with the performance requirements. Several experimental necks were tested and one, the GMR Polymeric Neck, demonstrated the feasibility of satisfying the requirements. However, additional efforts are required to assure proper performance of this neck when used in conjunction with a total dummy structure under a wider range of test conditions.

INTRODUCTION

In the early Fifties, anthropometric dummies were used by the Air Force to evaluate the injury potential of pilot ejection from high-speed aircraft. Because the response of the spine was being analyzed in these tests, the dummies had complex geometries for the spinal column. In particular, the neck structures were designed to simulate the articulation of the cervical spine. During ejection, the wind blast produced large differential forces between the head and the torso which resulted in relative rearward rotation of the head and hyperextension of the neck. In several tests, the neck failed structurally. Necks were then designed with sufficient strength to prevent structural failure in this severe exposure environment.

References p. 288

One of these early neck designs, which was also used in early automotive tests, is shown in Fig. 1. It consisted of five steel segments held together by a steel cable. Adjacent segments articulated on spherical surfaces with the resisting forces being frictional and controlled by the tension in the cable. This neck structure had many deficiencies. There was no positive control of the frictional forces between segments. Consequently, it was impossible to set the neck up in a repeatable fashion. The strength of the neck depended on the steel cable. Individual strands of wire were frequently severed during neck bending. This progressively reduced the strength of the neck until complete failure occurred. Adjacent segments frequently "bottomed out," i.e., metal-to-metal contact occurred. This produced large magnitude, short duration acceleration spikes which were superimposed on the head acceleration data.

Fig. 1. Steel cable neck.

The split ball and socket neck design, shown in Fig. 2, was an attempt to control the frictional resisting forces and minimize the large magnitude, short duration acceleration spikes caused by the metal-to-metal contact of adjacent segments. The relative stiffness between adjacent segments was individually adjustable by means of a set screw, spring, and nylon disk. Tightening the set screw which was threaded in the base of the segment compressed the spring which forced the nylon disk to bear harder against the ball of the adjacent segment. This increased the frictional force between the nylon disk and the ball, resulting in increased resistance to articulation of the joint. Metal-to-metal contact between adjacent segments was prevented by inserting a sandwich-type rubber disk between adjacent segments. This rubber insert consisted of an annulus of foam rubber situated between two annuluses of relatively hard rubber. The main mechanical deficiencies of this design were the time required to set up the frictional resistance for each joint and the extreme care required to maintain neck alignment during the process of setting the dummy in the test vehicle.

Fig. 2. Split ball and socket neck.

In an effort to obtain repeatability of the frictional resisting forces, a pin jointed neck structure was designed, Fig. 3. Two axes of the pin joints were perpendicular to the midsagittal plane of the dummy. They allowed forward and rearward rotation of the head relative to the neck structure, and forward and rearward rotation of the neck structure relative to the torso. The third pin joint axis was contained in the midsagittal plane and allowed lateral rotation of the head and the upper part of the neck relative to the torso. Resistance to articulation was controlled by frictional forces developed by clamping the clevis onto the male insert. Because of the frictional resisting forces, this neck structure required careful handling when a dummy was being positioned in a car for testing in order to prevent misalignment.

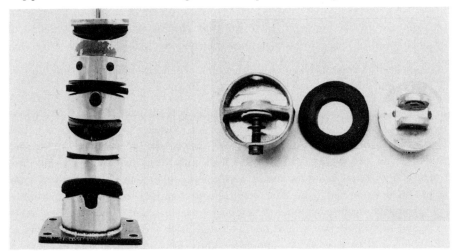

Fig. 3. Pin jointed neck.

To eliminate the necessity for careful handling of the neck structure during dummy positioning, numerous types of "rubber neck" structures were designed and used. These monolithic structures were elastic and therefore returned to a repeatable initial position once the dummy was properly oriented in the car. They eliminated metal-to-metal contacts and the high degree of uncertainty of resistance forces offered by the friction joints. Since no restrictions were placed on the bending resistance of the neck nor on the damping properties, rubber necks of different geometries and material properties were designed and used.

With numerous neck structures available, the inevitable question of which neck structure had the "proper" response characteristics was raised. Before a "proper" neck can be identified, the term "proper" must be defined. To define this term, the role of the neck in a collision environment must be analyzed.

In a collision environment, the neck controls the dynamics of the head; its trajectory, orientation, and velocity. These factors determine which structures in the car's interior will be impacted by the head, and the location, direction, and magnitude of the resulting impact forces on the head. This, in turn, produces the head acceleration history which is the basis of evaluating occupant protection. Consequently, to have meaningful evaluation of automotive occupant protection systems, "proper" response characteristics for dummy necks must be defined as human response characteristics.

The objectives of this paper are to:

1. Review a set of necessary performance requirements for neck structures based on human volunteer and cadaver response data and recommend modified performance requirements. These requirements are necessary but not sufficient to describe humanlike neck response.

2. Compare the response characteristics of commercial and experimental neck structures to these necessary performance characteristics.

3. Indicate the areas, such as head trajectory, which need additional study in order that conditions for these parameters may be specified.

NECESSARY PERFORMANCE REQUIREMENTS FOR NECK STRUCTURES

Mertz and Patrick Response Data — A set of necessary performance requirements for flexion and extension of mechanical necks was described by Mertz and Patrick (1). Their performance requirements were based on human volunteer and cadaver data obtained from sled tests. The subjects were exposed to dynamic environments which produced various degrees of flexion or extension of the neck. The subjects were seated in a rigid chair which was mounted on a sled. They were restrained by a lap belt and a criss-cross pair of shoulder straps. This prevented gross torso motion relative to the seat. The sled was accelerated to a specified velocity during which the head of the subject was restrained. After a short period when the sled traveled at a constant velocity, it was decelerated at a prescribed rate. When he was forward facing, this produced forward rotation of the subject's head relative to his torso, flexing his neck. It produced rearward relative rotation of his head and neck extension when he was rearward facing.

Neck response was determined by finding the relation between the moment about the occipital condyles of the forces acting on the head and the angular position of the head relative to the torso. (The occipital condyles are the articular surfaces of the skull which bear on the first cervical vertebra.) In flexion, the chin frequently impacts the chest. When this happened, the moment of this contact force about the occipital condyles was included with the resisting moment produced by the neck.

Because of the complexity of neck structure in the human, no attempt was made to distribute the forces to the various muscles, ligaments, and bony contacts. Only the total reaction was computed. If the neck is thought of as a beam, an extremely simplified model, it is evident that specifying the applied moment on one end does not give an unique beam configuration since shear and axial loads will also affect the final shape. However, if the test environment is maintained, the bending moment, shear, and axial loads are coupled and a single parameter may be adequate. In any case, the moment-angle relations found remain necessary conditions for humanlike neck response.

The volunteer used in the dynamic tests was representative of a 50th percentile adult male in height, weight, age, and range of neck motion. No atypical characteristics were noted. The volunteer was used to obtain non-injurious response characteristics. Cadavers were used to extend the data into the injury region. In some of the flexion tests, a 3-lb. weight was attached to the subject's head.

The response envelopes recommended by Mertz and Patrick are the areas enclosed by the solid lines shown in Figs. 4 and 5. These envelopes were synthesized from static volunteer, dynamic volunteer, and cadaver data and were considered to represent a tensed individual. Mertz and Patrick (1) imposed the following conditions:

"Two necessary conditions required for the response of mechanical simulation of the human neck of a 50th percentile adult male are:

1. The relationships between the equivalent moment and angular displacement of the head relative to the torso for loading and unloading must lie within the response envelopes shown ... for hyperflexion and extension. To evaluate the mechanical neck, it must be mounted between an appropriate dummy chest and head, and the testing must be done in a dynamic environment.

2. To insure adequate damping, the ratio of the area between the dynamic loading and unloading curves to the area between the loading curve and the abscissa

must not be less than 0.5 for that portion of the particular response curve lying below constant plateaus (45 and 22.5 ft. lbs.) of the flexion and extension response envelopes, respectivelely."

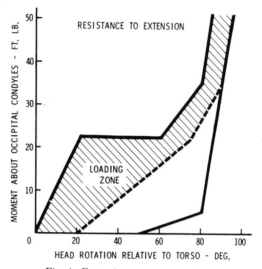

Fig. 4. Extension response envelope.

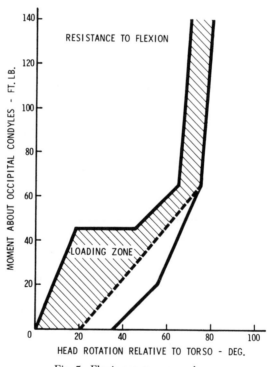

Fig. 5. Flexion response envelope.

Recommended Performance Requirements — In February of this year, Mertz formulated the "GMR Performance Requirements for Mechanical Necks" as a part of a general statement on dummies which was distributed to interested parties by General Motors Research Laboratories. This statement was an attempt to take into account the differences between dummies and humans and the problems of mechanical and experimental reproducibility. These requirements introduced loading corridors, shown as the shaded areas in Figs. 4 and 5, rather than response envelopes. The requirement then was that only the loading portion of the curve must lie within the shaded area of the torque-angle space. Figs. 6 and 7 show the recommended loading corridors separately for extension and flexion, respectively. The return path is controlled by required hysteresis ratio but is not as severely regimented as the loading requirement.

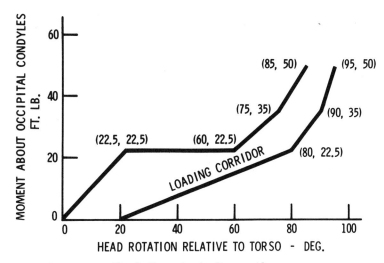

Fig. 6. Extension loading corridor.

The method of determining the hysteresis ratio was simplified to include all of the torque-angle data in the relevant quadrant. Upper limits were set on the ratio to avoid totally damped structures. (The response envelope prevented total damping in the earlier recommendation.) Loading conditions were also specified. For a valid test, resisting torque of 40 to 50 ft.-lbs. for extension and 120 to 140 ft.-lbs. for flexion had to be produced in the tests.

It has been repeatedly emphasized that the recommendations of Mertz and Patrick were necessary but not sufficient conditions to ensure humanlike response. In its testing, GMR has attempted to require some of the unquantified (even unidentified) sufficiency conditions by duplicating the original biomechanical test conditions on the original apparatus. This was easily achieved for extension; but in many of the original flexion tests, the subject had a 3-lb. weight attached to his head at various locations. Such an encumberance on dummy tests seemed unwelcome and

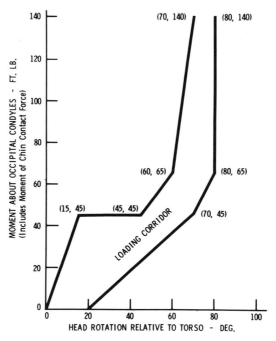

Fig. 7. Flexion loading corridor.

unnecessary. Compensation for the added weight can be made by using the empirical relation developed by Mertz and Patrick (2):

$$\frac{T}{WD} = 2.13 + 0.12a + 0.15a^2$$

where:

T – torque produced at the occipital condyles in ft.-lbs.
W – head weight in lbs.
D – distance from the center of gravity of the head to the occipital condyles in ft.
a – plateau sled deceleration in G's

Using this relation, it was found that the most severe weighted cadaver runs which produced 140 ft.-lb. resisting torque would require a plateau deceleration of 19 G for an unweighted dummy head. If the 10-in. stopping distance is maintained, an impact velocity 32 fps is required. Thus the GMR flexion tests were run under these conditions rather than those indicated in the original data.

It is evident that some viscoelastic material or damping device will be required to produce humanlike response in a dummy neck. All such devices are, of course, velocity sensitive, and require a common velocity experience to assure reproducible results. To achieve this and to assure the loading experienced by the dummy neck is comparable to that of the cadavers, an impact velocity parameter was introduced. Hence the following performance requirements are offered.

PERFORMANCE REQUIREMENTS FOR MECHANICAL NECKS

The neck, when subjected to the dynamic tests specified in this paragraph, has the following response:

1. The moment-angle, relationships during loading lie within the corridors shown in Figs. 6 and 7 for extension and flexion, respectively.

2. The quotient of the area between the loading and unloading curves and the area between the loading curve and the abscissa is to be 0.3 to 0.5 for flexion and 0.4 to 0.6 for extension.

It is necessary for a single structure to satisfy both flexion and extension requirements, without modification or adjustment.

Details regarding the required test procedure are as follows:

1. The neck is mounted between the head and chest structure of the dummy. The chest is reclined 15 deg. to the vertical, and the anterior-posterior axis of the head is horizontal. The chest is restrained to minimize its rotation. The initial head position is maintained until the onset of the prescribed loading.

2. The following loading is required:

    Flexion. The chest experiences a horizontal change in velocity of 32 ± 5 ft./sec. producing flexion of the neck. The maximum moment produced at the occipital condyles is at least 120 ft.-lbs.

    Extension. The chest experiences a horizontal change in velocity of 22 ± 5 ft./sec. producing extension of the neck. The maximum moment at the occipital condyles produced is at least 40 ft.-lbs.

3. For the purpose of these tests, the axis of the occipital condyles is defined to be perpendicular to the midsagittal plane of the head. The axis is 0.75 in. posterior and 2.45 in. inferior to the center of gravity of the head.

Trajectory Requirements — While necessary and sufficient performance requirements for mechanical necks may lie somewhat in the future, as data become available, new requirements should be made. Head trajectory is an area where additional data is required to improve neck designs. The head trajectory will determine whether head impact occurs in a collision environment and, if so, to what object. It would therefore seem imperative for mechanical necks to provide a humanlike head trajectory as well as proper resisting forces.

Melvin (3), who has made an exhaustive study of the literature, cites a paucity of trajectory data. He does present a mean human volunteer trajectory from Snyder, Chaffin, and Schultz (4).

References p. 288

There are two major difficulties in determining the head trajectory relative to the torso. If the torso has any rotation, then a point on the torso must be identified in order to have a unique trajectory. Selecting such a surface landmark on a human is not difficult, but for dummy work an analogous point must be defined.

For autogenous ranges of motion, the torso can be adequately restrained such that its rotation is negligible. However, total restraint is very difficult in a forced or a dynamic environment. The first thoracic vertebra was used by Mertz and Patrick as the torso reference point when the torso was subject to rotation. Once such a point is defined, its motion must be adequately monitored. It is very difficult to separate the surface motion from the motion of the underlying structure in volunteers; Mertz and Patrick monitored the torso motion by tightly strapping a custom molded acrylic plate to the back of the subject.

The high-speed movie of the most severe flexion test (Run 79) of the original Mertz and Patrick data was analyzed for the trajectory of the center of gravity of the subject's head relative to a back target. This trajectory is shown in Fig. 8. The back target was strapped firmly to the subject directly on the surface marker of the first thoracic vertebra. The relative position of this marker and its estimated motion relative to the target are also indicated in the figure. The data indicate that the position of the center of gravity of the head cannot be uniquely related to the head angle relative to the torso. For instance, a head angle of -13 degrees occurred for a cg position of (-0.2, 1.4) as well as for a position of (-0.35, -0.2) as indicated in Fig. 8.

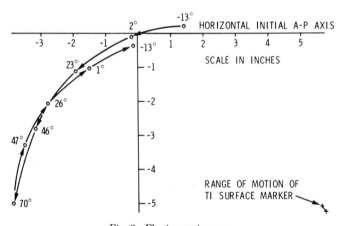

Fig. 8. Flexion trajectory.

Tests were conducted two years later to determine the autogenous trajectory of the head's center of gravity relative to a fixed point on the torso for the same volunteer, LMP. This time the back target was strapped on tightly directly on the surface marker of the second thoracic vertebra. This position was used instead of the first thoracic vertebra location used in the earlier dynamic tests because there was no palpable motion of this point when the head was rotated in the midsagittal plane. Slight motion was noted for the marker when it was positioned over the first thoracic vertebra. A typical result is shown in Fig. 9. The lack of a unique relation between the head angle relative to the torso and the position of the head cg is reinforced. Also, it is noted that the cg of the head is elevated relative to the torso as the head is rotated rearward from its initial position. This is because the cg of the head is located superior and anterior of the occipital condyles.

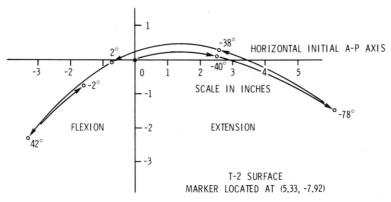

Fig. 9. Voluntary trajectory.

Because of the problems encountered in determining the uniqueness of the trajectory of the cg of the head, a requirement has not been specified in this paper.

PERFORMANCE OF COMMONLY USED NECKS

Neck Configurations — Four commonly used dummy neck configurations were evaluated against the recommended performance requirements. Two of these were standard commercial necks; a Sierra 1050 and an Alderson VIP-50A. A third configuration was obtained by inverting the Alderson neck, that is, the rubber section was placed at the base of the neck and the metal spacer was placed on top of it. Some investigators (5) have indicated that this was a "more humanlike" configuration. The fourth configuration tested was a GM Proving Ground solid rubber neck. This neck is used in the GM Hybrid I dummy. The necks are shown in Fig. 10.

In order to have a comparative evaluation of the necks, they were tested in a common dummy. The GM Hybrid I dummy was selected. This dummy consists of a slightly modified Alderson VIP-50A torso and limbs, a Sierra 1050 head, and the rubber neck with necessary adaptive hardware. Sierra and Alderson necks were also tested in their normal configuration.

The tests were conducted on the WHAM I sled at Wayne State University. Figs. 11 and 12 show the facility and the seat used with the dummy in place in the flexion configuration. This apparatus is described in detail by Mertz (6). Briefly the sled can be accelerated pneumatically up to 40 mph. A hydraulic snubber is capable of decelerating the sled up to a maximum rate of 25 G. The maximum usable snubber stroke is 22 inches. The seat frame was constructed of steel angles. It had a plywood bottom and back.

Fig. 10. Proving Ground (left), Alderson (middle), and Sierra (right) rubber necks.

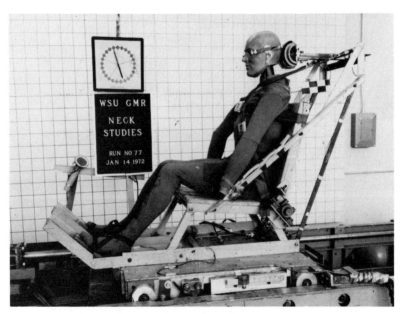

Fig. 11. WHAM I.

The dummy was tightly strapped into the chair by a lap belt, two crossing shoulder straps, and a horizontal belt across the chest and under the arms (see Fig. 12). The head of the dummy was kept in position during the acceleration phase by the headrest.

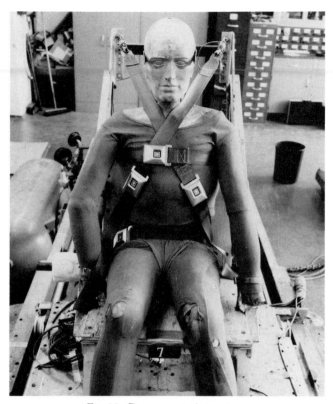

Fig. 12. Dummy restraint system.

Linear and angular acceleration data were collected from accelerometers. Angular position data were obtained by high-speed photography. Angular velocity data were obtained by differentiating the position data.

The Sierra head with accelerometers in place weighed 10.0 lbs. Head weights were not taken in accordance with the SAE recommended Practice J963, since that specification includes part of the neck. The mass moment of inertia about the center of gravity was 0.22 in.-lb.-sec^2. The neck skin was cut away to avoid interference with the various necks. The Alderson head with accelerometers in place weighed 7.37 lbs. Its mass moment of inertia about its center of gravity was 0.132 in.-lb.-sec^2. For both heads, the occipital condyles were taken as 0.75 in. posterior and 2.45 in. inferior to the center of gravity as found by Mertz and Patrick (7).

Results — Comparative Tests — Figs. 13 through 16 show the response characteristics of the four neck configurations in extension. In the background, the required loading corridor is plotted. The ordinate is the moment of force about an axis through the occipital condyles. The abscissa is the head angle relative to the torso. The similarities of the four rubber necks can be readily seen. All are too stiff and all lack damping.

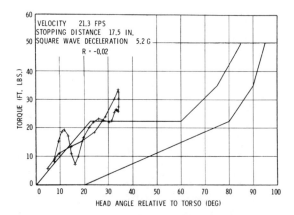

Fig. 13. Extension response characteristics of an Alderson (Standard Position) neck in a GM hybrid dummy.

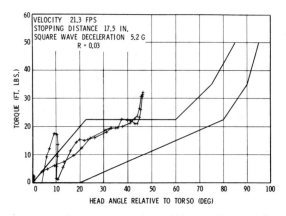

Fig. 14. Extension response characteristics of an Alderson (Inverted Postion) neck in a GM hybrid dummy.

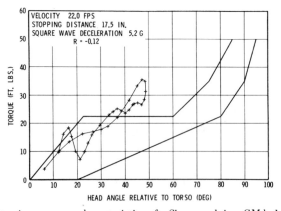

Fig. 15. Extension response characteristics of a Sierra neck in a GM hybrid dummy.

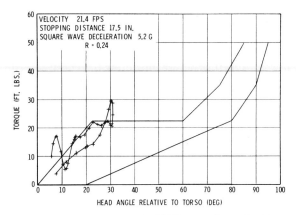

Fig. 16. Extension response characteristics of a GM Proving Ground rubber neck in a GM hybrid dummy.

Figs. 17 through 20 show the response characteristics of the same four necks in flexion. These plots differ from the extension plots in two aspects. The required values of the parameters are different and the torque plotted on the ordinate includes the moment of the chin-chest interaction.

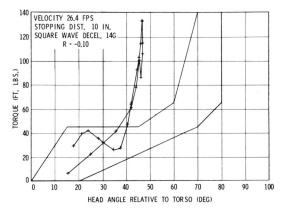

Fig. 17. Flexion response characteristics of an Alderson (Standard Position) neck in a GM hybrid dummy.

The GM Hybrid neck mount is designed to give the dummy head a proper orientation when the seat back is reclined 24° from the vertical. The seat back was reclined only 15° in these tests, and the dummy appeared to look down at about 9°. Therefore the total angulation in each of the four figures was foreshortened about 9°. This would not affect the basic conclusions except for the Alderson neck in the inverted position (Fig. 18). Rotation of all four necks was stopped by the impact of the chin onto the chest. On all but the inverted Alderson, this limited the rotation to less than the required amount. If the 9° offset were included, the inverted Alderson neck would marginally have the correct range of motion. The inverted Alderson has a greater range of motion than the standard Alderson because placing the flexible rubber section lower in the neck gives the head path a greater radius of curvature and allows the head to rotate further before impacting the chest.

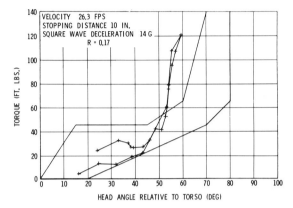

Fig. 18. Flexion response characteristics of an Alderson (Inverted Position) neck in a GM hybrid dummy.

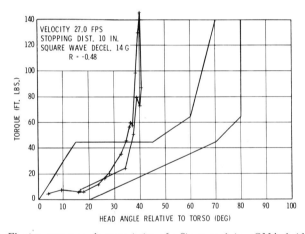

Fig. 19. Flexion response characteristics of a Sierra neck in a GM hybrid dummy.

The chin-chest impact also produced large sudden rises in the resisting torque for three of the four necks. The exception was the GM Hybrid (Fig. 20) which not only had a low maximum torque, but also had an unusual flat top on the torque-angle plot. Because of these differences, additional tests were made of this configuration which confirmed both items. The GM Hybrid neck was stiffer than the other necks and therefore produced a softer impact of the chest. This, coupled with sliding of the chin on the chest, probably produced the flat area on the curve.

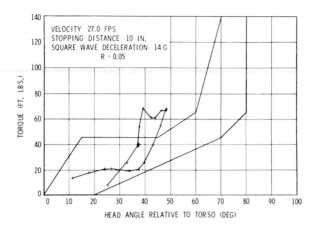

Fig. 20. Flexion response characteristics of a GM Proving Ground rubber neck in a GM hybrid dummy.

The sled velocity in these tests was only 27 fps which is lower than the specified mean of 32 fps. Since all the necks had obvious deficiencies based on the extension and flexion tests that were made, additional tests at the higher velocity were not considered necessary.

Results — Standard Configurations — Fig. 21 shows the extension response characteristics of the Alderson neck with a standard Alderson torso and head. In Fig. 13, this neck exhibited a larger angle and torque when exposed to a similar acceleration environment. However the Hybrid dummy uses a Sierra 1050 head which is heavier than an Alderson head (10.0 lb. as compared with 7.37 lb.). Hence while the sled acceleration level was the same, the loading on the neck was significantly different.

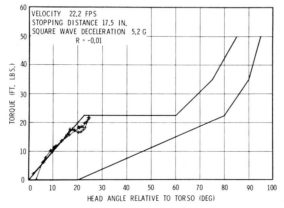

Fig. 21. Extension response characteristics of an Alderson (Standard Position) neck in an Alderson dummy.

Fig. 22 shows the extension response characteristics of a Sierra neck in a standard Sierra dummy. It also exhibited smaller angles and torques than it did in the Hybrid configuration. Fig. 15, although it used the same head in both cases. This can be explained by considering the initial positions of the head. The dummy head appeared to be looking up at about 10° when in the test position while it looked down at 9° in the Hybrid configuration. (No special adapters were used in either case.) Hence, the standard configuration eliminates 19° from the top of the arc where the moment produced by a constant horizontal acceleration of the head would be nearly constant. Its initial position is such that the moment arm is constantly and rapidly reducing in length. Therefore, the initial position of the head in Fig. 22 causes lower torque-angle loading of the neck than in Fig. 15.

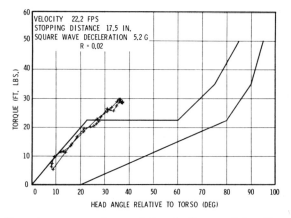

Fig. 22. Extension response characteristics of a Sierra neck in a Sierra dummy.

Both curves in Figs. 21 and 22 lack the oscillation during loading that can be observed in Figs. 13 and 15. The exact cause of this oscillation was not identified, but it would appear to be related to the sharp spike on the onset of sled deceleration (12 G peak to 5 G plateau). It was observed during the program that the rubber cushion on the sled which impacted the snubber for decleration had deteriorated badly. When it was replaced, the peak of the sled deceleration pulse was reduced to 10 G. In later tests, with this reduced sled deceleration spike, the rubber neck structures did not exhibit the large oscillation in loading observed earlier.

Figs. 23 and 24 show the results of two tests of an Alderson neck in an Alderson dummy in flexion. A third test of this configuration gave results very similar to Fig. 24. In the latter two tests, there apparently was a hard contact which did not occur in the first test. This presents a possible repeatability problem in addition to the other deficiencies of the neck. There was no significant change in angulation compared to the Alderson neck-Hybrid dummy configuration Fig. 17. (Both initially had a similar 9° forward offset.)

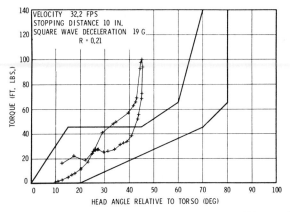

Fig. 23. Flexion response characteristics of an Alderson neck in an Alderson dummy, run 152.

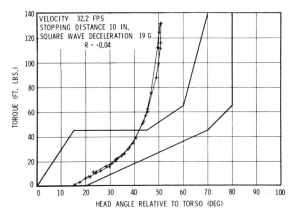

Fig. 24. Flexion response characteristics of an Alderson neck in an Alderson dummy, run 154.

Fig. 25 shows the response characteristics of a Sierra neck in a Sierra torso. It did not experience the repeatability problems of the Alderson dummy, and the change in dummy configuration did improve its range of rotation as compared to Fig. 19. Nineteen degrees of this difference can be accounted for by the change in initial position of the head (9° forward to 10° rearward). If the neck had been mounted with the head erect for this seat back angle, the head would have had the required range of motion relative to the torso in flexion.

Load comparisons of Figs. 23, 24, and 25 to Figs. 17 and 19 are not clear since the head rotation was terminated by different hard impacts. Also the stock configurations had a higher initial impact velocity as discussed earlier.

PERFORMANCE OF EXPERIMENTAL NECKS

Several institutions are actively involved in the development of humanlike neck structures. A number of these were tested and are reported here. Others, which have not been tested against the proposed performance requirements, are discussed briefly in the concluding section.

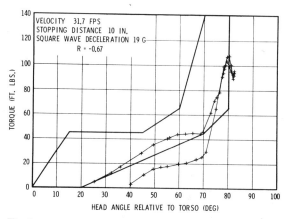

Fig. 25. Flexion response characteristics of a Sierra neck in a Sierra dummy.

MetNet Neck — The MetNet Neck was developed several years ago at GMR Laboratories to indicate maximum loads (axial and bending moment) and maximum deflections (axial and angular) experienced by the neck in simulated crashes. The neck, Fig. 26, consists of five split ball and socket joints with load carrying disks. The split balls and the load carrying disks are fitted with a crushable material (MetNet) which deforms permanently under load. The deformation of these resistive elements indicates the deflection of the neck and the load carried by the member.

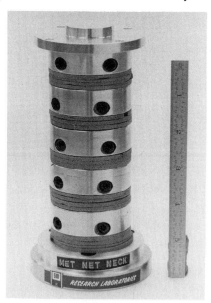

Fig. 26. MetNet neck.

Fig. 27 shows the extension response characteristics. The damping ratio, as expected, was nearly 1.0. The flexion response is indicated in Fig. 28. The chin-chest impact of the Hybrid dummy occurs as usual. Analysis of the crushable resistive elements indicated that the actual resistive torque in the top joint of the neck was only 22 ft.-lb. This means 120 ft.-lb. of the total 142 ft.-lb. resistive torque was due to the chin-chest interaction.

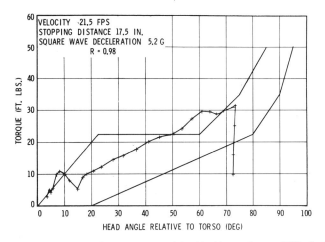

Fig. 27. Extension response characteristics of the MetNet neck in a GM hybrid dummy.

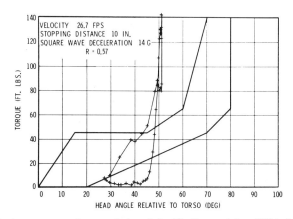

Fig. 28. Flexion response characteristics of the MetNet neck in a GM hybrid dummy.

GMR Ball Jointed Polymeric Neck — The GMR ball jointed polymeric neck (Fig. 29) was developed to have humanlike response characteristics in a crash environment. The basic structure is similar to that of the MetNet neck. However the resistive elements are of polymeric materials. They can be geometrically shaped to have different properties in flexion and extension. Various materials and shapes of resistive elements have been tested. Two different sets have satisfied the performance

References p. 288

requirements in extension. Shown in Figs. 30 and 31 are extension results using elements of polyvinylchloride and polyurethane, respectively. Multiple tests indicates no repeatability problems.

Fig. 29. GMR ball jointed polymeric neck.

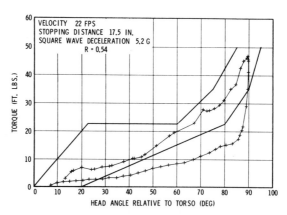

Fig. 30. Extension response characteristics of a GMR neck (polyvinylchloride washers) in a GM hybrid dummy.

Flexion characteristics had to be determined in a Sierra dummy due to the frequently mentioned chin impact problem with the GM Hybrid I dummy. Fig. 32 shows the flexion results for a set of polyurethane elements which satisfied the performance requirements. However, two more tests did not produce high enough torques due to twisting of the dummy's head during sled deceleration. This is a minor mechanical problem of the neck design not related to the dynamic response characteristics.

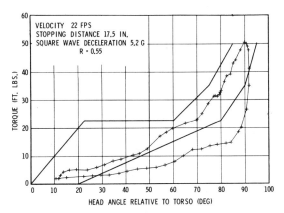

Fig. 31. Extension response characteristics of a GMR neck (polyurethane washers) in a GM hybrid dummy.

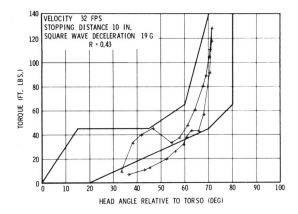

Fig. 32. Flexion response characteristics of a GMR neck (polyurethane 750 washers) in a Sierra dummy.

General Motors Proving Ground Monolithic Neck — The Safety Research and Development Laboratories of the GM Proving Ground have developed monolithic neck structures. Monolithic structures are very desirable from a users point of view since they are easy to set up and highly repeatable. A rubber neck developed by this group was discussed earlier. It was too stiff and lacked damping. Materials with better damping characteristics and different stiffnesses are being investigated. Fig. 33 shows the flexion characteristics of a polyacrylate model tested in a Hybrid II dummy. The Hybrid II employs different neck and head mounts that eliminate the chin-chest early impact problem of the GM Hybrid I dummy.

References p. 288

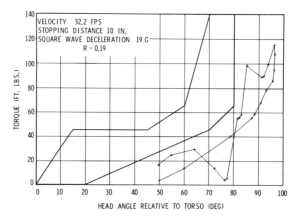

Fig. 33. Flexion response characteristics of GM Proving Ground polyacrylate neck in a GM hybrid II dummy.

Alderson Research Laboratories Viscously Damped Neck — This neck consisted of four articulative joints. The top joint had viscous resistance to motion. It allowed flexion-extension motion in the sagittal plane. The viscous resistance to motion could be controlled independently for each direction by adjustable orifices. Rubber stops limited the motion rearward. Impact of the chin on the neck provided a forward stop. Friction washers on the mounting bolts provided resistance to small static loads. The second and third joints allowed axial rotation and rotation in the lateral plane, respectively. The bottom joint was a short rubber block that provided a degree of universal motion.

Extension and flexion response characteristics for preferred orifice settings are shown in Figs. 34 and 35, respectively. The characteristics of the neck were changed

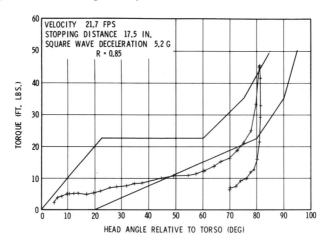

Fig. 34. Extension response characeristics of an Alderson Laboratories viscously damped neck in an Alderson dummy, run 91.

drastically by adjusting the orifices. This neck has been redesigned. The flexion-extension pivot point has been moved to a lower position and a spring return has been added to the joint to attenuate the damping. Tests have not been conducted on this new design.

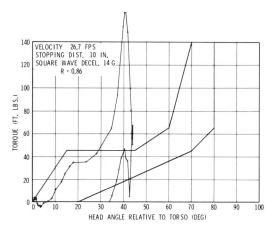

Fig. 35. Flexion response characteristics of an Alderson Laboratories viscously damped neck in an Alderson dummy, run 87.

Other Experimental Necks — A number of other necks are under development by various institutes that are conceptually promising. To the authors' knowledge, they have not been fully evaluated against the recommended performance requirements. GM Research Laboratories has a prototype that would replace the ball joint-disk arrangement with a cable-disk arrangement, thus greatly reducing the cost and complexity of the present structure. The General Tire Corporation is developing viscously damped resistive elements for the present GMR ball joint neck. The Highway Safety Research Insititute (3) at the University of Michigan has developed a neck that is similar in principle to the GMR ball joint neck. However, it uses a cheaper, commercially available universal joint in place of the ball and socket and, consequently, does not allow head rotation about a vertical axis. Its resistive elements are butyl rubber, and the desired characteristics are obtained by geometric shaping of the elements. The Safety Systems Laboratory of the National Highway Traffic Safety Administration is developing a neck that depends on the stretching of elastic cords over cam surfaces for resistance.

CONCLUSIONS

Performance requirements for mechanical necks have been proposed which are based on biomechanical data from volunteers and cadavers. These requirements also include consideration of mechanical reproducibility and repeatability in manufacturing and testing. Any dummy head-neck-chest system which meets these requirements will have significant necessary biomechanical characteristics.

References p. 288

While these conditions are necessary, they are not sufficient. One critical area needing additional study is the head trajectory. Trajectory requirements may need the load conditions specified as well. No unique relation between head angle and position can be specified. Other conditions needing study for a more humanlike neck are response characteristics in the lateral plane, torsional twist characteristics, and elongation-compression response to axial loads.

No known commercially available or commonly used neck satisfied these necessary requirements. The GMR Polymeric neck has demonstrated the feasibility of satisfying the requirements by meeting them using multiple configurations. Additional efforts are required to assure proper performance of the neck when used in conjunction with a total dummy structure under a wider range of conditions. Also, its durability must be established before this neck can serve as an adequate compliance device.

The test procedure described by Mertz and Patrick and modified here is not desirable for the qualification of mass produced necks which are of a type that meet the recommended requirements. Simplified test procedures are needed. However, before meaningful performance on a simplified system can be specified, its traceability to the test procedures and response characteristics of Mertz and Patrick or other biomechanical data must be documented.

REFERENCES

1. H. J. Mertz and L. M. Patrick, "Strength and Response of the Human Neck," Proceedings of the Fifteenth Stapp Car Crash Conference, SAE 1971.
2. H. J. Mertz and L. M. Patrick, "The Effect of Added Weight on the Dynamics of the Human Head," U.S. Army Natick Laboratories, Natick, Mass., 1971. Report on Contract No. DAAG-17-67-C-0202.
3. J. W. Melvin, "Improved Neck Simulations for Use with Anthropomorphic Dummies," Contract No. 329570, Highway Safety Research Institute, The University of Michigan, Ann Arbor, Michigan, 1972.
4. R. G. Snyder, D. B. Chaffin, and R. K. Schultz, "Link System of the Human Torso," AMRL, Wright-Patterson Air Force Base, Contract F33615-70-C-17777, 1971.
5. "Test Procedure and Performance of Anthropomorphic Test Devices," Safety Systems Laboratory, National Highway Traffic Safety Administration, 1971.
6. H. J. Mertz, "The Kinematics and Kinetics of Whiplash," Ph.D. Thesis, Wayne State University, Detroit, Michigan, 1967.
7. H. J. Mertz and L. M. Patrick, "Investigation of the Kinematics and Kinetics of Whiplash," Proceedings of the Eleventh Stapp Car Crash Conference, SAE 1967.

DISCUSSION

S. W. Alderson *(Alderson Biotechnology Corporation)*

Are the motion ranges well defined in the neck?

H. J. Mertz

They are not unique. The motion of any multiple link system will not be unique.

PROGRESS IN THE MECHANICAL SIMULATION
OF HUMAN HEAD-NECK RESPONSE

M. P. HAFFNER and G. B. COHEN

U.S. Department of Transportation, Washington, D.C.

ABSTRACT

A mechanical neck has been constructed which simulates $\pm G_x$ human head motion in the automotive crash environment. The present design simulates head motion in the sagittal plane by means of a two pivot model. Pre-loading torques, as well as linear and non-linear torque elements, can be accommodated at each pivot location. The proposed method of validation is the matching of displacement-velocity response of the head as predicted by a two degree-of-freedom computer simulation with existing human data. Simulations have been performed using Daisy sled volunteer data; comparisons of head angular displacements and angular velocities show favorable correlation.

INTRODUCTION

This paper describes the development of a mechanical analog of the human head-neck whose kinematic response in the automotive crash environment can be correlated with human response at certain levels of exposure. It is important to emphasize that a correlation between human and surrogate response is being sought, and not necessarily a one-to-one relationship, although this would certainly be desirable and convenient. Also, it should be expected that a mechanical test device, if it is to be relatively simple, repeatable, and usable, will most likely allow correlation within only a certain dynamic exposure range, and then perhaps only for occupants exhibiting a specific muscular activity state prior to impact.

The opinions, findings, and conclusions expressed in this publication are those of the authors and not necessarily those of the National Highway Traffic Safety Administration.

References pp. 314–315

The methodology of the current work consists of the examination of the kinematic response of the human head-neck to $\pm G_x$ impact, the synthesis of a mechanical design capable of demonstrating the primary features of this response, and the validation of the mechanical analog thus produced. Essential to the design process is the development of a mathematical simulation which can itself predict mechanical head-neck response. This simulation provides the link which permits rational comparisons to be made between human and surrogate responses.

The difficulties associated with the development of a general neck analog are numerous. The bony structure of the neck and the neck musculature are highly asymmetric. Seven anatomically different cervical vertebrae link the head to the thorax at T1; each articulation possesses its characteristic ranges of motion and translational and rotational degrees of freedom. Forces opposing head-neck motion are generated both by intervertebral ligaments and by extensive muscular connections between the head and upper thorax. Simulation of these resistive forces is complicated by the fact that the muscular tissue is capable of various active states. The result is that the muscle behaves as a viscoelastic system with parameters which are functions of muscle activity. For example, Moffatt, et al (1) have documented a strong functional relationship between isometric muscle tension and dynamic viscoelastic muscle properties. The muscle parameters are also known to be non-linearly dependent upon muscle length and velocity (2).

These observations indicate that a unique kinematic response of the human head-neck to a specified dynamic input does not exist. However, a specific response must be chosen for simulation.

The literature has been searched for dynamic response data for the human head-neck. Mertz and Patrick (3,4) instrumented a volunteer and several cadavers, and measured accelerations on the head and neck for both $+G_x$ and $-G_x$ exposures. Their emphasis was upon correlation of computed torques at the occipital condyles with the probability of neck injury. Tarriere and Sapin (5) present $-G_x$ human impact data, and model the neck-thorax link as a single pivot at C7-T1 restrained by a spring and damper in series. Ewing, et al, (6) and Ewing and Thomas (7) present the results of human $-G_x$ sled runs made under Army-Navy-Wayne State University auspices. They consider the head-neck input to be the timed-based accelerations at T1; outputs are the linear and angular accelerations, velocities, and displacements of the head anatomy.

Chandler and Christian (8) document photographically a series of human volunteer $-G_x$ exposures on the Holloman Daisy Track. Volunteers were seated in production automobile seats and wore both Type 1 and Type 2 automotive restraints. Major emphasis has been placed in the present research upon the analysis of these films for the deduction of head-neck kinematics. Selected early Daisy runs (9) have also been utilized, primarily for the determination of head kinetics.

The data sources investigated are summarized in Table 1.

TABLE 1

Human Volunteer Data Sources

Reference	Sled Type	Restraint	Pulse Shape & Duration	Peak g's (Sled)	Sled Accel.	No. of Volunteers
3	Decelerator	N/A	140ms	1.3 & 3.2	+G_x	1 (2 Runs)
4	Decelerator	Lap Belt & Dbl. Shoulder Harness	100ms	2.0- 4.4 (Relaxed) 1.5- 9.6 (Tensed)	-G_x	1 (46 Runs)
6,20	Accelerator	Lap Belt, Shoulder Harness, Chest Safety Strap	250ms	3-10 g	-G_x	17 Volunteers (197 Runs)
8	Daisy Sled Decelerator	Automotive Restraint Type 1 Type 2	100ms	7.6-14.5 10.0-14.8	-G_x	(14 Type 1 Runs) 19 Volunteers (18 Type 2 Runs)
9	Daisy Sled Decelerator	Lap Belt & Dbl. Shoulder Harness	65ms	33.7-34.4	-G_x	Runs 674 & 675

MECHANICAL DESIGN CONSIDERATIONS

While mathematical simulation and formal parameter deduction efforts were underway, a parallel effort was directed at combining the essential elements of head-neck geometry, mass distribution, and static human measurements into a practical mechanical design. The intent was, however, to build into the prototype a high degree of flexibility, so that the results of the dynamic simulations could be incorporated into the design as they became available. The development of criteria for design of the prototype neck structure is described below.

The head and neck were analyzed as a sub-system with dynamic inputs at T1; dynamic outputs were characterized as linear and angular motions of the head anatomy. The head was considered to be a rigid body for the time frame under consideration, while the neck was assumed to be a non-rigid viscoelastic linkage. Response of the head anatomy was assumed to occur in the mid-sagittal plane for $\pm G_x$ impact.

In order to gain insight into the head kinetics during -G_x impact, films of Daisy Runs 674 and 675 (February 1960) were examined. Pertinent data for these runs is shown in Table 2.

References pp. 314–315

TABLE 2

Data for Daisy Runs 674 and 675

Run No.	Sled g Peak	Onset (g/sec) (Nominal)	Peak Subject g's	Entrance Velocity (ft/sec)	Duration Subject-X (sec)
674	33.7	1200	37.4	45.3	.066
675	34.4	1200	39.8	46	.065

Fig. 1 illustrates motion of the head relative to a ground mounted camera during Run 675. Figs. 2 and 3 show the motion of the head relative to the sled during the loading phase (flexion angle increasing) and the unloading phase (flexion angle decreasing) respectively for Run 674. Fig. 4 is the superposition of Figs. 2 and 3. Since the volunteers in these early runs did not wear skull caps with photometric tracking targets, two targets were inserted at each head position by means of a template cut to match the head outline.

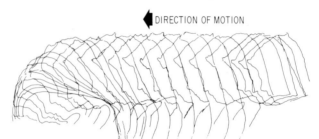

FORWARD HEAD MOTION DURING SLED DECELERATION PULSE

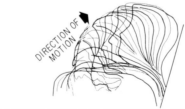

HEAD MOTION AFTER SLED DECELERATION PULSE *(SLED STOPPED)*

Fig. 1. Kinetics of head-neck motion – Daisy Run 675.

The unloading and loading trajectories are seen to superimpose for small head angles from the vertical, but soon diverge. The unloading trajectory loops down at maximum flexion and returns under the loading trajectory.

The trajectory shape is largely a function of the interaction of the thorax with the upper restraint. During the loading the upper thorax exhibits considerable forward displacement, with the result that the head and neck trajectory is high and flat. Maximum forward torso displacement occurs before the head reaches full flexion,

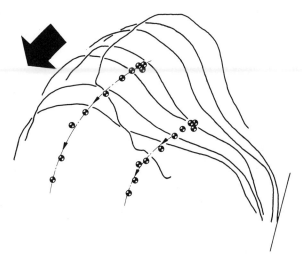

Fig. 2. Head kinetics during loading – Daisy Run 674.

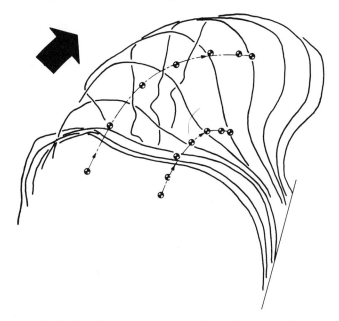

Fig. 3. Head kinetics during unloading – Daisy Run 674.

with the result that as the torso begins to move backward towards the seat, the head is pulled down and into the sternum. Subsequently, the upper torso moves back into the seat as the head rotates back into its initial position.

A technique for the location of instantaneous centers of rotation of rigid bodies in planar motion was applied to the unloading portion of Runs 674 and 675. The goal was the deduction of the range of effective centers of rotation of the head for application to $-G_x$ mechanical neck design. The loading phase did not prove useful for this purpose, since the torso translates and rotates substantially in the sled coordinate system, and its motion tends to obscure the kinetics of the neck. Fig. 5 illustrates the graphical location of centers of rotation for the head anatomy during the unloading phase of Daisy Run 674.

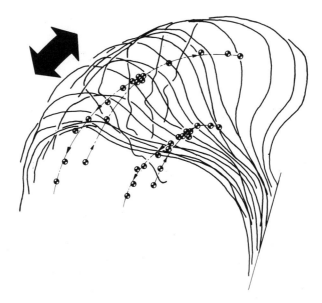

Fig. 4. Superposition of loading and unloading head kinetics — Daisy Run 674.

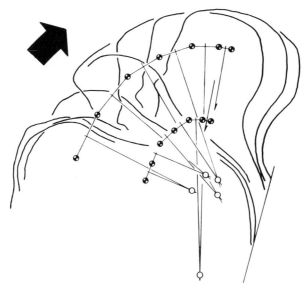

Fig. 5. Construction of instant centers of rotation for unloading — Daisy Run 674.

With the influence of the torso largely isolated, the centers of rotation of the head anatomy are found to plot closely to one another. These instant centers fall below the estimated location of the transverse plane through the spinous process of T1*.

A similar construction for the unloading phase of Run 675 is shown in Fig. 6.

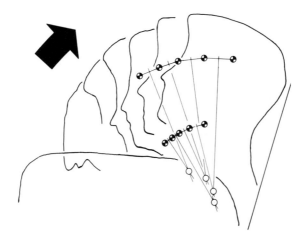

Fig. 6. Construction of instant centers of rotation for unloading – Daisy Run 675.

Further examination of Figs. 5 and 6 reveals that head motion cannot be characterized as simple one degree-of-freedom motion. Translation of the head anatomy relative to T1 as well as rotation can be observed, as has also been noted by Ewing. Therefore, it is desirable that the mechanical neck should provide the capability for translation of the head, as well as rotation, during -G_x impact.

A concurrent goal of the design effort was the incorporation of realistic +G_x (whiplash) kinetics. Little published data on human head-neck response to +G_x acceleration exists in a form that is useful for design purposes. However, some film data of human experiments conducted at Case Institute of Technology is available. Fig. 7 shows tracings from a film of a typical volunteer exposure. It should be noted that the effective centers of rotation of the head anatomy are considerably higher than those observed for -G_x impact. An important contributing factor to this result is the significant capacity for extension ($\sim 30°$) at the atlanto-occipital articulation.

Another important consideration was that of muscular pretensioning. Upon examination of the responses of the sled volunteers, it became clear that isometric muscular pretensioning prior to impact was likely to be a significant factor in determining response, and that capability for providing preloading in the mechanical neck should, if possible, be incorporated. A typical example of the effect of

*Hertzberg & Daniels (10) data for head length, stature, and cervicale height were utilized to estimate the position of T1.

pretensioning may be seen in Fig. 8, where the volunteer's response to a 3.2g, $+G_x$ impact is seen to lag behind that of two cadavers exposed to the same nominal impact. More quantitative information was deduced from (11). According to Gadd, et al, males of medium stature can resist loads up to 79 lbf applied in the P-A direction against the occiput. Assuming a distance of 6.5 inches between transverse planes intersecting C7-T1 and occiput, and isometric muscular torque of 503 in-lbf about C7-T1 is required to resist flexion.

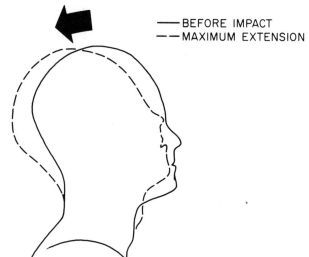

Fig. 7. Head kinetics during $+G_x$ impact.

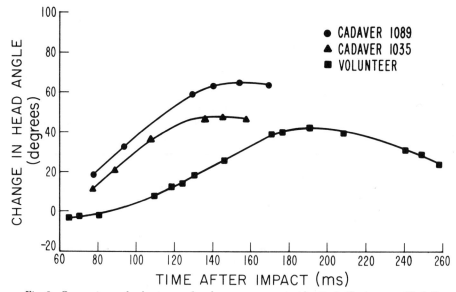

Fig. 8. Comparison of volunteer and cadaver exposures to identical $+G_x$ impacts (Ref. 3).

The net result of this investigation was the formulation of the following design criteria:

1. A higher effective denter of rotation for $+G_x$ exposure than for $-G_x$ exposure.

2. Provision for simulation of head translation as well as rotation relative to the thorax for $-G_x$ impact.

3. Provision for varying neck stiffness in both flexion and extension.

4. Provision for the incorporation of linear and non-linear stiffness and damping elements.

5. Provision for preloading capability to simulate the vehicle occupant aware of impending impact (e.g., a sled volunteer).

6. Relative simplicity of design.

7. Fidelity of response within limitations of 6 above.

8. Compatibility with an existing dummy; ease of use in the testing environment.

9. Repeatability.

10. Ease of modification to permit matching to new bio-mechanical data.

11. Suitability for mathematical simulation.

A concept was sought which would meet the criteria established and yet retain relatively straightforward mechanical design. It was decided that for the first prototype, the main effort would concentrate upon obtaining humanlike response to both $\pm G_x$ exposures, leaving the incorporation of lateral flexion and rotation capability to a later date. Thus, the first design, shown in Fig. 9, operates in the mid-sagittal plane only.

The neck is designed to be used with a properly ballasted Alderson VIP-50A head; its base attaches directly to the existing VIP-50A thorax. The neck is of a two pivot design, these pivots being located at the equivalent of the anatomical locations of C7-T1 and C4-C5 respectively. Resistive torques developed about these pivots are generated by calibrated lengths of elastic shock cord guided by rotating sectors. The capability for adjustment of elastic cord diameter and length, and of sector profile, yields considerable flexibility in the generation of torque characteristics at the pivots.

Torsional preloading at each pivot (if desired) is achieved by pretensioning the elastic cords using adjustments which are accessible with the head-neck-thorax complex fully assembled. Damping characteristics are at present inherent in the elastic cords; provision has been made for the incorporation of additional damping elements.

The neck is so designed that when the base is impacted from the front ($-G_x$ impact), the restraints on the moving elements of the neck allow relative motion at both pivots, to permit simulation of both rotation and translation of the head. Thus in -G$_x$ impact, the head-neck is a two degree-of-freedom system, and may be mathematically modeled as such. When the base of the neck is impacted from the rear (whiplash simulation), the design is such that the bulk of the response occurs at the upper articulation. A higher effective center of rotation for +G$_x$ impact is thus achieved.

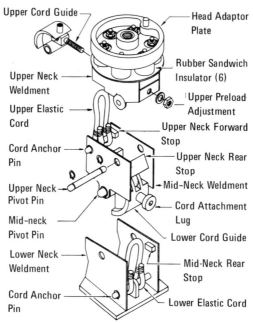

Fig. 9. Two pivot mechanical neck.

Fig. 10 shows a three-quarter rear view of the prototype neck "skeleton." The upper part of the neck is shown in full extension, this extension having taken place about the upper pivot. Six rubber sandwich type isolation mounts separate the head attachment at the top of the neck and the main structure. There is thus no direct metallic path within the neck through which high frequency ringing may be transmitted. Fig. 11 is a view of the neck skeleton from the front quarter. The upper part of the neck remains in full extension, but the head attachment plate and isolation mounts have been removed.

PROTOTYPE NECK: -G$_x$ IMPACT TESTS

Preliminary modeling indicated that the torsional stiffness at C7-T1 lay in the range of 600-800 in.-lbf/radian; preload capability up to 200 in.-lbf about C7-T1 was deemed achievable in the design. Because of space limitations, a material with extremely high energy storage capability per unit volume was sought for use as the elastic element at the lower pivot. The material finally selected for this purpose was military specification (MIL-C-5651) elastic shock cord, composed of multiple strands of natural rubber tightly encased in two woven coverings. Considerable static testing of this material was conducted to document its performance, and methods were investigated for clamping the ends of this cord at high tensile loads and accomplishing this in the very limited space available. It was found that heavy-duty looping rings would permit clamping of the shock cord within minimum volume and with adequate strength. Fig. 12 shows a rear view of the SSL head-neck (without head backplate) as mounted on the VIP050A thorax. Fig. 13 further ducuments the flexion test configuration.

Fig. 10. Rear view of prototype skeleton in extension.

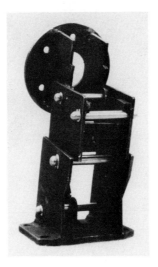

Fig. 11. Front view of prototype skeleton in extension (head adaptor and isolation mounts removed).

References pp. 314–315

Fig. 12. Rear view of prototype neck mounted on VIP-50A thorax.

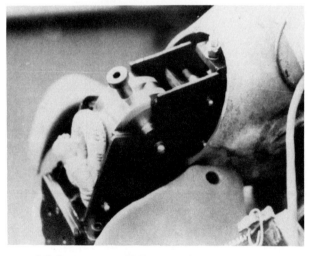

Fig. 13. Prototype in full flexion about C7-T1 pivot. Lower elastic shock cord shown detached from lower sector.

The geometric and inertial properties of the head-neck assembly in the test configuration have been measured; results are given in Table 3 and compared with those chosen as representative of the 50th percentile male by Tarriere and Sapin.

TABLE 3

Selected Geometric and Inertial Properties
of the Prototype in -G_x Configuration

	SSL Head-Neck as Ballasted for Test	Tarriere & Sapin "50th Percentile" Head-Neck
Total Moving Mass	12.87 lbm	13.45 lbm
Vertical Height of Assembly c.g. above C7-T1	5.4 in	5.8 in
Moment of Inertia about C7-T1 (C4-5 pivot fixed)	1.29 lbf-in-sec^2	1.50 lbf-in-sec^2

Sixteen -G_x sled runs have been made to date, with sled peak accelerations up to 34g, corresponding to a simulated entrance velocity of 46 ft./sec. Tests have been made both in Type 2 configuration and with the thorax securely restrained. Fig. 14 is a typical time lapse photograph of a test in the latter type of restraint. Peak sled g's were 29.2; change in velocity was 42.4 fps. No degradation of the neck structure was observed during these tests. Triaxial head, chest, and sled accelerations were recorded, and high speed films were made for the bulk of these experiments.

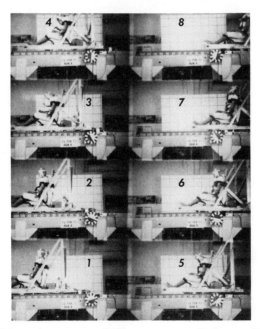

Fig. 14. Typical dynamic exposure of SSL prototype to -G_x impact. Peak sled g = 29.2.

PROTOTYPE NECK: +G$_x$ IMPACT TESTING

A parallel experimental program of dynamic +G$_x$ exposure of the mechanical neck on the SSL sled has also been conducted. Figs. 15 and 16 document the test configuration. Static measurements of the torque developed by the upper cord used

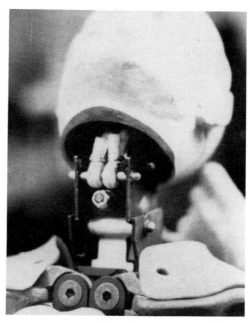

Fig. 15. Prototype neck in place on VIP-50A thorax; head held in hyperextension to expose upper shock cord assembly.

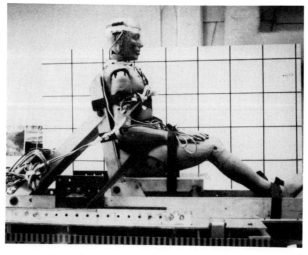

Fig. 16. Test configuration for +G$_x$ impact.

in the test configuration were made; the resulting torque curves for the two preload settings are shown in Fig. 17 (future whiplash experiments will be run with zero preload to simulate an unaware occupant). "Preload II" settings was used in the sled tests.

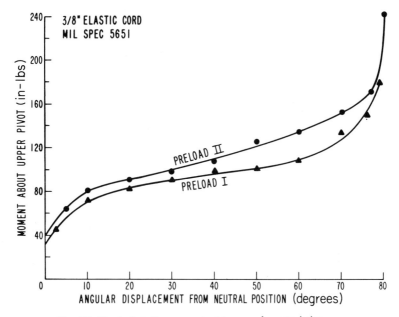

Fig. 17. Typical static upper pivot torque characteristics.

Measurements were made of the location of the c.m. of the head-neck relative to the C4-C5 pivot, moment of inertia, and weight of the moving mass prior to sled testing. Results for the whiplash configuration are shown in Table 4.

To date, the prototype in the $+G_x$ configuration has been subjected to seven dynamic exposures (SSL sled runs 1014-1020). Peak g's ranged between 8 and 14, with corresponding simulated entrance velocities of 11.6 and 16.9 fps. The primary

TABLE 4

Selected Geometric and Inertial Properties
of the Prototype in $+G_x$ Configuration

Moving Mass of Head-Neck	11.10 lbm
Radial distance from Pivot to c.g. of moving portion of head-neck assembly	3.90 in
Moment of inertia of moving mass about C_4 - C_5 pivot	.63 lbf-in-sec^2

References pp. 314–315

dummy instrumentation consisted of triaxial accelerometers in the head and chest. Fig. 18 is a time-lapse photograph of run 1016 (8g peak). Throughout the test series, no degradation of the prototype was noted.

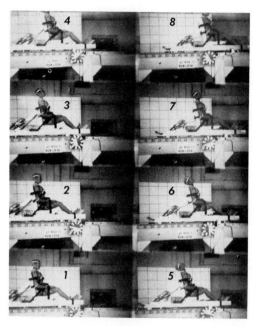

Fig. 18. Typical exposure of prototype neck to $+G_x$ impact (Run 1016, 8g sled peak).

STATIC AND DYNAMIC BENCH TESTING

A bench test has been developed which permits the acquisition of considerable neck performance information. This technique permits the measurement of static or dynamic resistive torques at the neck pivots as a function of both angle and time. In use, the neck is so mounted that the pivot being investigated is co-linear with the sensitive axis of a torque transducer. Angular deflection is monitored by a precision linear potentiometer. Display of torque versus angle is effected directly on a storage oscilloscope, and the time base is established by blanking the trace at known time intervals.

The prototype neck, instrumented for determination of lower pivot torque-angle characteristics, is shown in Fig. 19. Fig. 20 shows typical dynamic and static torque-angle loops for the lower pivot. Blanking at intervals of 10 msec provides the time base for the measurements. Each vertical division represents a torque of 100 in.-lbf, while each major horizontal division represents 10 degrees of rotation.

A comparison of static versus dynamic loading has been made (Fig. 21) with the goal of quantifying the dynamic torques generated by the cord upon rapid elongation. In general, little dynamic stiffening has been observed. In this respect, results are similar to those reported by Gadd, et al for human ligament generated dynamic torques in a dissected cervical vertebral section.

Fig. 19. Prototype neck mounted for dynamic bench test.

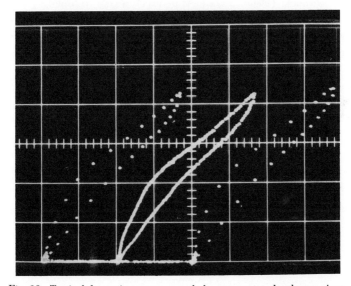

Fig. 20. Typical dynamic torque — angle loops generated at lower pivot.

Fig. 21 also illustrates directly the capability of the prototype neck to produce various levels of torsional preloading at the pivots. The four loops shown, read left to right, document four discrete preload settings at the lower pivot.

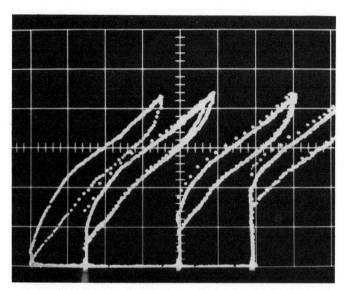

Fig. 21. Family of torque-angle loops for various preload settings at lower pivot.

SIMULATION AND VALIDATION

For validation of the mechanical neck model, a mathematical model of the mechanical head-neck (Fig. 22 and Appendix) was developed. The equations of motion are essentially those of a two degree-of-freedom inverted compound pendulum with dynamic inputs at the lower pivot. In addition, resistive torques are accommodated at each pivot and the mechanical stops of the mechanical neck are modeled.

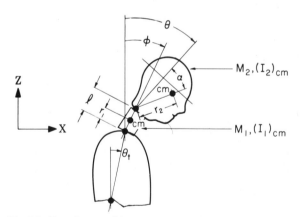

Fig. 22. Two degree of freedom model of the head and neck.

The inputs found necessary for the mathematical model (in addition to the torso angular displacement and velocity) are the translational accelerations (in the X and Z directions) of the first thoracic vertebra (T1).

These inputs to the math-model were obtained from films of human volunteers on the Holloman Daisy Track. The specific runs chosen for consideration were those in the 4500 and 4600 series reported by Chandler and Christian (8).

As previously mentioned, the volunteers were restrained in Type 2 automotive restraints facing forward in regular production automobile seats (Fig. 23). Each volunteer was subjected to nominally similar 15 mph, 12g ($-G_x$) decelerations. The objective of the original tests was to study the responses of subjects with widely varying anthropometrics. The initial head orientation of the subjects was not controlled.

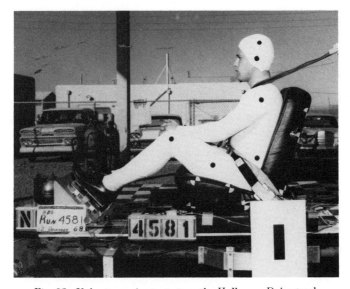

Fig. 23. Volunteer prior to test on the Holloman Daisy track.

Some of the early simulations indicated that the head motion was sensitive to the initial conditions; especially the head angle. Therefore, it was decided to use the Chandler and Christian data of subjects in Type 2 restraints to investigate the influence of the initial head orientation upon head motion. Shown in Fig. 24 is a linear regression analysis of the initial head angle from the vertical with the ratio of the peak velocity of head target No. 1 to the sled impact velocity. Target No. 1 is the upper-most tracking spot on the skull cap worn by each volunteer.

All of the data was found to lie within two standard deviations of the regression line. In addition, a correlation coefficient of -0.724 indicated a significant correlation at the p=0.01 level. Based upon this analysis, a decision was made to remove the variability of initial head orientation by selecting runs for simulation in which the subjects had the same initial head angles. The runs that were selected are numbers 4570, 4573, and 4581. Table 5 lists the percentile groups of the subjects and abstracts of the test data.

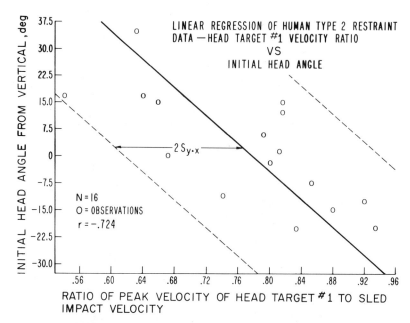

Fig. 24. Human type 2 restraint data — head target #1 velocity ratio vs. initial head angle.

TABLE 5

Subject Anthropometric Percentiles and Test Data

Daisy Run No.	Weight Percentile	Sitting Height Percentile	Sled Impact Velocity fps	Sled Peak G decel.	Initial Head Angle from Vertical	Ratio of Peak Velocity Target #1 to Sled Impact Velocity
4570	41	24	21.5	12.4	16	0.54
4573	72	48	21.7	12.4	15	0.82
4581	41	22	20.4	10.8	15	0.66

Digitizing of the 16mm films was accomplished using a free cursor digitizer on a projected image of the test. Although the digitizer accuracy is ±0.008 inch, the grain size of the film coupled with operator error and the image size gave an overall measuring accuracy of ±0.4 inch when the data was scaled. The two targets provided head orientation; T1 was approximated as the intersection of a line horizontal to the top of the shoulders and a line tangent to the back of the neck. The torso angle was defined as the angle from the vertical to the line connecting the lower torso target and T1.

The digitized data was smoothed and differentiated using an eleven point floating least squares fit to the displacement date. Accuracy of the smoothed derivatives was ascertained by comparing the integrated values with the displacement data. Good results were obtained in smoothing and calculating derivatives of the data (Figs. 25-27) with the exception of the vertical T1 accelerations which could not be distinguished from system error. Hence, they were not used in the mathematical simulations.

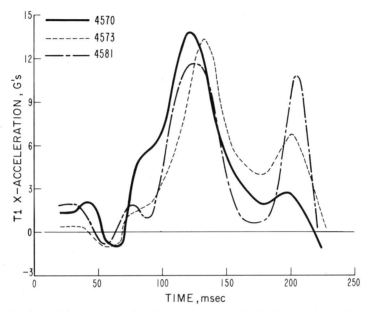

Fig. 25. Inertial acceleration of the first thoracic vertebra in the negative x direction.

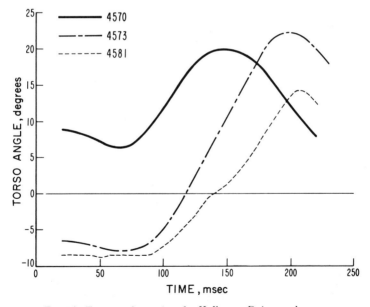

Fig. 26. Torso angle vs. time for Holloman Daisy track runs.

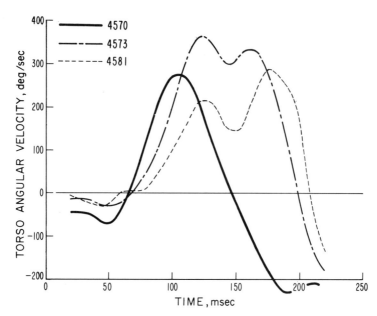

Fig. 27. Torso angular velocity vs. time for Holloman Daisy track runs.

By using as inputs to the math-model the information derived from the film data (namely, the T1 accelerations and the angular displacement and velocity of the torso versus time), a basis for comparing human and surrogate responses was established. As a first approximation, τ_1 and τ_2, the resistive torques at the lower and upper pivots, respectively, were assumed to be generated by a pre-loaded linear spring in parallel with a linear damper, i.e.,

$$\tau_1 = \tau_{p1} + k_1 \left(\phi - \phi_o - (\theta_t - \theta_{to}) \right) + c_1 (\dot{\phi} - \dot{\theta}_t), \tag{1}$$

$$\tau_2 = \tau_{p2} + k_2 (\theta_r - \theta_{ro}) + C_2 \dot{\theta}_r, \tag{2}$$

One should note that the time variation of the torso angle is included, as the neck pivots relative to the torso; also, the pre-loading torques, τ_{p1} and τ_{p2}, are achievable in practice because of the mechanical stops built into the design.

The approach taken was to attempt a simulation giving reasonable agreement with the head angular displacements and velocities under the following conditions:

1. The calculated inputs from the film data were used for the loading portion of the head-neck motion.

2. The masses, moments of inertia, c.m. locations, etc., were those measured from the mechanical design.

3. All parameters were kept constant with the exception of the preloading torques, spring constants, and damping coefficients.

Examination of Figs. 28-33 shows that favorable agreement has been obtained for the angular velocities and displacements. It was found that the same set of parameters, i.e., τ_{p1} = 5 ft.-lbf, k_1 = 95 ft.-lbf/rad, c_1 = 0.012 ft./lbf.-sec./deg. for the lower pivot and τ_{p2} = c_2 = 0, and k_2 = 3 ft.-lbf/rad for the upper pivot gave good agreement for Runs Nos. 4570 and 4573. However, a satisfactory simulation for Run 4581 could not be obtained until k_1 was lowered to 60 ft.-lbf/rad and the damping coefficient c_1 set to zero.

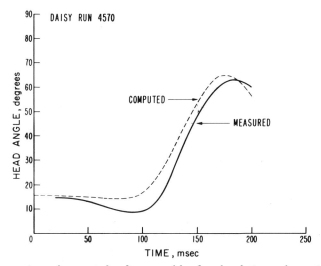

Fig. 28. Comparison of computed and measured head angle relative to the vertical vs. time.

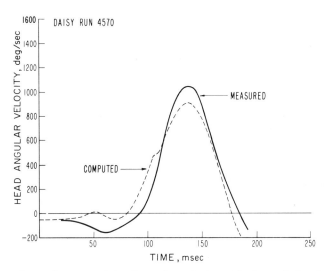

Fig. 29. Comparison of computed head angular velocity with that calculated from measured data.

References pp. 314–315

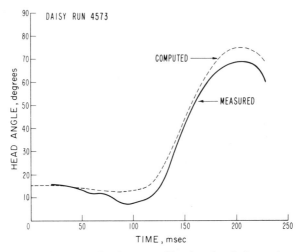

Fig. 30. Comparison of computed and measured head angle relative to the vertical vs. time.

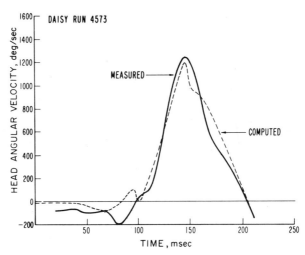

Fig. 31. Comparison of computed head angular velocity with that calculated from measured data.

The fact that a different set of parameters was found to give good agreement with the data for Run 4581 is not as surprising as the fact that 4570 and 4573 could be simulated well with the same pivot torque parameters. This is especially true, as their anthropometrics are markedly different and the masses, c.m. locations, etc. in the math-model were not varied.

With the exception of early portions of the head angle and head angular velocity curves, very good agreement was obtained using a simple viscoelastic model for the pivot torques.

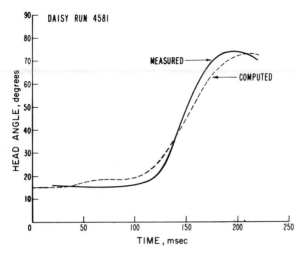

Fig. 32. Comparison of computed and measured head angle relative to the vertical vs. time.

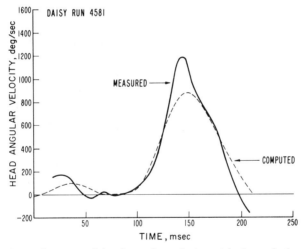

Fig. 33. Comparison of computed head angular velocity with that calculated from measured data.

It is important to emphasize that only three runs were simulated, and then only for the loading portion of the head motion. Emphasis was placed upon matching the loading portion of the head motion as it is suspected that chin-chest impact occurred.

SUMMARY AND CONCLUSIONS

Results obtained in the simulation indicate favorable correlation between the mechanical head-neck response and observed human response. The parameters deduced from the simulations can readily be incorporated into the prototype neck. In addition, sled testing has verified the integrity of the design concept. Further sled

References pp. 314–315

testing of an updated prototype will enable validation of mechanical neck performance in $-G_x$ impact. Validation of $+G_x$ performance and incorporation of $\pm G_y$ (lateral flexion) and rotational degrees of freedom awaits the analysis of additional data.

Simulations of head kinematic response have been performed for head motion during the loading phase of three subjects with the same initial head orientation. Additional runs must be simulated in which the effects of different initial head angles, sled decelerations, muscular tension, and body anthropometrics may be studied to determine the range of parameters necessary to obtain good agreement with the observed head motions.

Simulation of the unloading phase is not feasible at present, since unloading kinematics are dependent upon the nature of chin-chest impact. Dummy thorax construction will influence the unloading kinematics.

Even a humanlike head-neck structure will respond inaccurately if the dummy provides unhumanlike inputs at the base of the neck during impact. It is suggested that experiments be conducted to compare human and surrogate response at T1 to the same impact conditions.

It should be recognized that a neck analog which "matches" the responses of sled volunteers will be biased in favor of muscular, tensed subjects. The "worst" case for a typical vehicle occupant is likely to occur under conditions in which he is unaware of impending impact. The literature indicates that under such conditions in which the musculature is relaxed, stiffness and damping components are greatly reduced. In this case, kinematic response may be expected to be more severe. The use of a simple "relaxed" neck, then, is a possibility to be considered.

The thorax exerts a significant influence upon head kinematics; therefore, it is recommended that the design of the head, neck, and torso be integrated in the "next generation" of dummies.

REFERENCES

1. C. A. Moffatt, E. H. Harris, and E. T. Haslam, "Experimental and Analytic Study of Dynamic Properties of the Human Leg", ASME 69-BHF-4, March 1969.
2. G. L. Gottlieb, G. C. Agarwal, "Dynamic Relationship between Isometric Muscule Tension and the Electromyogram in Man", J. Applied Physiology 30: pp. 345-351, 1971.
3. H. J. Mertz and L. M. Patrick, "Investigation of the Kinematics and Kinetics of Whiplash", 11th Stapp Car Crash Conference, SAE paper no. 670919, New York.
4. H. J. Mertz and L. M. Patrick, "Strength and Response of the Human Neck", 15th Stapp Car Crash Conference, 1971, SAE, New York.
5. Tarriere & Sapin, "Biokinetic Study of the Head to Thorax Linkage", 13th Stapp Car Crash Conference, pp. 365-380, SAE, New York.
6. C. L. Ewing et al., "Living Human Dynamic Response to $-G_x$ Impact Acceleration", 13th Stapp Car Crash Conference, pp. 400-415, SAE, New York.
7. C. L. Ewing, D. J. Thomas, "Human Head and Neck Response to Impact Acceleration", NAMRL Monograph 21, August 1972.

8. R. F. Chandler, R. A. Christian, "Crash Testing of Humans in Automobile Seats", 1970 International Safety Conference Compendium, SAE paper no. 700361.

9. Selected Test Data and Facilities; Holloman Air Force Base Daisy Decelerator, First Interim Report, NSL 67-218 (Rev 2), May 1967, Northrop Systems Laboratories.

10. H. T. E. Hertzberg, G. S. Daniels, "Anthropometry of Flying Personnel – 1950, WADC Technical Report 52-321.

11. C. W. Gadd, A. M. Nahum, C. C. Culver, "A Study of Responses and Tolerances of the Neck". 15th Stapp Car Crash Conference, SAE, New York, 1971.

12. R. G. Snyder, D. B. Chaffin, and R. K. Schutz, "Joint Range of Motion and Mobility of the Human Torso", 15th Stapp Car Crash Conference, SAE paper no. 710848, November 1971.

13. D. R. Wilkie, "The Relation between Force and Velocity in Human Muscle", J. Physiology, London 110: pp. 249-280, 1950.

14. J. V. Basmajian, "Muscles Alive (Electromyography)", Second Ed. Baltimore, Md.: Williams & Wilkins, 1967.

15. S. W. Alderson, "The Development of Anthropomorphic Test Dummies to Match Specific Human Responses to Accelerations and Impacts", 11th Stapp Car Crash Conference, SAE, New York, 1967.

16. R. O. Godby, S. B. Browning, D. S. Belski, E. R. Taylor, "Anthropometric Measurements of Human Sled Subjects", Tech. Doc. Report ARL-TDR-63-13; DDC Report AD 407668, 1963.

17. F. Latham, "Linear Deceleration Studies and Human Tolerance", Institute of Aviation Medicine, Royal Air Force, DDC Report AD 141044, 1957.

18. H. T. E. Hertzberg, Ed., "Annotated Bibliography of Applied Physical Anthropology in Human Engineering", WADC Technical Report 56-30.

19. W. T. Dempster, "Space Requirements of the Seated Operator", WADC Technical Report 55-159, 1955.

20. C. L. Ewing et al, "Dynamic Response of the Head and Neck of the Living Human to $-G_x$ Impact Acceleration", 12th Stapp Car Crash Conference, pp. 424-439, SAE, New York.

NOMENCLATURE

c	damping coefficient
g	gravity
I	Iyy moment of inertia about the center of mass
k	spring constant
l	distance between pivots
m	mass
r	distance from the segment pivot to the center of mass
T	kinetic energy with respect to an inertial frame
t	time
U	potential energy
x,y,z	inertial coordinate system
α	angle from vertical ref. line to c.m. of head
θ	head angle with respect to the vertical
θ_r	angle of head segment relative to lower neck segment
θ_t	torso angle
ϕ	angle of lower neck segment wrt vertical
τ	resistive pivot torque

τ_p preloading torque
$(\ )_0$ initial condition
$(\ )_{1,2}$ segment number 1 or 2
$(\dot{\ })$ derivative wrt time

APPENDIX

The model for the mechanical head-neck is as shown in Fig. 22. Using a Lagrangian formulation, the equations of motion are determined by differentiating the kinetic and potential energies, i.e.,

$$\frac{d}{dt}\frac{\partial T}{\partial q_i} - \frac{\partial T}{\partial q_i} + \frac{\partial U}{\partial q_i} = Q_i \quad \dot{\tau} = 1,2 \tag{1}$$

where $q_i = \phi$, $q_2 = \theta_r$ and $Q_i = -\tau_i$, the resistive torques. The kinetic and potential energies are respectively:

$$\begin{aligned} T = \ & 1/2\,(m_1 + m_2)\,(\dot{X}^2 + \dot{Z}^2) + 1/2\,[I_1 + I_2 + m_1 r_1{}^2 \\ & + m_2\,(l^2 + r_2{}^2) + 2\,m_2 r_2 l\,Cos\,(\theta_r + \alpha)]\,\overset{\circ}{\phi}{}^2 + \\ & (m_1 r_1 + m_2 l)\,(\dot{X}\,Cos\,\phi - \dot{Z}\,Sin\,\phi)\,\overset{\circ}{\phi} + \\ & m_2 r_2\,(\dot{\phi} + \theta_r)\,[\dot{X}\,Cos\,(\phi + \theta_r + \alpha) - \dot{Z}\,Sin\,(\phi + \theta_r + \alpha)] \\ & + [I_2 + m_2 r_2{}^2 + m_2 r_2 l\,Cos\,(\theta_r + \alpha)]\,\dot{\phi}\theta_r \\ & + 1/2\,(I_2 + m_2 r_2{}^2)\,\dot{\theta}_r{}^2 \end{aligned} \tag{2}$$

$$U = m_1 g r_1\,Cos\,\phi + m_2 g\,[l\,Cos\,\phi + r_2\,Cos\,(\phi + \theta_r + \alpha)] \tag{3}$$

Performing the prescribed operations, the equations of motion are found to be

$$\begin{aligned} [I_1 + I_2 + &m_1 r_1{}^2 + m_2\,(l^2 + r_2{}^2) + 2\,m_2 r_2 l\,Cos\,(\theta_r + \alpha)]\,\ddot{\phi} \\ & + [I_2 + m_2 r_2{}^2 + m_2 r_2 l\,Cos\,(\theta_r + \alpha)]\,\ddot{\theta}_r = \\ - (m_1 g r_1 &+ m_2 g l)\,[\ddot{X}\,Cos\,\phi - (1 + \ddot{Z})\,Sin\,\phi] - m_2 g r_2\,[\ddot{X}\,Cos\,(\theta + \alpha) \\ & - (1 + \ddot{Z})\,Sin\,(\theta + \alpha)] + m_2 r_2 l\,Sin\,(\theta_r + \alpha)\,(2\dot{\phi} + \dot{\theta}_r)\,\dot{\theta}_r - \tau_1 \end{aligned} \tag{4}$$

and

$$\begin{aligned} (I_2 + m_2 r_2{}^2)\,\ddot{\theta}_r &+ [I_2 + m_2 r_2{}^2 + m_2 r_2 l\,Cos\,(\theta_r + \alpha)]\,\ddot{\phi} = \\ - m_2 g r_2\,[\ddot{X}\,Cos\,(\theta + \alpha) &- (1 + \ddot{Z})\,Sin\,(\theta + \alpha)] - 2\,m_2 r_1 \varrho \sin\,(\theta_r + \alpha)\,\dot{\phi}^2 - \tau_2 \end{aligned} \tag{5}$$

where X and Z are given in g's while θ is the head angle with respect to the vertical (i.e., $\theta = \phi + \theta_r$).

DISCUSSION

R. A. Wilson (GM Proving Ground)

During the runs on the sled with the dummy did you also measure head accelerations? The reason I ask is the transition from the one pivot to the other looked rather abrupt and I'm sure this would evidence itself in head acceleration spikes. This might be a problem.

M. P. Haffner

I think you have a good point. Essentially, the prototype in its present form awaits modification to meet the requirements of our latest parameter studies. We will be improving the joint stop characteristics.

G. B. Cohen

I should also mention the fact that we did not make comparisons in these particular three runs of the linear head accelerations as John Melvin had previously. The primary reason for this was the fact that the film data we had was very good, and we obtained reasonable results for first thoracic vertebra accelerations. However, we did not know where the center of mass of the subject's head was, not did we feel that we could make any reasonable estimate which would give us reliable results.

K. R. Trosien (Wayne State University)

I'm very curious about all of the dummy neck development. All of the necks are based on models or experiments where the input is to the first thoracic vertebra. Is anyone giving any consideration to input to the first cervical vertebra, which is the primary mode of loading during head impact?

M. P. Haffner

We have not considered a neck model based on dynamic inputs at C1 because of the obvious difficulties of obtaining human volunteer data for correlation. Our current model reflects the fact that in the $-G_x$ environment in the Holloman series studied, where impact of the head with hard surfaces external to the body does not occur, head-neck dynamic input can reasonably be taken at T1.

K. R. Trosien

Some of the things that have to be considered then are the responses of an absolutely rigid neck, in the vertical direction, subjected to head impact loading; and also the very complex motions such as those encountered in a vehicular environment where there are windshield, heater and instrument panel impacts. I'm curious if anyone of the groups doing this work is looking at their mathematical models from the other end of the neck. When you obtain a model that works fine with low level human volunteer inputs, will it be useful for analyzing windshield impacts?

M. P. Haffner

Again, our model is based upon human volunteer experiments not involving head impact. For this case, we have found it possible to assume the head as a rigid body, and have gotten good agreement between our model and observed head angular displacement and angular velocity. A more complex model incorporating realistic head impact response characteristics will probably be necessary for the windshield impact case.

H. E. VonGierke *(Aerospace Medical Research Laboratory)*

There are several biodynamic data sets available from upward ejection tests on the motion of the head in response to Z acceleration, or at several angles to the Z axis. It would be interesting to compare the dynamics of your model, which describes the motion in the XZ plane, to inputs from directions other than X; for example, the Z direction and angles to the Z axis for which data are available.

W. Goldsmith *(University of California)*

I would like to ask a question on how your angular velocity and angular acceleration were derived from your photometric data? I assume a graphical or numerical differentiation was used. Is that correct?

G. B. Cohen

The data itself was measured using a free curser digitizer. Smoothing was accomplished by means of an eleven-point floating least-squares fit of the least squares fit to the data. The angular velocities were calculated using the coefficients of the least squares fit to the displacement data. The angular velocity was again smoothed using the same technique and from the new coefficients, angular acceleration was computed.

W. Goldsmith

In trying to obtain two differentiations from a photometric measurement, I found it expedient to try to integrate backwards to see if the proper results are obtained. The feature that would worry me in a smoothing process is the fact that short duration peaks of acceleration or even velocity might be wiped out.

G. B. Cohen

Two things were done here. Number one, the displacement data was double-differentiated and secondly, the differentiated data was, in turn, integrated backwards. This was one of the criteria used in differentiating the raw data. Now as to the effect of short-duration pulses; integration of the equations of motion is almost tantamount to a smoothing operation, as you are integration the accelerations. Therefore, it was felt that the very short-duration pulses would be smoothed, resulting in a average output.

M. P. Haffner

I think there's another point to add here too, and that is that we have put a lot of effort into the extraction of quantitative information from film. Reliability of the higher derivatives is difficult to establish, and frequency response is poor. The availability of complete data sets based on transducer data will improve things considerably.

C. L. Ewing *(Naval Aerospace Medical Department)*

I'd like to know how you measured the acceleration at the first thoracic vertebra?

G. B. Cohen

The first thoracic vertebra, as I mentioned previously, was approximated as the intersection of two lines; one horizontal at the top of the shoulders and another parallel to the back of the neck. This was indeed an approximation; however, we had no other means without high speed X-ray or photometric targets. I should also mention that the gross results of the photometric analysis provided a scaled accuracy to within half an inch, which we thought was acceptable.

Y. K. Liu *(Tulane University)*

I will not make any comment on the mathematical model. Mr. Haffner gave a very good account concerning the complexity of the effects of neuromusculature. He pointed out that in fact it depends on activation levels, strain, strain rate, and so on. In fact, it was a very accurate description, but he left the impression that the problem is impossible to do. After two years, we are successful in doing this, and I would like to read one of the conclusions which is germain to all the previous papers that have noted certain spikes, in the head acceleration time curve. "The initial low level of head acceleration is due to the resistance of the passive elements of the neuromusculature. The neuro-loop governs this passive response by the level of anticipation. The overshoot of the head acceleration at the end of the input pulse is due to the energy stored in the passive elements of the neuromusculature, as well as the rising neural signal level caused by the stretch reflex." These are quantitative results.

M. P. Haffner

I would like to make a comment concerning the stretch reflex feedback mechanisms. I question whether within the duration of the dynamic event those feedbacks can occur in time to significantly change the kinematics. I would be interested in getting your opinion on this.

Y. K. Liu

I would like to disagree. The neuro stretch reflex time constant is in the neighborhood of 60 to 100 milliseconds, and all the events that have taken place in either plus G_x or minus G_x are way over that. For low levels of acceleration, the neuromusculature is in fact the predominant feature of the problem.

PRELIMINARY DISCUSSION OF AN APPROACH TO MODELING LIVING HUMAN HEAD AND NECK TO -G$_x$ IMPACT ACCELERATION

E. B. BECKER

Naval Aerospace Medical Research Laboratory Detachment, New Orleans, Louisiana

A joint Army-Navy-Wayne State University study of the dynamic response of the living human head and neck to -G$_x$ impact acceleration was initially presented at the Twelfth Stapp Car Crash Conference (1). In the course of this study a number of volunteer subjects underwent -G$_x$ impact acceleration known to be well within previously established tolerance limits. The motion of the subject's head and neck in the midsagittal plane were monitored with inertial instrumentation and high speed photography.

These collected data are being studied in order to relate the motion of the T1 vertebra to subsequent head motion. It is hoped that this study will eventually yield the following:

1. A limited set of parameters for each subject quantifying the transformation of motions sustained at the subject's T1 vertebra during -G$_x$ impact acceleration to subsequent motion of the head.

2. Correlations between these sets of parameters and more easily measured anthropometric quantities such as total body weight, sitting height, etc.

3. A mechanical analog of the human head and neck embodying these sets of parameters.

   At the onset of this study the high speed films of the impact experiments were reviewed. A mechanical linkage was then selected as a context for the analysis.

   This mechanical linkage (shown in Fig. 1) consists of a number of kinematic and dynamic components. The kinematic components consist of:

1. A point fixed relative to T1.

References p. 329

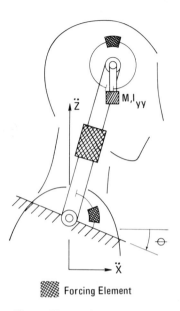

Fig. 1. The mechanical linkage.

2. A platform that passes through this point and that is fixed parallel to the T1 anatomical coordinate x axis defined in reference 4.

3. A second point fixed in the head.

4. A link of variable length, L1, that connects these two points and that pivots freely about both of these points, hereafter referred to as "hingepoints".

5. A third point, also fixed in the head, which is the position of the head center of gravity.

6. A link of constant length, L2, connecting the head center of gravity and the second hingepoint.

The dynamic components consist of:

1. The head mass and the head moment of inertia for rotation about an axis normal to the midsagittal plane.

2. A forcing element, such as might be composed of springs and dashpots, etc., operating on the angle formed by L1 and T1 platform. This element produces torques about the first hingepoint as a function of the time history of this angle.

3. A forcing element operating on the length of L1. This element produces tension in L1 as a function of the time history of the length of L1.

4. A forcing element operating on the angle formed by L1 and L2. This element produces torques about the second hingepoint as a function of the time history of this angle.

The assumption implicit in the selected locations of these forcing elements is that the force developed across any of the variables is an exclusive function of that variable and its derivatives. This assumption greatly simplifies the mechanism and will be discarded only at such time as it is demonstrated that this configuration is not sufficiently complex to account for the observed motions.

The kinematic parameters of this model are determined by the anatomical locations of three points: The two hingepoints and the head center of gravity. Mass distribution information for the human head is available from a number of sources (2,3) and likely positions for the two hingepoints are taken from a kinematic analysis of the high speed films.

In this analysis all the data reduced as described in reference (4) from the high speed films of a particular subject undergoing $-G_x$ acceleration are pooled. Two points, one fixed in the head and one fixed relative to T1 are selected so as to satisfy the following criterion: The variation of the distance between the two points, as measured over the range of body orientations observed in the photographic data, is a minimum. These two points are defined as the hingepoints. The position of the head hingepoint selected for subject 004 is shown in Fig. 2 in a coordinate system fixed relative to T1 and therefore fixed relative to the T1 hingepoint for all the photographic data available on that subject. The arc in this figure is the circle whose radius is the averaged distance between the two points and whose center is the T1 hingepoint.

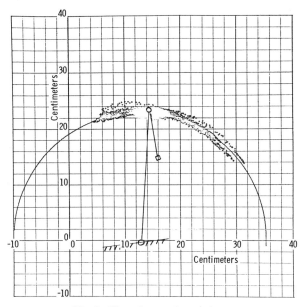

Fig. 2. Hinge locus subject of 004.

As a result of examination of the results of this kinematic analysis and of the similar results obtained for other subjects, the length of L1 was made constant and equal to the average distance between the hingepoints as measured in the kinematics analysis and the forcing element presumed to be operating across this length was discarded. The linkage now has only two degrees of freedom, as shown in Fig. 6.

Once the kinematic parameters for a particular subject are available it is possible to find parameter fits for the forcing elements. Since it is hoped that basic forms for these elements will apply to the whole range of subjects with individual characteristics being accounted for by numerical values, the elements used to fit the data will initially be held to a minimum of complexity.

The values obtained for the kinematic and forcing parameters for subject 004 are shown in Figs. 3, 4 and 5. The response of this mechanism to T1 inputs taken from one run are shown with the subject's own dynamic response in terms of angular

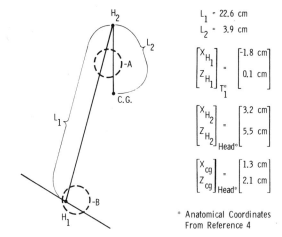

$$L_1 = 22.6 \text{ cm}$$
$$L_2 = 3.9 \text{ cm}$$

$$\begin{bmatrix} X_{H_1} \\ Z_{H_1} \end{bmatrix}_{T_1^\circ} = \begin{bmatrix} -1.8 \text{ cm} \\ 0.1 \text{ cm} \end{bmatrix}$$

$$\begin{bmatrix} X_{H_2} \\ Z_{H_2} \end{bmatrix}_{\text{Head}^\circ} = \begin{bmatrix} 3.2 \text{ cm} \\ 5.5 \text{ cm} \end{bmatrix}$$

$$\begin{bmatrix} X_{cg} \\ Z_{cg} \end{bmatrix}_{\text{Head}^\circ} = \begin{bmatrix} 1.3 \text{ cm} \\ 2.1 \text{ cm} \end{bmatrix}$$

° Anatomical Coordinates
From Reference 4

Fig. 3. The kinematic parameters.

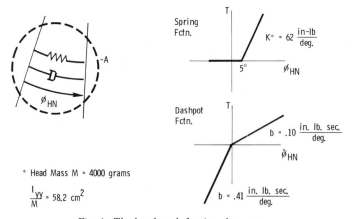

Spring
Fctn.

$$K^\circ = 62 \frac{\text{in-lb}}{\text{deg.}}$$

$5°$ ϕ_{HN}

Dashpot
Fctn.

$$b = .10 \frac{\text{in. lb. sec.}}{\text{deg.}}$$

$\dot{\phi}_{HN}$

° Head Mass M = 4000 grams

$$\frac{I_{yy}}{M} = 58.2 \text{ cm}^2$$

$$b = .41 \frac{\text{in. lb. sec.}}{\text{deg.}}$$

Fig. 4. The head-neck forcing element.

displacement, velocity and acceleration for the two angular degrees of freedom in Figs. 6-12.

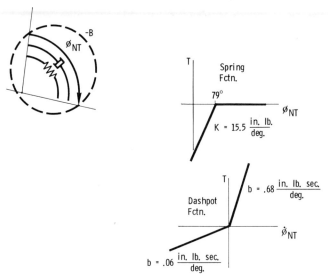

Fig. 5. The neck-torso forcing element.

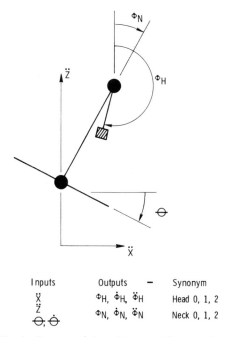

Inputs	Outputs	—	Synonym
$\ddot{X}$	$\Phi_H, \dot{\Phi}_H, \ddot{\Phi}_H$		Head 0, 1, 2
$\ddot{Z}$			
$\Theta, \dot{\Theta}$	$\Phi_N, \dot{\Phi}_N, \ddot{\Phi}_N$		Neck 0, 1, 2

Fig. 6. Context of the subject-model comparison.

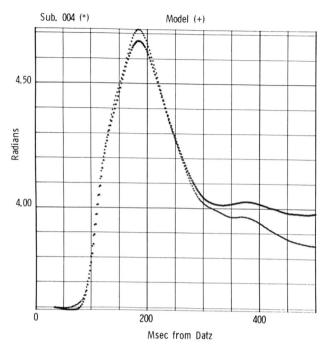

Fig. 7. Head 0.

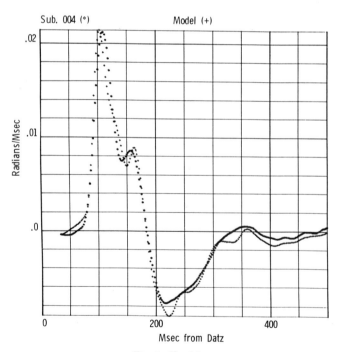

Fig. 8. Head 1.

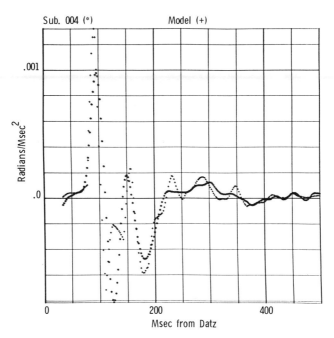

Fig. 9. Head 2.

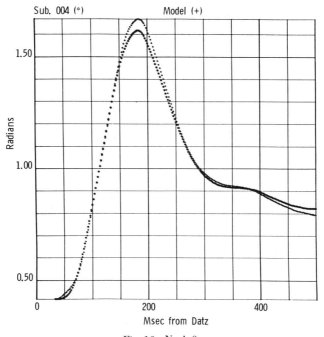

Fig. 10. Neck 0.

References p. 329

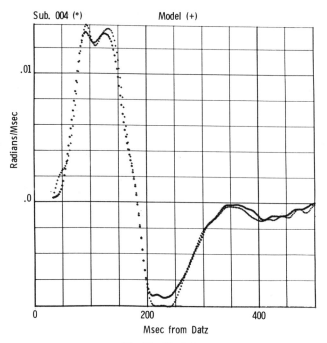

Fig. 11. Neck 1.

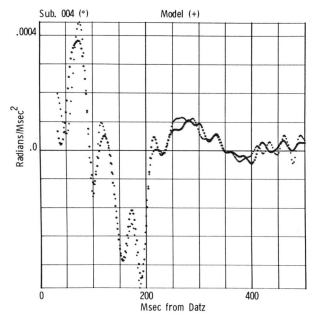

Fig. 12. Neck 2.

The linkage-subject agreement for this set of parameters is good for all five of the runs examined so far, although the elbow of the head-neck spring function seems to be increasingly displaced for sequential runs. It is as though the subject was able to rotate his head further back on his neck as the testing went along. The total increase in this articulation seems to be within 5 to 10 degrees. Such an increase in articulation may have actually taken place although this has not been shown conclusively at this time.

This approach to modeling living human head and neck response to -G_x impact acceleration shows considerable promise both as a tool for further analysis and as a head-neck mechanism such as might be employed in anthropomorphic dummies.

ACKNOWLEDGEMENTS

This work was funded by the U. S. Navy Bureau of Medicine and Surgery and the Medical and Dental Program of the Office of Naval Research.

Opinions or conclusions contained in this report do not necessarily reflect the views or endorsement of the Navy Department.

REFERENCES

1. C. L. Ewing, D. J. Thomas, G. W. Beeler, Jr., L. M. Patrick and D. B. Gillis, "Dynamic Response of the Living Human to -G_x Acceleration," Proceedings of the Twelfth Stapp Car Crash Conference, pp. 424-439, New York: Society of Automotive Engineers, Inc., 1968.
2. L. Walker, E. Harris and U. Pontius, Letter Report of Data from ONR Contract N00014-69-A-0248-0001, with Tulane University. "The Evaluation of Some Physical Measurements of the Head and Head and Neck of Human Cadavers," November 16, 1971 (Unpublished).
3. E. B. Becker, "Measurement of the Mass Distribution Parameters of Anatomical Segments", accepted for publication for the 16th Stapp Car Crash Conference, August 11, 1972. In press.
4. C. L. Ewing, D. J. Thomas, "Human Head and Neck Response to Impact Acceleration," NAMRL Monograph 21. Pensacola, Florida. August 10, 1972.

SESSION IV

MOBILITY AND KINEMATICS

Session Chairman
R. R. McHENRY

Cornell Aeronautical Laboratory
Buffalo, New York

NEW ADVANCES IN VOLITIONAL HUMAN MOBILITY SIMULATION

D. B. CHAFFIN and R. G. SNYDER

University of Michigan, Ann Arbor, Michigan

ABSTRACT

The use of computerized kinematic models of the human body has greatly increased in the last few years. This paper describes the results of research conducted to quantify the configurations of the human body most often chosen by people when reaching with one hand about their immediate environment. In performing this research basic size and volitional mobility data of the human torso were developed. These data have been used in constructing a computerized kinematic model, the output of which is a linkage representation of the body which is either displayed on a CRT or drawn by a computer driven X-Y plotter.

It is proposed that with this type of model a designer of crash dummies can begin to develop future dummies that better represent the size and mobility of various body segments, particularly the torso. In addition, the effects of various gross anthropometric variations (e.g., body weight and stature) on volitional body configurations and specific segment dimensions can be predicted for occupant packaging design evaluations. Also, the "most preferred" configurations of the body for given right hand positions becomes useful data for the design of restraint systems which not only then could minimize injury potential but maximize volitional function, thereby adding to their social acceptance and the resulting user safety.

INTRODUCTION

This paper is concerned with demonstrating how a more basic understanding of human volitional motion can assist a designer to assure that man can better function

---

Much of the research reported was developed under an Air Force Contract, AMRL, TR-71-88.

References p. 343

in his physical environment. It is the contention that protection of man's health and safety must not only concern itself with the study of man's impact response characteristics, but also must assure that man can function in a normal manner within the physically restrictive environment required for his protection.

In other words, the crash impact models reported in these proceedings and elsewhere are excellent for simulating specifically structured impact situations. They could, however, result in sub-optimum recommendations as to safety if the designs change normal driver functions. One example that one often hears is that a recommended shoulder belt rubs against the driver's neck. This in turn causes the driver to attempt to readjust it from time to time while driving, raising a potential hazard due to his diverted attention. In a more general sense, any time a restraint system interferes with normal function (e.g., reaching to common controls, changing postures, looking in different directions) it is classified as a "nuisance" by the user. It therefore engenders a poor attitude on the part of the user, which can lead to his rejecting the restraint system or vehicle design completely.

Therefore, it is the thesis of this paper that to better protect a person the designer of a vehicle must study both the impact protection afforded by his designs (via impact models) and the interaction of his designs with normal volitional functions. It is the latter category of simulation models (i.e., Volitional Mobility Models) that are discussed in the following. It should also be noted that the data resulting from the study of human volitional mobility are also contributing directly to the physical parameter values used in the many different crash impact models.

Background of Volitional Mobility Models — Several groups have concerned themselves with human volitional mobility. The BOEMAN development is noteworthy (1). This four year project by Boeing Aircraft has as a general objective to predict pilot body configurations when reaching for various controls in the cockpit. Excellent CRT display characteristics of both the pilot and the cockpit have been achieved. Further validation of the model over a larger workspace with program logic simplications to decrease the computer execution costs is currently underway.

Another computer based biokinematic model was developed by Kilpatrick at The University of Michigan to not only predict body configurations, but also to predict normal motion times for occupational situations (2, 3, 4). This model is faster in execution, but does not have the graphical display capabilities of the BOEMAN model. It should also be noted that a British group has a model that appears to be similar in the execution speed and graphical display capability of the Kilpatrick model (5).

One common aspect of the above three models is that they assume that the body is composed of some predetermined set of solid links which rotate about specific articulations. Thus input anthropometry data defines the lengths of the links. Assumed heuristic or empirically derived relationships then describe the spatial form of the links given some hand position data as input. This approach has been recently altered by another group at The University of Michigan (6, 7). These researchers have assumed that the body configurations can be predicted just as accurately and with a much simpler program logic by simply regressing the coordinates of predetermined body reference markers onto a specified group of anthropometric dimensions and hand coordinates. The end result of such an approach is an Empirical Prediction Model. In other words, it is a set of prediction equations for the coordinates of the body reference markers. Each equation has a known prediction accuracy. Such an approach also provides a model that can be programmed to run on a computer having less than 8K of direct address computer memory. It is this latter model that has been used for the demonstrations described in this paper.

The prediction accuracies of all the above models has not been well stated by the investigators, or it has been determined over a small set of feasible body positions. An estimate of the mean prediction error of these models appears to be about 2.0 inches for shoulder and elbow positions over a wide range of hand positions (6). Thus if it is assumed that either the shoulder or elbow could be feasibly located in a volume having a radius of 16 inches (including a wide range of hand positions and anthropometric variables), then the models have the capability of explaining approximately 80% of the potential body configurations variance. Since unknown variables (primarily behavioral) account for about half of the prediction error, the accuracy of the existing models is believed to be quickly approaching the universal accuracy limit of models based on physical input data (i.e., hand positions and anthropometry). Improved accuracy will need to rely on other data (i.e., fatigue states, alertness, motion dynamics, motor response requirements, etc.)

Review of U-M Empirical Model — The empirical prediction model developed at The University of Michigan was developed by the following procedure:*

1. Twenty-eight young males were selected to match on stature and weight U.S. flying personnel.

2. Seventy-two anthropometric dimensions were obtained on each individual.

3. A set of eight X-rays were obtained with each subject in various standardized but extreme postures.

4. The X-rays were measured for coordinates of bone reference mark positions (e.g., estimated centers-of-rotation of spinal and shoulder links).

5. The X-rays were also measured for coordinates of body surface reference markers.

6. Vectors were then computed between body surface markers and bone reference markers.

A more complete description of this model is given in references 6 and 7.

References p. 343

7. The Vector magnitudes and directions determined in Step 6 were then regressed onto body position reference axes for each major body segment and anthropometric variables.

8. Tables and graphical illustrations of bone and surface marker relative configurations were constructed for design purposes, see Fig. 1.

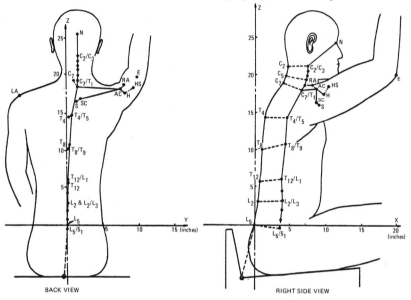

Fig. 1. Representative torso links for 50 percentile male.

9. The subjects then demonstrated thirty-five different body positions while the body surface marker locations were photographed from four views.

10. These surface marker coordinates were then regressed onto the right elbow coordinates and a selected set of anthropometric dimensions.

11. The surface marker coordinate predictions from the resulting regressions were validated by repeating some extreme positions and comparing these to the regression equation values.

12. The resulting regression equations were then programmed to form a total body configuration model which accepts as input data the right arm locations (the elbow position is predicted based on the earlier Kilpatrick model) and a set of basic anthropometry, and outputs a graphical display of the resulting body configurations most often chosen by the sample population.*

*The development of the computerized body configuration model was performed by Marvin Bolton, a research assistant in the Engineering Human Performance Laboratory at The University of Michigan.

DEMONSTRATIONS OF EMPIRICIAL PREDICTION
MODELS OF VOLITIONAL MOBILITY

Since the intent of this paper is to demonstrate the technology of volitional mobility modeling for design reference, the practical aspects of running the model are of great concern. Fig. 2 displays the computer system used for the model demonstrations discussed later in this paper.

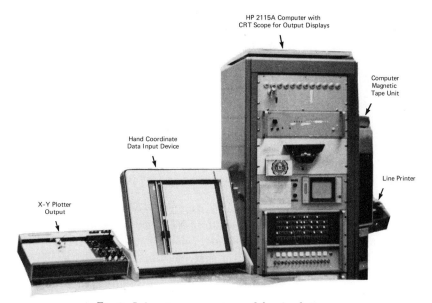

Fig. 2. Laboratory computer used for simulations.

To enter data about the right hand locations of interest (i.e., the "reach to" points) and X-Y encoder device has been utilized. Fig. 3 displays this device. A scale drawing of any workspace is inserted into the X-Y encoder device to assist the designer in locating the desired hand positions and seat reference point or H point. For this paper vehicle dimensions from the *SAE Controls Reach Study* (Fixture II for light truck and passenger cars) were assumed with three specific reach motions (8). These motions are meant to illustrate the right hand moving from the top of the steering wheel to:

Point 1 – Rear view mirror

Point 2 – Radio

Point 3 – Control under dash to right of steering column

Simply by moving the cursor (shown over point 1 in Fig. 3) to any point of interest in both views, the necessary positional input data about the physical task is supplied.

References p. 343

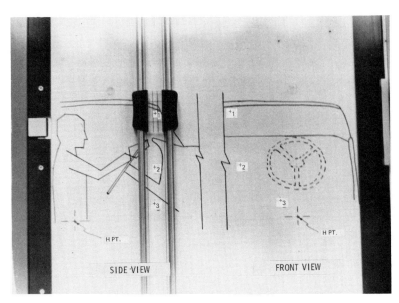

Fig. 3. Right hand coordinate input device and task reach points.

Data output is by the X-Y plotter for hard copy or by display on the CRT, as shown in Fig. 4, for more on-line evaluations. As can be seen, both right-side and front views are given. The left-side shoulder coordinates are estimated, but as the present model assumes that the left arm is at the side of the body (hand in lap when seated) the left-side is not displayed.

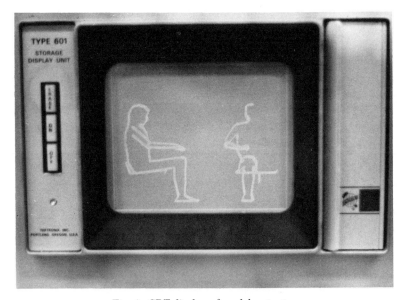

Fig. 4. CRT display of model output.

All the models discussed earlier assume sequential-static positions are representative of the postures during normal dynamic motions. As a systematic investigation of this assumption has not been made over the large variation in motion parameters (speed, location, sequences, etc.) it must be accepted as reasonable based on a small scale evaluation by the Boeing research group (1).

A typical output from the sequential-static motion is displayed in Fig. 5. The task is "reaching to mirror" (point 1 in Figure 3). What is clearly evident is that with a 50 percentile male's anthropometry as input, significant torso and shoulder assistance is predicted for this type of reach.

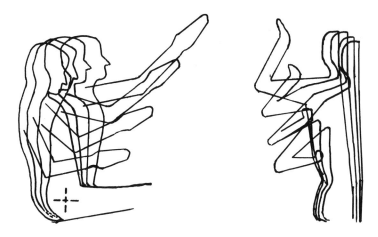

Fig. 5. Illustration of sequential static motion output when reaching to mirror (point 1).

Comparison of Body Anthropometry and Seat Horizontal Motion — Another question that can be easily evaluated is in regards to whether an adjustment that is a specific dimension of the environment is sufficient to compensate for a large anthropometric variation. An example chosen for demonstration is a seat track with five inch horizontal seat travel and no seat rise. Five and 95 percentile male dimensions were assumed for both the seat full forward and the seat full rear respectively. Each of three reach tasks were simulated, with the terminal postures displayed in Figs. 6, 7 and 8. What is apparent is that the five inch horizontal adjustment has been adequate to compensate for the male anthropometric variation as evidenced by the fact that equal torso assistance (same general torso postures) were predicted at the extreme points in each reach motion.

Comparison of Shoulder Belt and Torso Motion Interference — Another general question that can be evaluated for early design reference is in regards to whether some physical object or restraint would possibly interfere with normal body movements. For this paper a shoulder belt was assumed to be present. Its tightness was adjusted such that when the 50 percentile male sat with his right hand on the steering wheel it

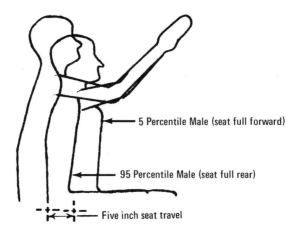

Fig. 6. Comparison of two seat positions when reaching to rear view mirror (point 1).

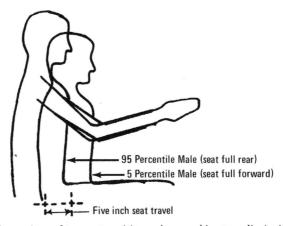

Fig. 7. Comparison of two seat positions when reaching to radio (point 2).

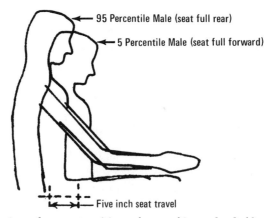

Fig. 8. Comparison of two seat positions when reaching under dashboard (point 3).

would allow the suprasternale reference point to move forward four inches. In other words, it was a "normally" adjusted belt. Once again all three reach tasks were simulated with the terminal body postures depicted in Figs. 9, 10 and 11.

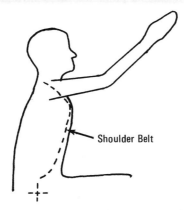

Fig. 9. 50 percentile male reaching to mirror (point 1).

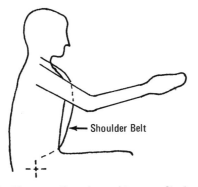

Fig. 10. 50 percentile male reaching to radio (point 2).

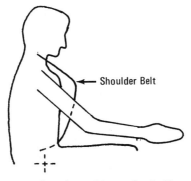

Fig. 11. 50 percentile male reaching under dashboard (point 3).

References p. 343

What is clearly evident from the simulations is that the significant torso motion when reaching to the mirror (point 1), as mentioned earlier in regards to Fig. 5, could cause significant interference with the shoulder belt. The other positions were not nearly as critical, due to the smaller predicted torso movement.

It must be quickly added that the predicted interference between the shoulder belt and the front of the torso and shoulder in reaching to the mirror may not be a major functional problem to the person since he simply may extend his arm further (i.e., rotate the shoulder and straighten the elbow more). Yet one must still ask, "In doing this is the *normal* motion pattern changed?" The answer from the simulation is, "Yes!" Thus the user may be motivated to not accept the shoulder belt since it interferes with his normal function. It might also be of note that a vehicle with the seat further back or with a smaller than 50 percentile male would result in even greater interference with normal movement, and even potential limitation of the mirror reach capability with a tighter belt. It should also be noted that the extreme reaches were not used in the simulations. Obviously, reaches to controls far under the dashboard, to the glove compartment, or to the opposite door would reveal significantly more interference problems which could be evaluated by the models discussed in this paper.

SUMMARY AND RECOMMENDATIONS

It is believed by these authors that computer based biokinematic volitional mobility models provide a powerful tool for evaluating initial design concepts. The ease of interacting with the models in a graphical mode should facilitate their acceptance.

Various biokinematic approaches have been utilized by different groups. Some evaluations of their prediction accuracies have been made, but more development and validation is needed. It is specifically proposed that additional development of simulations and validation studies of various seat designs and two-handed activities for both men and women be beneficial to the enhancement of vehicle design. It is expected that such an effort would result in major advances in the design concepts necessary to maximize a person's normal motor functions in the vehicle. In addition, the basic human anthropometry and mobility data gathered in the experiments necessary to develop the models, also would serve to enhance both the quality of existing crash impact computerized models and dummies.

ACKNOWLEDGEMENTS

The authors wish to express their sincere gratitude to Mr. Marvin Bolton, a research assistant, who not only programmed the volitional mobility model, but also ran the simulations reported in the paper. Also, thanks go to Miss Susanne Catchell, a Ph.D. student who assisted in the formulation of the tasks to be simulated. It should also be mentioned that the mobility data used for the model was formulated under

Contract F-33615-70-1777 for the Anthropology Branch, 6570th Aerospace Medical Research Laboratories.

REFERENCES

1. P. W. Ryan, "Cockpit Geometry Evaluation – Phase II-A," Final Report to JANAR, under ONR Contract N00014-68-C-0289, November 1971.
2. D. B. Chaffin, K. E. Kilpatrick, and W. M. Hancock, "A Computer-assisted Manual Work-design Model", AIIE Trans., Vol II (4), pp. 348-354, December 1970.
3. K. E. Kilpatrick, "A Model for the Design of Measured Work Stations", Ph.D. Dissertation, Department of Industrial Engineering, The University of Michigan, 1970.
4. K. E. Kilpatrick, "Computer Aided Workplace Design", MTM Journal, Vol. XIV (4), 1969.
5. M. C. Bonney and N. A. Schofield, "Computerized Work Study Using the SAMMIE/AUTOMAT System, MTM Journal, Vol. XVII (3), 1972.
6. D. B. Chaffin, R. K. Schutz, and R. G. Snyder, "A Prediction Model of Human Volitional Mobility", presented at SAE Annual Meeting, Detroit, 1972, and accepted SAE Transactions.
7. R. G. Snyder, D. B. Chaffin, and R. K. Schutz, "Joint Range of Motion and Mobility of Human Torso", presented at 15th Stapp Car Crash Conference, Coronado, California, 1971.
8. R. Roe, "Test Fixtures and Procedures, SAE Controls Reach Study", presented at SAE Annual Meeting, Detroit, Michigan, 1972.

DISCUSSION

J.N. Silver (GM Proving Ground)

Could techniques and tools such as these be used to perhaps map the preferred orientation of the skeleton system sitting in an automotive seat with respect to pelvic and thoracic vertebra orientation *and* with respect to belts *as well as to indicate* how these segments might be oriented for various size occupants?

D.B. Chaffin

I, quite frankly, see no particular *technical* problems at all in *developing and modeling such* data. We have been able to say, in *the case of the* hard seat, which removes one variation, that you can gather *these* data. The data shows that people are consistent. That's a relative thing though and consistency has to be looked at in terms of how much prediction accuracy you can *develop by using* a model as compared to the *unpredictable* variation in the position that a person would take. I think that a person does follow, with a reasonable constraint of the immediate environment that *he is* in, consistent *positional* patterns, and these patterns can be predicted in the manner that we described.

J.N. Silver

I'm thinking more specifically in terms of predicting the skeleton orientation rather than the orientation of the outer surfaces.

D.B. Chaffin

There is another *development of this project that I should have shown. This displays the motion of the interior skeletal points relative to the external points.* Regarding the interior points, the accuracies hold at the same level that we have for the surface *points.*

T.L. Black *(GM Design Staff)*

I'm sure you'll be glad to know that the SAE reach information will be published soon. You will probably want to check that against your model and see how it compares.

How do you know how far to move the subject forward when you're making a comparison between large and small people? Do you use a flat track or a curved track?

D.B. Chaffin

We used a flat track for this simulation. We obviously could have selected many different *modes but* at this point we decided to try a *simple* flat track with a 5 inch travel. All we could do was demonstrate *that this adjustment* seemed to be adequate for the *5 to 95 percentile* male. It would have been interesting to *use* female *anthropometric* data. I would speculate that *the 5 inch travel* would probably not be adequate for the female because of her smaller size. For male subjects, what we're saying is the 5 inch travel did not seem to *require extra torso motions for the small versus the large man when performing the specific reach motions indicated.*

VALIDATION OF A THREE-DIMENSIONAL MATHEMATICAL MODEL OF THE CRASH VICTIM

J. A. BARTZ

Cornell Aeronautical Laboratory, Inc., Buffalo, New York

ABSTRACT

A forty-degree-of-freedom, three-dimensional mathematical model of the crash victim, either vehicle occupant or pedestrian, has been validated by comparing predictions of the model with the results of a variety of experiments. These experiments included static bench tests and pendulum drop tests for checking the adequacy of the air bag submodel, but the validation effort centered on experiments with anthropometric dummies in impact sled tests and a full-scale automobile crash test. For this study, inputs to the digital computer program were based on detailed measurements of the dummy characteristics, as well as measured properties of the contact surfaces and restraints. A typical set of results of the dummy measurements is presented. These measurements include segment weights, segment moments of inertia obtained with a torsional pendulum, contact surface and link dimensions, joint torque characteristics obtained from static torque measurements, and material properties of the contact surfaces obtained from static load-deflection measurements. The mathematical model and digital computer program, previously reported in the literature, are briefly summarized. Predictions of the computer simulation using the measured inputs are compared with experimental results. These experiments consisted of an automobile crash test and a number of impact sled tests, including some with air bag restraint systems. The generally good agreement between the predictions and the experimental results is discussed, as well as probable sources of some differences noted in the comparisons. It is concluded that the generality and detail incorporated in the developed computer model, coupled with the demonstrated good prediction accuracy and relatively low computation cost, make it a valuable engineering tool for application in continuing research efforts to enhance the safety of the crash victim.

References p. 375

INTRODUCTION

In the development of planar computer simulations of the motor vehicle crash victim at CAL, (1-7) comparisons of predictions of the computer program with appropriate experimental results have been found essential to assess the assumptions and logic applied in the model development as well as to check the coding and execution of the computer program. McHenry and Naab have reported the results of a validation of one of these planar models, including detailed measurements of the properties of an anthropometric dummy used in the experimental effort. (2, 3).

During the recent development of a three-dimensional mathematical model of the crash victim at CAL*, (8, 9) the need for assessment of predictions of the computer program with experimental results on a continuing basis was made clear by the increased complexity of the three-dimensional model. The purpose of this paper is to summarize this validation effort, with emphasis placed on the measurement of those properties of an anthropometric dummy that were required for validation of the model. This experimental effort closely followed the basic approach described in References 2 and 3. The object of the experimental program was to obtain values of all the parameters that define the simulated crash victim from a set of measurements that would not be prohibitively time consuming or expensive. For this reason, the total number of required measurements was reduced wherever possible, and only simple static or quasi-static measurements were made.

In the following paragraphs, the mathematical model and the computer program are briefly reviewed to provide the background and motivation for the experimental study.

MATHEMATICAL MODEL – COMPUTER PROGRAM

Both the mathematical model and the computer program are documented in some detail in References 8 and 9. Consequently, only those aspects of the model and program that are directly applicable to the validation study are presented here.

The body dynamics model, which contains the equations of motion and joint constraint of the dynamical system, is formulated for an arbitrary number of segments and joints, but is presently programmed as illustrated in Fig. 1. It consists of fifteen rigid body segments connected by fourteen joints, both ball-and-socket and pinned types, resulting in a forty degree-of-freedom system. Muscle tone is simulated by torques acting at the joints. The joint torques consist of viscous, coulomb friction, and spring components, and include nonlinear springs for joint stops.

The body contact model, Fig. 2, consists of an ellipsoidal contact surface corresponding to each body segment. The simulation includes treatment of contact

*Developed under the joint sponsorship of the National Highway Traffic Safety Administration and the Motor Vehicle Manufacturers Association.

between body segments and vehicle contact surfaces, as well as segment-segment contacts. In the program, normal contact forces are generated as a function of penetration of contacting surfaces, and forces from sliding friction oppose relative motion between contact surfaces.

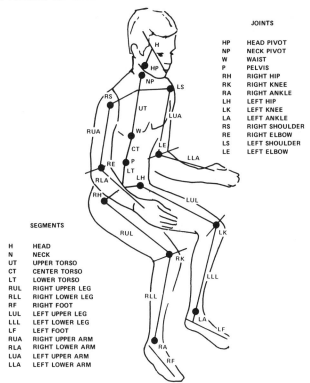

JOINTS

HP	HEAD PIVOT
NP	NECK PIVOT
W	WAIST
P	PELVIS
RH	RIGHT HIP
RK	RIGHT KNEE
RA	RIGHT ANKLE
LH	LEFT HIP
LK	LEFT KNEE
LA	LEFT ANKLE
RS	RIGHT SHOULDER
RE	RIGHT ELBOW
LS	LEFT SHOULDER
LE	LEFT ELBOW

SEGMENTS

H	HEAD
N	NECK
UT	UPPER TORSO
CT	CENTER TORSO
LT	LOWER TORSO
RUL	RIGHT UPPER LEG
RLL	RIGHT LOWER LEG
RF	RIGHT FOOT
LUL	LEFT UPPER LEG
LLL	LEFT LOWER LEG
LF	LEFT FOOT
RUA	RIGHT UPPER ARM
RLA	RIGHT LOWER ARM
LUA	LEFT UPPER ARM
LLA	LEFT LOWER ARM

Fig. 1. Body dynamics model.

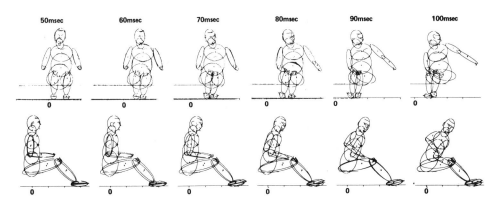

Fig. 2. Contact model – graphics display model.

References p. 375

Occupant restraints, both belt type and inflatable restraints are included in the simulation. The air bag model includes the processes of gas dynamic deployment, inflation, and deflation, with contact forces computed as a function of the deformed air bag geometry and gas dynamics.

The graphics display model, developed for interpretation and diagnosis of results, provides several optional levels of output. The plotter graphics display, Fig. 2, produces two orthogonal views of the ellipsoidal contact model, with certain human characteristics and features added to improve comprehension of results.

The computer program is written entirely in Fortran IV for simplicity of use on other digital computer facilities. The program includes a variable step exponential integrator, developed as part of the research effort to minimize computation costs. With this program, a 200 millisecond simulation of a crash event on the IBM 370/165 digital computer at CAL consumes about 60 seconds cpu time, at a cost of about $50.

With this brief description to serve as both the background and motivation for the experimental effort, the dummy measurement program will now be summarized.

ANTHROPOMETRIC DUMMY MEASUREMENTS

A Sierra 292-1050 (50th percentile male) anthropometric dummy (serial number 2004), a representative 1971 model crash test dummy, was selected for this study. The following dummy properties were measured:

1. Inertial properties; segment centers of gravity, masses, and principal moments of inertia,

2. Link dimensions and contact surface dimensions,

3. Static joint torque characteristics,

4. Static force-deflection and frictional characteristics of the dummy contact surfaces.

To obtain these detailed measurements, the dummy was disassembled as shown in Fig. 3. The breakdown into fifteen segments, to correspond to the body dynamics model, was as follows:

Head,

Neck (joint),

Upper torso (rib cage assembly with shoulder joints, upper torso foam rubber padding, upper torso skin),

Center torso (lumbar joint, viscera sac attached with tape),

Lower torso (pelvis with hip joints, lower torso foam rubber padding, lower torso skin),

Left and right upper leg segments,

Left and right lower leg segments,

Left and right feet,

Left and right upper arm segments,

Left and right lower arm segments (including corresponding clenched hands locked tightly to the lower arm segments).

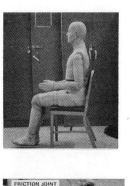
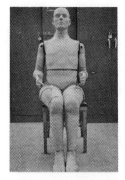

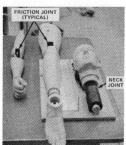
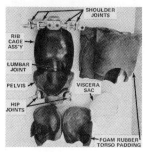

Fig. 3. Sierra 292-1050 (50th percentile) dummy.

In addition, properties of the assembled head and neck (joint) and the assembled torso were measured, to allow ease of input of the measured data into other computer programs. In all cases, spacers, fasteners, and mounts were fastened securely to the corresponding dummy segment. All measurements were referenced to the following coordinate system. Consider an erect standing position; for all body segments, x is positive in the posterior-anterior (forward) direction, y is positive to the right and z is positive in the superior-inferior (downward) direction, except for the feet, for which x is positive in the upward direction, and z is positive in the forward direction. The measurements were performed in the following manner.

INERTIAL PROPERTIES

Centers of Gravity — The centers of gravity of the segments were located by balancing each segment on a knife edge, as indicated in Fig. 4, in planes defined by the x, y and z axes. For long segments, such as the limbs, it was found that the center of gravity was within about one-tenth of an inch of an imaginary line connecting joint centers. For these segments, the z axis was defined as the line parallel to the line connecting the joint centers, and passing through the center of gravity.

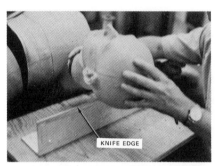

DETERMINATION OF SEGMENT CENTER OF GRAVITY

MEASURING INSTRUMENTS FOR LINK DIMENSIONS
AND CONTACT SURFACE DIMENSIONS

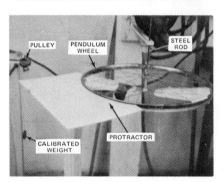

STATIC CALIBRATION OF THE TORSIONAL PENDULUM

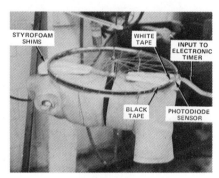

DETERMINATION OF MOMENT OF INERTIA
ON THE TORSIONAL PENDULUM

Fig. 4. Measurement of dummy characteristics.

Masses — Segment masses were determined by weighing each of the previously mentioned segments, except the assembled torso, on a Friden model 1367 scale (0-70 lbs.), which was checked with calibration weights and found to be accurate to within one ounce. The assembled torso was weighed on a Fairbanks Morse Model 1118 scale (0-2000 lbs.) to an estimated accuracy of 1/4 pound. In the summary of inertial data presented in Table 1, note that the sum of the masses of the individual torso segments is within 2 ounces of the mass of the assembled torso. It should also be noted that the measured masses are in quite good agreement with values measured at the National Bureau of Standards on another Sierra 292-1050 dummy (10). The difference between the present measurements and the NBS values for a given segment are no more than the difference between corresponding left and right segments in either set of measurements. These differences, on the order of several percent, are therefore traceable to manufacturing variations.

TABLE 1

Measured Inertial Properties

Body Segment	Formulation Symbol	Program Symbols	Segment Weight lbs	Segment Moment of Inertia, lb-sec²-in			Segment Weight lbs	Segment Moment of Inertia lb-sec²-in y Axis
				x Axis	y Axis	z Axis		
Head	H	1	10.13	0.259	0.311	0.200	10.28	0.301
Neck (Joint)	N	2	3.13	0.0400	0.0400	0.00660	3.17	—
Upper Torso	UT	3	31.75	2.32	1.65	1.33	—	—
Center Torso	CT	4	9.38	0.325	0.314	0.149	—	—
Lower Torso	LT	5	35.75	1.78	1.14	1.71	—	—
Right Upper Leg	RUL	6	17.56	0.727	0.703	0.154	17.24	0.707
Right Lower Leg	RLL	7	6.94	0.435	0.437	0.0165	7.46	—
Right Foot	RF	8	2.75	0.0383	0.0434	0.0132	2.8	—
Left Upper Leg	LUL	9	17.31	—	—	—	17.81	0.740
Left Lower Leg	LLL	10	7.00	0.445	0.447	0.0215	7.28	—
Left Foot	LF	11	2.75	—	—	—	2.8	—
Right Upper Arm	RUA	12	5.13	0.167	0.165	0.0132	5.73	0.169
Right Lower Arm	RLA	13	4.63	0.255	0.269	0.0115	4.88	0.269
Left Upper Arm	LUA	14	5.38	0.160	0.167	0.0149	5.78	0.174
Left Lower Arm	LLA	15	4.94	—	—	—	4.63	0.250
			164.5					
Head Plus Neck (Joint)	H & N		13.25	0.412	0.484	0.207	13.45	0.477
Torso	UT, CT, LT		76.75	14.25	13.21	3.83	77.19	13.208

Moments of Inertia — Segment moments of inertia about the x, y and z axes of each segment were determined on a torsional pendulum. (2, 3) Essentially, the pendulum consists of a bicycle wheel suspended on a 1/4" steel rod, as shown in Fig. 4. The moment of inertia of a segment about a given axis was determined from

$$I = k (p_2 + p_1) (p_2 - p_1) \qquad (1)$$

where I is the moment of inertia, k is the torsional spring constant of the pendulum, p_1 is the oscillation period of the unweighted pendulum and p_2 is the period of the oscillation with the segment mounted on the pendulum. The spring constant k of the pendulum was determined statically by applying a series of known couples to the pendulum within its linear torque-angular displacement range. This was done by suspending pairs of calibrated weights from two cords which were connected tangentially to the wheel periphery and passed over pulleys mounted on bearings, as shown in Fig. 4. Angular deflections were measured with a protractor at approximately equal increments of static torque up to a total angular deflection of about 40°.

Care was taken to approach the equilibrium position of the wheel for any given torque from both angular directions, to avoid hysteresis errors. The pendulum torque-angular deflection characteristic was found to be linear to about one percent, and the measured spring constant was within two percent of the value determined in a previous calibration.

Segment moments of inertia were determined by securely fastening the segment under study to the wheel with tape or wire, making certain that the rod center line was colinear with the segment axis under consideration. For the long segments (limb segments), a short piece of threaded rod was epoxied to a metal joint part with the rod center line on the segment z axis. After the epoxy cured, the segment was attached to a threaded hole at the bottom of the pendulum wheel, resulting in a rigid attachment. For the heavier segments, styrofoam shims were used in positioning. Every effort was made to reduce added weight, to minimize error in measurement. The period of oscillation was determined with a photodiode sensor connected to an electronic timer, or counter. Adjacent pieces of white and black tape were fastened to the periphery of the wheel, facing the sensing element of the photodiode. During pendulum oscillation, the change in illumination sensed by the photodiode triggered the counter, which was set to record ten periods of oscillation. Measurements of the period of oscillation of the unweighted pendulum were made both before and after the segment measurement, with care taken to add the mass of the fasteners (tape, wire or shims) at approximately the correct radial distance from the center of the wheel. A check on the accuracy of the determination of segment center of gravity (using the knife edge) was made on the pendulum in the following manner. After the required pendulum measurements, selected segments were purposely displaced from the pendulum center line, and oscillation again induced. It was found that, with the segment center of gravity displaced from the pendulum center line, a precession mode of oscillation of the pendulum occurred. Pendulum oscillation was carefully observed during measurements of moments of inertia to be certain that no precession was present, as a check on the accuracy of the center of gravity determined by the knife edge.

Results of the moment of inertia measurements are summarized in Table 1, along with National Bureau of Standards (NBS) measurements on the same type of dummy. (10) The maximum difference between the present measurements and the reported NBS values for a given segment is four percent. It can be seen from the table that this is the same range of differences between the values for corresponding left and right segments, in either the present or NBS measurements. Again, it appears that the differences in the measurements are attributable to small manufacturing variations. Because of the generally good agreement between corresponding right and left segments, all possible moment of inertia measurements on both right and left segments were not made.

For most segments, it is estimated that the moments of inertia determined by the present technique are accurate to a few percent. The exceptions are the moments of inertia about the z axes of the less massive long segments. These are estimated to be accurate to about ten percent, based on examination of Equation (1). Note that as the moment of inertia to be measured becomes small, $p_2 \to p_1$, and the accuracy of measurement decreases. To increase the accuracy of these measurements, a considerably less massive pendulum would be required. However, the effort to achieve greater accuracy is hardly worthwhile. For the long segments, a relatively large percentage error in the very small value of the moment of inertia about the z axis will have only a small effect on the moment of inertia tensor as a whole. For this reason, the tabulated values were considered satisfactory for the validation experiments.

It should be noted that experimental error is inherently present in moment of inertia measurements when the principal axes of the body in question are assumed (as was done in the present research), rather than experimentally determined. To assess the effect of this experimental error, a computer run with the mathematical model was made in which the moments of inertia about the long axes of selected body segments were decreased by 20% from the measured values. This corresponds to misalignment of 0.1 radians $\approx$ 5.2° of the assumed long principal axis. This angular misalignment is about 1° for these segments. It was found that the 20% change in the inputted moments of inertia was typically reflected as only a several percent change in the elements of the direction cosine matrices of these segments after a rather complicated motion. Based on this result, and from the result of hand calculations, it was concluded that the effect of the experimental error in question is negligible, and that more refined measurements were not justifiable.

Link Dimensions and Contact Surface Dimensions — These caliper and scale measurements, using instruments shown in Fig. 4, are separated into link dimensions, Fig. 5, and contact surface dimensions, Table 2. For convenience of inputting the link dimensions into other computer programs, they were developed into graphical form, Fig. 5, and referenced to the segment centers of gravity. It should be noted that the effective locations of the neck and torso joints are defined by neck and torso joint characteristics, and consequently are not included in this section. In the interests of economy, not all possible link measurements were made. It was found in making sample comparisons between corresponding left and right segments that variations in dimensions were typically several percent, comparable to variations in the inertial properties of corresponding left and right segments.

Note that the measured center of gravity of the assembled torso segments is within 0.2 inches of the value calculated from the measured masses and centers of gravity of the individual torso segments. Likewise, the measured center of gravity of the assembled head and neck segments is within 0.2 inches of the value calculated from the measured masses and centers of gravity of the individual head and neck (joint) segments.

References p. 375

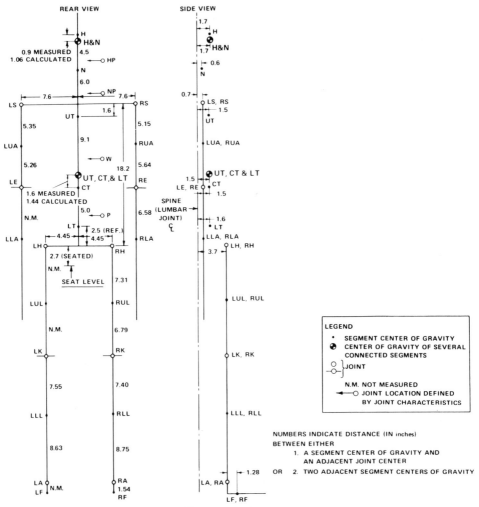

Fig. 5. Link Dimensions.

The contact surface measurements, to define the ellipsoidal contact model, represent the half-depth, half-width, and half-height of the individual segments, using the midpoint of the segment, not the center of gravity, as the reference. It should be noted that the ellipsoid dimensions corresponding to the z axes of the neck and torso segments are somewhat arbitrary, but with the exception of these dimensions, the tabulated values are estimated to be accurate to about one-tenth of an inch. Locations of the centers of the contact ellipsoids relative to the segment centers of gravity are presented in Table 2 along with the contact surface dimensions.

Static Joint Torque Characteristics — The general procedure in measuring the static joint torque characteristics, illustrated in Fig. 6, consisted of disassembling the dummy limbs into selected pairs of segments, with each pair connected by a common joint. The heavier segment was then rigidly mounted, with the z (long) axis of the less massive segment pointing vertically upward, for the sake of establishing a convention. Measurements on the elbow, knee, and ankle were made in this configuration. For the shoulder measurements, the rib cage assembly was removed from the dummy and mounted rigidly, to obtain joint characteristics and motion limits that were unaffected by interference between segments. In a similar manner, for the hip and

TABLE 2

Contact Surface Dimensions

Body Segment	Formulation Symbol	Program Symbol	a_i, Inches	b_i, Inches	c_i, Inches	Z_{CG}, Inches
Head	H	1	3.99	3.10	4.59	0
Neck	N	2	2.57	2.28	3.28	0
Upper Torso	UT	3	4.41	6.78	4.49	0
Center Torso	CT	4	4.91	6.35	7.03	-2
Lower Torso	LT	5	4.94	6.94	7.60	0
Right Upper Leg	RUL	6	3.03	3.77	12.63	-2.6
Right Lower Leg	RLL	7	2.37	2.25	9.07	-0.45
Right Foot	RF	8	1.51	1.83	5.23	+0.95
Left Upper Leg	LUL	9	2.94	3.71	12.16	-2.6
Left Lower Leg	LLL	10	2.34	2.20	9.07	-0.45
Left Foot	LF	11	1.53	1.77	5.21	+0.95
Right Upper Arm	RUA	12	2.06	1.65	6.80	0
Right Lower Arm	RLA	13	1.27	1.07	8.50	0
Left Upper Arm	LUA	14	2.08	1.63	6.96	0
Left Lower Arm	LLA	15	1.33	1.14	8.25	0

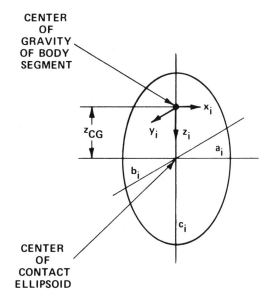

lumbar spine measurements, the lumbar spine-hip joint assembly was removed and mounted rigidly to avoid interference between segments. The experimental procedure consisted of tightening the joint to various known levels with a torque wrench and measuring the "breakaway" torque of the joint (i.e., the torque required to overcome the static coulomb friction of the joint). Torque was applied to the joint with a spring scale (Chatillion Instrument T, 0-60 lbs., 0.5 lb. divisions) over a known moment arm, usually the distance from the joint being studied to an adjacent joint. Prior to the measurements, the spring scale was calibrated with known weights in 5 lb. increments and found to be accurate to about ± 0.1 lb. over its entire range. In general, the joint was adjusted so that it would support 2 g's; that is, twice the mass of the less massive segment acting over the moment arm between the center of gravity of that segment and the joint. The loading-unloading characteristics of a knee joint, a typical result of these measurements, are shown in Fig. 7.

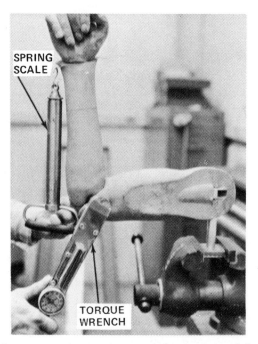

Fig. 6. Measurement of static joint torque characteristics of the dummy.

In this figure, for the data and curve labeled "load", the "breakaway" torque required to move the joint is plotted versus the angular deflection at which the measurement was made. The torque plotted is the sum of the torque measured with the spring scale and the torque produced by gravity acting on the less massive segment as it was displaced from its vertical position, calculated from the known segment mass and the moment arm of the torque. The data and curve labeled "unload" represent the static torque required to maintain a given angular deflection during the unloading cycle. The torque plotted is the sum of the torque measured with the spring scale and the calculated torque produced by gravity acting on the less massive segment, as it returned to the initial vertical position. Hysteresis in the range $\theta_a < \theta$ was caused by friction between compressed adjacent pieces of rubber on the two adjacent leg segments, as the knee was bent to negative values of θ_a. This conclusion was reached after repeated observations of the joint flexion.

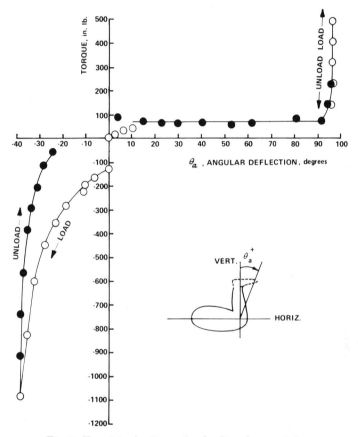

Fig. 7. Knee joint loading and unloading characteristics.

Additional measurements of joint torque characteristics with the dummy in a fully assembled condition were also performed, to compliment those measurements which have been discussed. The static joint torque characteristics of the head-neck-torso assembly were determined by reassembling the dummy and by flexing and twisting the head with respect to the upper torso, and by flexing and twisting the upper torso with respect to the lower torso.

The required flexural torques on the torso were produced by strapping the dummy torso to a rigid frame and applying force to the dummy with a jack through belts

attached to the dummy at the shoulder joints (with the arm segments removed), in
the manner indicated in Fig. 8. The applied force was measured with a Mintex pedal
force indicator, 0-200 lbs., 5 lb. divisions. The sensor was calibrated with known
weights and found to be accurate to about 5 lbs. over its range of measurement.

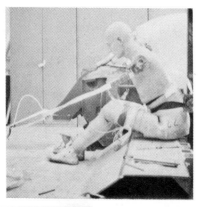

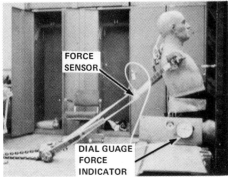

Fig. 8. Application of torque to the torso segments.

The required torsional loadings on the torso were produced by attaching cords to
the shoulder joints (with the arm segments removed), passing one cord over a pulley
located behind the dummy, attaching both cords to a bar located at shoulder height
in front of the dummy, and applying a force on the bar. This arrangement produced a
pure torsional couple on the torso.

Flexural and torsional torques on the head were produced in a similar manner. In
both cases, the flexural measurements consisted of applying a known force and
photographing the test setup and dummy in a direction perpendicular to the plane of
flexural motion.

The effective locations of the neck and torso joints, and the torque characteristics of these joints were determined in the following manner. Copies of the photographs were made for data reduction purposes. For the neck joints, lines were drawn coincident with the neck center line and parallel to the z axes of the head, neck, and upper torso for each photograph of the flexural measurements. It was found that the locations of the intersections of these three lines remained essentially fixed on the dummy during angular flexion, regardless of the direction of flexion of the head. Consequently, the intersection of the lines drawn through the head and neck was taken to be the location of the head pivot joint, and the intersection of the lines drawn through the neck and upper torso was taken to be the location of the neck pivot joint.

For the torso joints, the procedure was similar, except that no measureable flexion of the center torso relative to the lower torso (as they are defined by the inertial measurements) occurred. This is undoubtedly because of the rigid construction of the torso below the lumbar spine joint. Consequently, only the location of a waist joint between the upper torso and center torso could be determined by the photographic data reduction procedure. It was found that the location of the intersection of lines drawn coincident with the center line of the spine in the center torso and upper torso segments remained essentially fixed on the dummy during angular flexion, regardless of the direction of flexion of the torso. Consequently, this intersection was taken to be the location of the waist joint.

The flexural joint torque characteristics were determined by plotting the total torque (applied force plus appropriate segment masses, both acting on moment arms about the joint) versus angular deflection of the joint, determined in the photographic data reduction procedure.

For the torsional measurements, torsional angular deflections were measured with a protractor mounted on the dummy. It was found that the locations of the neck and torso joints in torsion, determined from visual inspection, were consistent with the locations of these joints as determined in the flexural measurements.

Typical results of these measurements on the torso and neck joints are shown in Fig. 9. It can be seen that for each of the joints, the flexion characteristics are reasonably symmetrical for the directions of flexion studied. The exception is the torso joint in the rearward direction. Its construction is such that disc slippage in the joint causes collapse of the joint for large rearward flexion. It can also be seen from the figure that each of the joints is essentially a linear (torsional) spring in loading, with a relatively small coulomb friction component, and that energy dissipation in unloading is quite large.

Additional measurements of the shoulder and hip joints were also made during this study. These measurements on the fully assembled dummy were made to compliment those made with the various joints isolated from the rest of the dummy. Typical results of these measurements are shown in Fig. 10. The joint stops for the fully assembled dummy are somewhat ambiguous, because of interference between segments. The values shown for the fully assembled condition are the angular deflections at which the increase of applied torque with angular deflection first

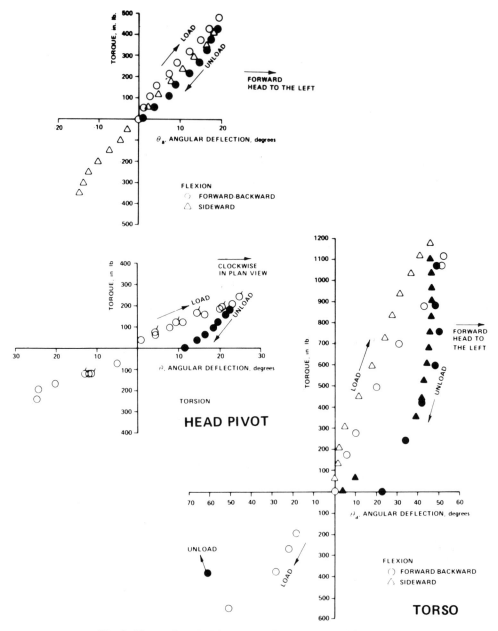

Fig. 9. Measured static joint torque characteristics of the dummy.

became highly non-linear. It can be seen from Fig. 10 that interference between segments does have important effects on the joint stops in certain directions.

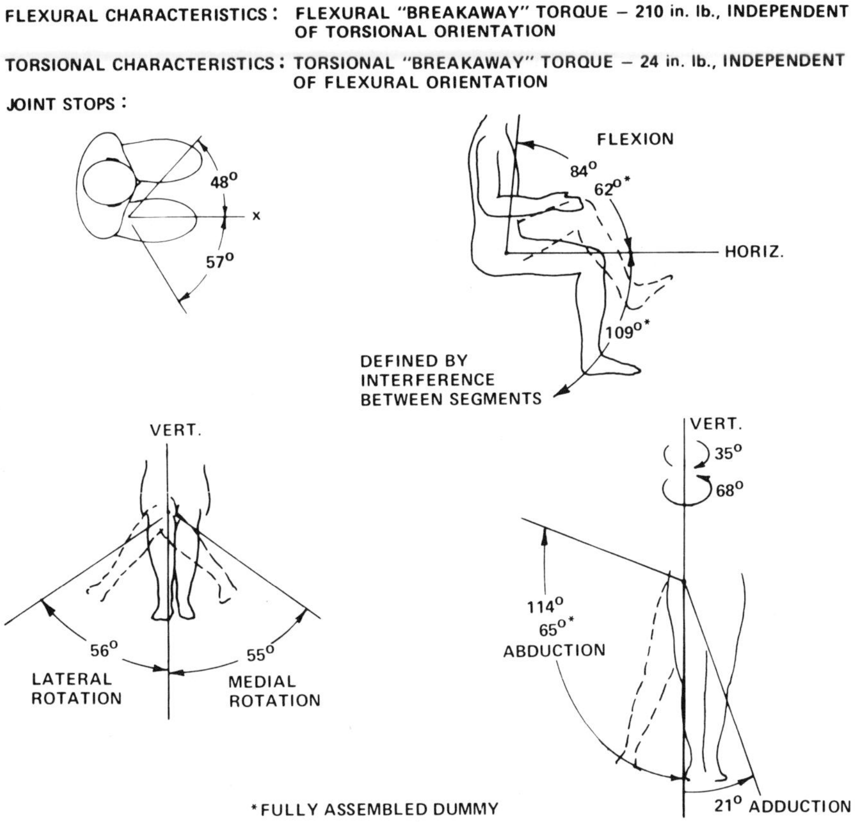

FLEXURAL CHARACTERISTICS : FLEXURAL "BREAKAWAY" TORQUE – 210 in. lb., INDEPENDENT OF TORSIONAL ORIENTATION

TORSIONAL CHARACTERISTICS : TORSIONAL "BREAKAWAY" TORQUE – 24 in. lb., INDEPENDENT OF FLEXURAL ORIENTATION

JOINT STOPS :

Fig. 10. Measured static joint torque characteristics of the dummy.

Following completion of the measurement of joint characteristics, the data were reduced to a form suitable for input to the program, Table 3.

It should be emphasized that the tabulated viscous torque coefficients are only approximate values. They are based on the observation that, during the joint measurements, after the coulomb "breakaway" torque was reached, continued application of that same torque caused a rotation of most of the joints at about 1 rad/sec, and increases in angular velocity above this value were roughly proportional to the applied torque. Consequently, these values should be considered to be first approximations to the viscous torque coefficients, probably indicating the correct order of magnitude of the coefficients.

The linear spring coefficients were obtained directly from the plotted torque-angular deflection characteristics. Only the neck and torso joints have a measurable linear spring contribution.

TABLE 3

Joint Torque Characteristics

Legend

C_0 — Coulomb Friction Torque Coefficient
C_1 — Viscous Torque Coefficient
K_1 — Linear Spring Coefficient
K_2 — Quadratic Spring Coefficient
K_3 — Cubic Spring Coefficient
R — Energy Dissipation Coefficient

	Flexion				
Joint	C_0, in. lbs. C_1, in. lb. sec/rad	K_1, in. lbs/rad	K_2, in. lbs/rad^2	K_3, in. lbs/rad^3	R
HP	10	1.22×10^3	0	0	0.63
NP	10	9.84×10^2	0	0	0.66
W	60	1.20×10^3	0	0	0.32
P					
RS, LS	185	0	2×10^6	0	1
RE, LE	74	0	3.84×10^4	0	1
RH, LH	229	0	2×10^6	0	1
RK, LK	198	0	5.50×10^4	0	1
RA, LA	25	0	2.06×10^6	0	1

	Torsion				
Joint	C_0, in. lbs. C_1, in. lb. sec/rad	K_1 in. lbs/rad	K_2, in. lbs/rad^2	K_3, in. lbs/rad^3	R
HP	30	4.59×10^2	0	0	0.30
NP	30	4.59×10^2	0	0	0.30
W	75	1.97×10^3	0	0	0.45
P					
RS, LS	42	0	2×10^6	0	1
RH, LH	280	0	2×10^6	0	1
RA, LA	6	0	3.10×10^5	0	1

The nonlinear joint stops were all represented by a quadratic fit, since they are all rather hard stops, and the torque-deflection characteristics near the stops were not defined by the measurements with sufficient precision (because of the small available range of angular deflection) to justify attempting a more exact data fit.

The energy dissipation coefficients for the joints were determined by measuring the energy dissipation in a loading and unloading cycle. This information was obtained from the torque-deflection plots by planimeter measurements.

STATIC FORCE-DEFLECTION AND FRICTIONAL CHARACTERISTICS OF THE DUMMY CONTACT SURFACES

Force-Deflection Characteristics — Rather than attempt to make every possible measurement of the force-deflection characteristics of the dummy contact surfaces, it was decided to concentrate on measurement of the characteristics of the surfaces most likely to be contacted, and to investigate the effect of varying the main parameters in the measurement of the force-deflection characteristics. Two of the most important parameters are the shape of the surface contacting the body segment, and the force-deflection characteristics of that contacting surface. Only rigid contact surfaces were used in the present study, in the interest of economy, but treatment of deformable surfaces is planned for the future. The following additional factors were considered, however: symmetry of the force-deflection characteristics on a given segment; variation of the force-deflection characteristics on a given segment with location of the applied force; and variation of the force-deflection characteristics among various segments.

The measurements were obtained by pressing various dummy segments in a hydraulic tester and recording the deformations with several height gauges, as shown in Fig. 11. In all cases, a line was marked on the dummy segment that referenced the skeletal structure of that particular segment. During the measurements, both the total segment deflection, and the deflection of both surfaces in contact with the press relative to the reference were measured. It should be noted that deflection of the segment relative to the rigid skeletal structure of the dummy is the quantity of most

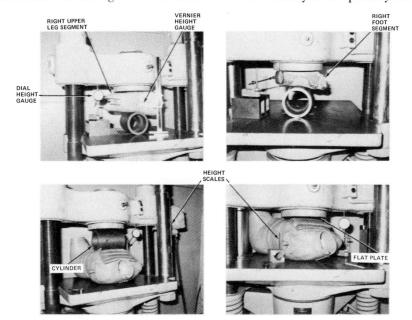

Fig. 11. Measurement of static force-deflection characteristics of the dummy contact surfaces.

interest. Care was taken to mark the reference line for each segment at a location at which distortion of the line relative to the skeletal structure, caused by distortion of the segment under load, was negligible.

In all cases, the segment to be tested was mounted in the press, a Southwark-Emery 0-100,000 lb. hydraulic tester, with the appropriate segment axis aligned to be parallel to the direction of force application, by applying a small (∿ 5 lb.) preload with the press. All deflections were then measured relative to this initial position. No attempt was made to account for the preload in the data reduction procedure.

The applied force was read to an estimated accuracy of ± 2 lbs. from the dial gauge on the press, which is graduated in 5 lb. increments. Height measurements were made with dial and vernier gauges, with redundant scale measurements taken to serve as a check on reading the gauges. The tests consisted of applying a load to the segment, measuring deflections, and rechecking the measured force to insure that slippage of the segment had not occurred. The loading was continued up to the point at which segment deformation suggested further loading might damage the segment. A series of measurements was then made during unloading of the segment. Reloading and (re)unloading measurements were also made, but in the interest of economy not for all segments, since the general characteristics of the reloading and (re)unloading curves were similar for all segments measured; the reloading curve in each case approached the original loading curve, and the (re)unloading curve approached the original unloading curve.

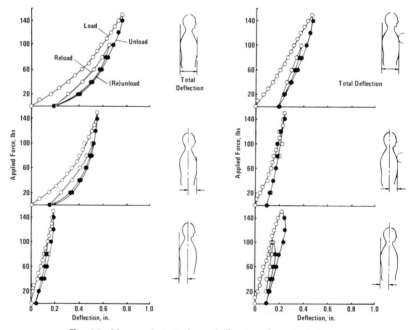

Fig. 12. Measured static force-deflection characteristics.

Typical results of these measurements are shown in Fig. 12. A distinct similarity in the characteristics is evident, in spite of vast differences in construction at the points of measurement. This can be seen in Fig. 13, in which some of these measurements were replotted on a log-log plot, to determine if the force-deflection characteristics have an asymptotic form

$$F = k \triangle^n \qquad (2)$$

where

F — applied force

$\triangle$ — deflection

k, n — constants

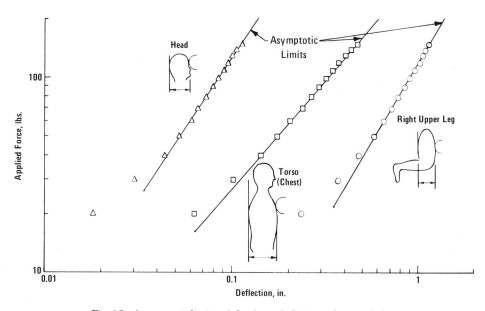

Fig. 13. Asymptotic limits of the force-deflection characteristics.

Total deflection measurements were chosen for this plot; plots using deflections measured relative to the skeletal structure give similar results. Three types of characteristics were considered, as illustrated in the figure: the head (stiffest), chest (medium stiffness), and upper leg (softest). Note that for each, the force-deflection characteristics have asymptotic limits. It therefore should be possible to extrapolate the loading characteristics beyond their measured range with reasonably good accuracy, probably up to the force level corresponding to damage of the segment.

In addition, it can be seen that the force-deflection characteristics represented by the asymptotic limits are quite similar. Note that the slopes of the curves shown are about equal, giving the exponent a nearly constant value. Only the relative stiffness, defined by variations in the coefficient k, varies appreciably; about a factor of 20 in deflection (for a given force level) for the curves shown. It was found that the variation in stiffness between segments greatly exceeds the variation in characteristics caused by varying the shape of the press surface, Fig. 11.

Frictional Characteristics — For analysis of the validation tests, the coefficients of static friction were measured for some of the contacts occurring in the tests. Both Teflon and polished aluminum seats and floors were used in these validation tests. Measurements of the coefficients of friction between the dummy lower torso and both a Teflon and polished aluminum seat were made by pulling on the lower torso parallel to the seat with a belt encircling the torso and attached to a dial force gauge (Chatillion, Model WT-10, 0-5000 lb., 20 lb. increments). The dummy was positioned with the feet elevated from the floor so that the only contact was between the lower torso and the seat. Prior to the test, the lower torso and seat surfaces were cleaned with methyl ethyl ketone and dried, to produce repeatable test conditions. Measurements of the force required to cause slippage of the torso on the seat were then made. The required force was found to be repeatable to about 5 lbs., or 3 percent. It can be seen in Table 4 that the Teflon surface, as expected, was quite slippery, but the polished aluminum-rubber torso combination has a coefficient of friction of 1.0.

TABLE 4

Measured Cofficient of Static Friction

(Contact Between Lower Torso of Dummy and Bench Seat)*

Seat Covering	Coefficient of Friction
Teflon	0.24
Polished Aluminum	1.0

All Contacting Surfaces Cleaned with Methyl Ethyl Ketone

Restraint Belt Characteristics — Characteristics of restraint belts used in some of the validation tests were determined by loading samples of the belts in the hydraulic tester. Typical results are shown in Fig. 14. Belt characteristics were generally repeatable, with typical variations in strength of about ten percent.

VALIDATION STUDIES

To illustrate the application of these measurements, the model validation studies will be reviewed. These studies are reported in detail (9), so that only representative comparisons are presented here.

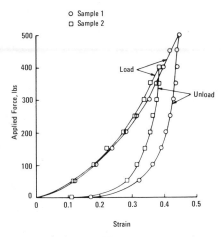

a. (1-1/2 in. Wide x 0.04 in. Thick Undrawn Nylon)

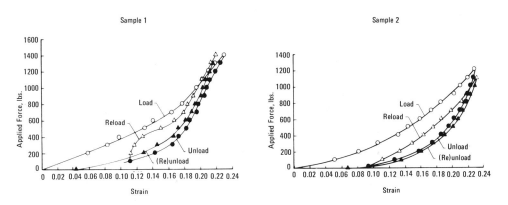

b. (1-1/4 in. Wide x 1/8 in. Thick Undrawn Nylon)

Fig. 14. Static loading and unloading characteristics of the restraint belts.

A critical assessment of the model was made by comparing its predictions with the results from a series of impact sled tests performed at CAL. In these tests, the Sierra 292-1050 dummy previously discussed was deliberately forced into a very general three-dimensional motion with four restraint belts located unsymmetrically with respect to the dummy and sled, as shown in Fig. 15. The resulting dummy responses, and the corresponding predictions of the computer program are presented in Figs. 16 and 17. The agreement is considered quite good, and differences between the experimental results and the simulation are explainable. For example, the difference in head rotation is believed to be the result of contact between the dummy chin and the belt loading the neck, a contact that was not simulated.

References p. 375

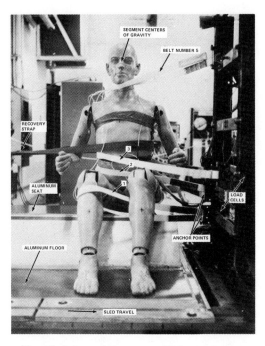

Fig. 15. Impact sled tests — initial conditions.

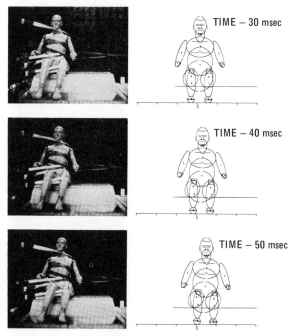

Fig. 16a. Validation of the model simulation of impact sled experiments.

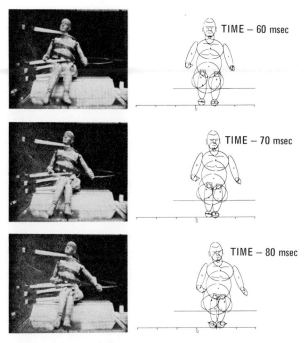

Fig. 16b. Validation of the model simulation of impact sled experiments.

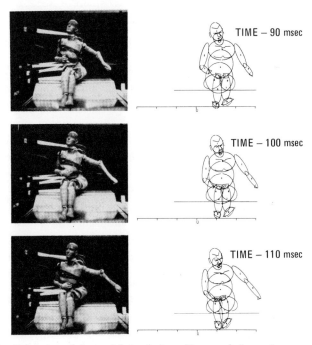

Fig. 16c. Validation of the model simulation of impact sled experiments.

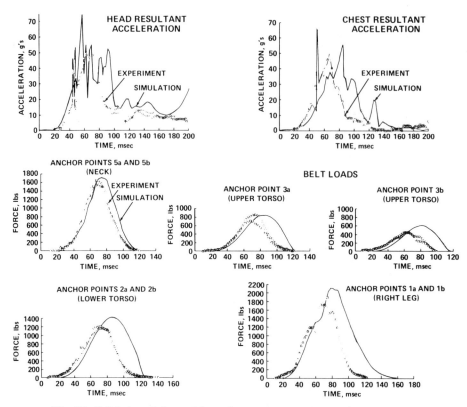

Fig. 17. Validation of the model simulation of impact sled experiments 30 mph.

To assess the air bag submodel, the program was used to simulate two replicate sled tests in which a 50th percentile Sierra 292-850 dummy was decelerated by an air bag deployed at the start of the sled acceleration pulse. The test setup prior to sled firing is shown in Fig. 18, and the program prediction of dummy and air bag responses during the event is illustrated in Fig. 19, using results from the plotter graphics program. It should be noted that not all of the measurements of the Sierra 292-850 dummy required for program input were available for this simulation. Consequently, the results of the measurements on the 50th percentile Sierra 292-1050 dummy, discussed in the previous section, were used to input the program instead. It is felt that differences in the properties of the two types of dummies should not have a large effect on the comparison. Nevertheless, the comparison should not be considered a test of model validity in the strictest sense.

The frontal component of acceleration of the head measured in these sled tests is compared to the program prediction in Fig. 20. Both the trend and the peak magnitude of the measured acceleration are predicted well, but the simulation predicts too rapid response of the dummy. It is believed that this lack of time correlation is caused by the rather massive air bag membrane requiring more time to deploy that the massless membrane of the air bag model. Considering the complexity of the simulation (forty-six degrees of freedom and twenty contacts and interactions between body segments, lap belt restraint, vehicle contact surfaces, and the air bag), the agreement is considered quite good.

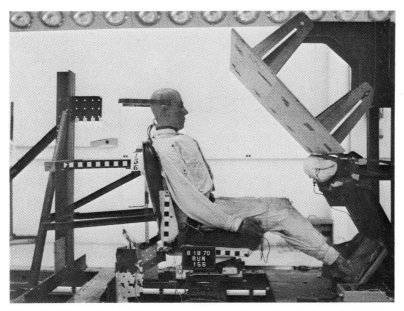

Fig. 18. Impact sled tests with air bags.

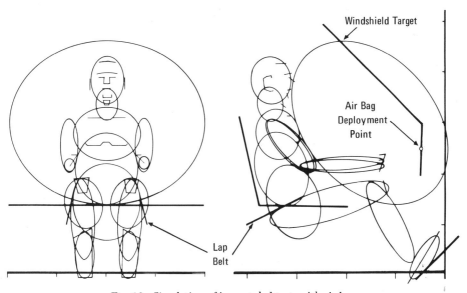

Fig. 19. Simulation of impact sled tests with air bags.

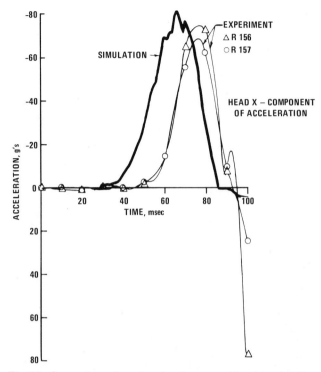

Fig. 20. Comparison of predicted and measured head acceleration.

To test predictions of the model in an actual crash environment, simulations of dummy responses in a full-scale car-to-car crash test were also performed. A concise summary of the test, consisting of a head-on laterally-offset crash between two vehicles, one containing two Sierra 292-1050 dummies, is presented in the photographs of Fig. 21. As a representative comparison, the program prediction of the dummy head displacement in the forward direction and the corresponding measured result, extracted from moving picture coverage of the crash, are presented in Fig. 22.

FUTURE EFFORTS

Although the program is at present considered a most useful tool, further generalizations will increase its utility. As an example, a pre-processor program to generate input data sets for the simulated crash victim* will automate the process of obtaining the required dimensional and inertial parameters. With this program, specification of the crash victim sex, height, and weight will result in the generation of the required inputs. The resulting crash victim model for a range of heights and weights is illustrated in Fig. 23.

---

*In development under the sponsorship of the Motor Vehicle Manufacturers Association.

Fig. 21a. Car-to-car test pre-test.

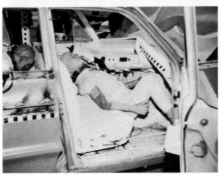

Fig. 21b. Car-to-car crash test post test.

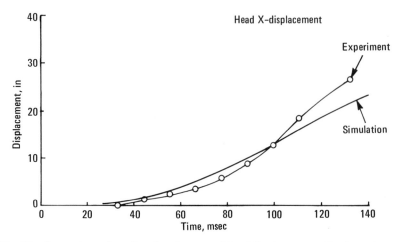

Fig. 22. Comparison of predicted and measured head displacement right front dummy.

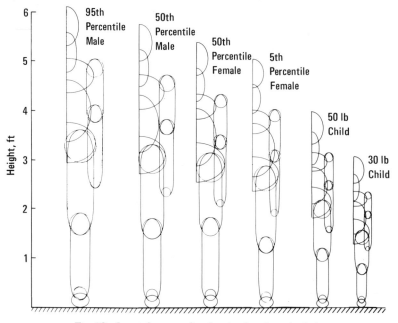

Fig. 23. Input data sets for the simulated crash victim.

SUMMARY

A detailed series of static measurements on an anthropometric crash dummy, plus a series of experiments ranging from static bench tests to impact sled tests and a full-scale automobile crash, have been used to test the validity of a three-dimensional mathematical model of the crash victim. The good prediction accuracy and relatively low computation cost, coupled with the generality and detail incorporated in the developed computer model, make it a valuable engineering tool for application in continuing research efforts to enhance the safety of the crash victim.

ACKNOWLEDGMENTS

The author gratefully acknowledges the contributions of Dr. John Fleck and Mr. Frank Butler in computer program development and Messrs. James Walunas and Earl Cluckey in performing the anthropometric dummy measurements. Technical guidance provided by Messrs. Raymond McHenry, Norman DeLeys, and Leonard Garr has also been most helpful.

REFERENCES

1. R. R. McHenry, "Analysis of the Dynamics of Automobile Passenger Restraint Systems, "The Seventh Stapp Car Crash Conference Proceedings (1963), Charles C. Thomas, Springfield, Illinois, 1965.
2. R. R. McHenry and K. N. Naab, "Computer Simulation of the Automobile Crash Victim – A Validation Study," CAL Report No. YB-2126-V-1R, July 1966.
3. K. N. Naab, "Measurement of Detailed Inertial Properties and Dimensions of a 50th Percentile Anthropometric Dummy," 10th Stapp Car Crash Conference (1966), Society of Automotive Engineers, New York, New York, 1966.
4. D. J. Segal and R. R. McHenry, "Computer Simulation of the Automobile Crash Victim – Revision No. 1," CAL Report No. VJ-2492-V-1, March 1968.
5. D. J. Segal, "Revised Computer Simulation of the Automobile Crash Victim," CAL Report No. VJ-2759-V-2, January 1971.
6. D. J. Segal, "Computer Simulation of Pedestrian Accidents," presented at the Third Triennial Congress of the International Association for Accident and Traffic Medicine, New York, New York, May 29 – June 1, 1969.
7. "Research in Impact Protection for Pedestrians and Cyclists," CAL Report No. VJ-2672-V-2, May 1971.
8. J. A. Bartz, "A Three-Dimensional Computer Simulation of a Motor Vehicle Crash Victim – Phase 1 – Development of the Computer Program," CAL Report No. VJ-2978-V-1, PB 204172, July 1971.
9. J. A. Bartz, "A Three-Dimensional Computer Simulation of a Motor Vehicle Crash Victim – Phase 2 – Validation of the Model," CAL Report No. VJ-2978-V-2, August 1972.
10. P. J. Brown and R. W. Armstrong, "Research Activity in Occupant Restraint Systems," Office of Vehicle Systems Research, National Bureau of Standards, U. S. Department of Commerce, December 1970.

DISCUSSION

D. A. Nagel (Stanford University)

Was the difference in the active and passive motion that you noted produced only by the active muscle contraction in the lower extremity or were you anticipating abdominal muscle contractions.

J. A. Bartz

No, the only parameter that was varied was the component of the torque at the knee joint, which was allowed to increase in a sinusoidal fashion.

A. I. King *(Wayne State University)*

Did you make validation studies on pedestrian impact and if so, how do you simulate fractures such as those of the pelvis and tibia?

J. A. Bartz

No, we have not made pedestrian validation studies. I should note that the model was not developed only for motor vehicle occupant simulations. There are no real restrictions for simulating pedestrians; only a few minor program modifications are necessary.

R. A. Potter *(GM Engineering Staff)*

I'd like you to comment on the importance you give to occupant segments moments of inertia. We saw in the Robbins and Roberts paper that there was, for instance, a factor of 4 difference, I believe, in the torso moment of inertia. One could speculate that the difference in response of these two volunteers was due to differences in moment of inertia, if the data are correct. SAE J963 does not have a moment of inertia specification for the dummy, nor does, I believe, the new RFP coming from Department of Transportation. Should we be concerned about the moment of inertia of the dummies and should it be specified precisely?

J. A. Bartz

We made one limited study in which we varied the moment of inertia. We were concerned about possible misalignment of the estimated long axis of the long body segments. If you consider the moment of inertia about the long axis of the arm segment for example, a very small misalignment causes a relatively large error in the moment of inertia. We assumed that we were misaligned and studied the effect of the error in moment of inertia for the nonplanar impact sled test condition which I discussed. We found that the dummy kinematics changed negligibly at about 200 milliseconds into the event. From this we concluded that our moment of inertia measurements were sufficiently accurate. We never did a parametric variation of the type you mentioned, however. We would very much like to do an extensive parametric investigation and answer some of these questions.

S. H. Backaitis *(NHTSA, Department of Transportation)*

I'm confused by a couple things. I think from the gross kinematic standpoint, the 3D model looks very believable. This morning we saw how much complexity is involved to attain an exact response between dummy and cadaver. It looks to me that your neck simulation is a rather simplified representation. Yet you claim to have a very good replication of the actual experiment. So the question arises, is this because your input, more or less, masks the entire event or is it truly that you can get good replication without having to worry about 5, 6, or 7 mass model systems in the head-neck complex?

J. A. Bartz

We have yet to perform a validation comparison in which the head acceleration is poorly predicted. We did go into considerable detail measuring the neck torque components. We also found that by increasing the viscous coefficients to a value that is probably more realistic we got improved agreement. I think that if the inputs are measured properly you should expect fairly good agreement.

Y. K. Liu *(Tulane University)*

I have two questions; one concerns a little bit of detail. In your moment of inertia measurements, let us assume your segments have mid-sagittal plane symmetry, how did you get the product moment of inertia terms? This is my first question.

J. A. Bartz

We didn't measure the product moment of inertia terms. We estimated the orientation of the principal axes. As I said previously, we questioned that procedure. In particular, we studied the most critical case, namely the long segments. In our parametric investigation we assumed a misalignment of 5 degrees of the orientation of the estimated principal axis with respect to the actual axis. This caused a negligible error at 100 milliseconds into the event.

Y. K. Liu

But, as in the head and neck problem, it can make quite a bit of difference in the equations of motion with respect to the center of mass whether you neglect the off-diagonal terms in the inertia matrix or not.

J. A. Bartz

No, we didn't neglect them. We felt that we knew the principal moments of inertia with sufficient accuracy from our measurements. Dynamically the head behaves essentially as a sphere so it wouldn't make much difference if it were misaligned. Also, we did make comparisons with corresponding National Bureau of Standards measurements. We're in agreement within 4% of all measurements taken in the sagittal plane including the head segment.

Y. K. Liu

You can assume knowledge of the principal axes only if you know the product moments of inertia.

J. A. Bartz

I think you can estimate the principal axes for these purposes if you can also estimate the resulting error. It is of course, possible to make a more refined set of measurements and determine not only the principal moments of inertia, but the principal directions as well. However, that's a much more elaborate set of measurements.

Y. K. Liu

I'll come to my second question. In your measurement for the time constant you indicated a voluntary effort test but that the situation which you face is a situation involving stretch reflex. Shouldn't the time constant be different for these two cases? Stark gave the voluntary effort time constant to be from 100 to over 300 milliseconds and that includes your particular case. I would like to know where you got your voluntary data and what is your assessment of the stretch reflex time constant if it is different?

J. A. Bartz

The time of response of a hundred milliseconds which I quoted came from the book *The Human Body in Equipment Design,* by Damon, Stoudt and McFarland.

Y. K. Liu

That's voluntary muscle contraction?

J. A. Bartz

Yes, I should have added that. I assumed that the simulated occupant anticipated the crash and began responding at the start of the crash deceleration pulse, but no attempt was made to include stretch reflex. As I said, this was an exploratory study designed mainly to demonstrate the program capability. If we had plots of torque vs. angular deflection, including strain effects, we could include the stretch reflex.

Y. K. Liu

Well, the stretch reflex, in fact, is the first item that comes into play, right? Long before any voluntary intervention can take place? This is just an observation.

J. A. Bartz

Well yes, unless the subject happens to anticipate the event.

D. H. Robbins *(HSRI, University of Michigan)*

How were the belt force deformation curves obtained for the sled test? Where you had a side impact was the occupant restrained from moving by belts?

J. A. Bartz

We used the static force-deflection characteristics for the webbing and the deformation characteristics for dummy contact surfaces as well. In other words, we accounted for dummy segment deformation as well as webbing elongation during loading.

D. H. Robbins

How was the correction obtained. Did you have to lump all of the properties including belt, dummy, anchorages, sewing strips, all of this together?

J. A. Bartz

Yes, to get an equivalent force-deflection curve.

C. K. Kroell *(GM Research Laboratories)*

I believe you mentioned that the relative difference between the active and passive joint torques decreased with increasing severity level. I'm not sure I really understand what you're saying there. Could you possibly elaborate on this a bit please?

J. A. Bartz

Well, I guess the easiest way would be to look at plots of torque versus time at say 15, 30 and 60 mph. I didn't bring those slides along. The magnitudes of both torques increased with increasing speed as you'd expect, but the relative difference actually decreased.

C. K. Kroell

Why would this be true in the passive case? Is it the damping element that you had in the passive case that we are seeing there?

J. A. Bartz

The value plotted is the resultant torque. The magnitude of the resultant torque includes viscous, coulomb and spring components.

KINEMATIC ANALYSIS OF
HUMAN VOLUNTEER TESTS

V. L. ROBERTS and D. H. ROBBINS

University of Michigan, Ann Arbor, Michigan

ABSTRACT

One of the primary difficulties in providing adequate simulation of human response during a crash has been the lack of data describing human response under similar circumstances. Because of the medical care problems which may arise from volunteer tests, the bulk of the testing has been performed under the sponsorship of governmental agencies who can provide a proper medical staffing and hospital care facilities should they become necessary.

This report contains the results of an analysis of high speed cinephotographic and electronic transducer records from a series of impact tests on human volunteers conducted by the 6571st Aeromedical Research Laboratories at Holloman Air Force Base, New Mexico. The volunteers were restrained by lap belts. Linear and angular displacement, velocity and acceleration data for the volunteers' heads are given and provide useful information regarding human head-neck response to impact.

INTRODUCTION

The following question has been asked many times recently, "How closely do the results of impact sled tests using anthropometric dummies reproduce each other?" The answer most often given is "not too well." This answer is not acceptable when it is necessary to comply with design or performance standards.

A companion question can also be asked. "How well does human volunteer impact test data agree from test to test." The answer which is developed in the following text is also "not too well." This answer serves to point out the extraordinary complexity of the human mechanism and the difficulty of replacing the human with a test device.

References p. 393

Recent tests conducted at Holloman Air Force Base have provided sufficient data on which to base an initial look at comparative human kinematics. McElhaney (1) reports an extensive analysis of the film and transducer records obtained during these tests. A supplemental report by Robbins (2) includes a large buulk of comparative anthropometric data taken on the test subjects.

TEST DESCRIPTION

Six male human volunteers were subjected to increasingly severe impacts until in the subjects' (subjective tolerance) or the medical monitor's opinion the testing should be terminated. The seat used on the sled was one designed for various ARL human impact tests and contained instrumentation capable of measuring all forces transmitted to it by the subject during impact. The seat pan was horizontal and the seat back was angled backwards 13° from vertical. The Type I lap belt restraint consisted of a 1-3/4 inch-wide webbed dacron belt rated at 6000 lbs. The belt angle at its attachment point approximated 50° to the horizontal at the initiation of the test.

The tests were conducted with sled pulse peak G-levels of 8, 12, and then 15 G's. At 15 G's, increasing complaints of post-run neck and pelvis pains were being reported by the subjects although none felt that they had reached their tolerance limit with this system. Following analysis of impact data, however, several items were noted by the medical monitor. First, there was a marked increase in severity of post-run neck and hip pain complaints and second, the mean lap belt load peak had risen from 760 lbs at 12 G's to 975 lbs at 15 G's with one subject's belt loading as high as 1163 lbs. Assuming a linear extrapolation to an 18 G level, it was felt that a proper safety factor of belt strength to belt load could not be maintained. The combination of these two items led the medical monitor to limit further increases in test severity.

A careful and detailed photometric analysi was performed on the high speed movie films taken during the tests. The basic measuring device used in this work was a Vanguard film analyzer (Model M-160W). Four-place accuracy in linear and angular measurements is obtainable with this instrument. The linear and angular displacements of the target points on the head, shoulder, hip and thigh were measured frame by frame. The film analyzer was coupled to an IBM 029 card punch unit and computer cards were automatically punched containing the displacement data. Several computer programs have been developed to analyze photometric data at HSRI, including routines to compute linear and angular velocities and accelerations with appropriate smoothing and filtering operations. The digitized displacement data was analyzed using these programs and the linear and angular head target velocities and accelerations computed. A sample set of analyzed data for the head is included as Fig. 1, 2 and 3.

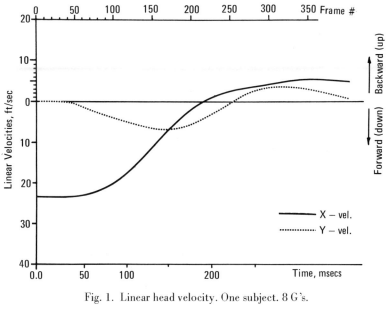

Fig. 1. Linear head velocity. One subject. 8 G's.

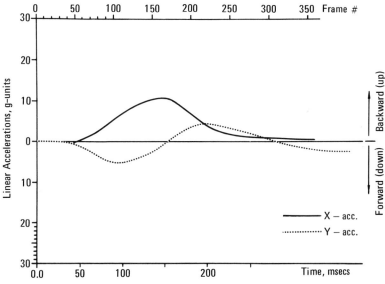

Fig. 2. Linear head acceleration. One subject. 8 G's

INCREASING EVENT SEVERITY — ONE SUBJECT

Fig. 4 shows one effect of an increase in test severity when a single subject receives 8, 12, and then 15 G's deceleration. Impact velocity was increased in these tests but the length of the deceleration stroke remained fixed. This is reflected by the fact that

References p. 393

all position curves are in phase with one another. The increase of head excursion with G-level is as expected although the increase from 12 to 15 G's is relatively small.

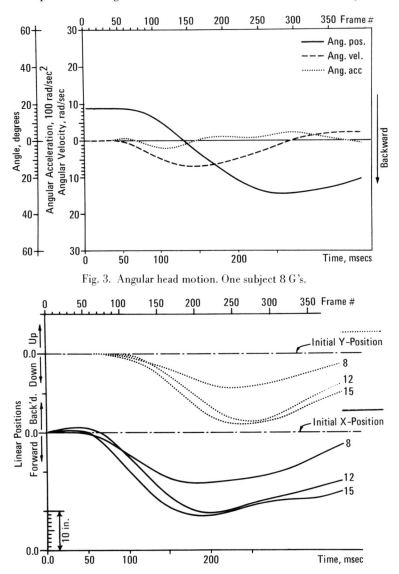

Fig. 3. Angular head motion. One subject 8 G's.

Fig. 4. Linear head position. Three G-levels.

COMPARATIVE SUBJECT ANTHROPOMETRY

Although test subject weight varied from 136 to 199 lbs. and stature from 169.2 to 180.5 cm., it was found that two of the subjects were very similar. They had weights of 155 and 157.5 lbs. and statures of 169.2 and 170 cm. respectively. An estimation of the weights and moments of inertia of various body segments based on formulas and data included in Reference 2 is given in Tables 1 and 2. Moment of inertia is based on a combination of mass and geometry. Because body geometry has been determined by classical and non-classical anthropometric measurements, it is not surprising that percentage differences between moments of inertia for the two subjects are greater than body element weights. This is particularly marked for the torso where it was necessary to define the "joint" between the upper and middle torso as an average of the locations of T-12, the bottom of the sternum, and the bottom of the rib cage. All in all the geometric and mass properties of the two subjects are quite similar.

TABLE 1

Body Segment Weights

Body Segment	Subject 144 (weight-lbs.)	Subject 128 (weight-lbs.)
head and neck	12.4	12.2
trunk	73.6	72.7
upper arms	9.7	9.5
forearms and hands	8.1	7.9
upper legs	31.6	31.1
lower legs and feet	21.0	20.7
Total body weight	157.5	155.0

TABLE 2

Body Segment Moments of Inertia

Body Segment	Subject 144 I (in lb. $\sec.^2$)	Subject 128 I (in lb. $\sec.^2$)
head and neck	.601	.667
upper torso	7.04	5.37
middle torso	.345	.379
pelvis-buttock mass	.549	.517
upper arms	1.079	1.170
lower arms and hands	2.060	2.100
upper legs	1.40	1.65
lower legs and feet	4.68	4.99

COMPARATIVE TEST RESULTS – TWO SIMILAR SUBJECTS

Fig. 5, and 6 show the head motions for the two subjects in the 12 and 15 G sled tests. In all cases subject 128 experienced substantially greater excursions. The reasons for this are not known. It may be related to muscle stiffening but is too large to be explained by the slight difference in sitting height between the two subjects.

References p. 393

Considering the similarity in size, one might also expect the test results to be similar. The complexity of the human system again precludes duplication.

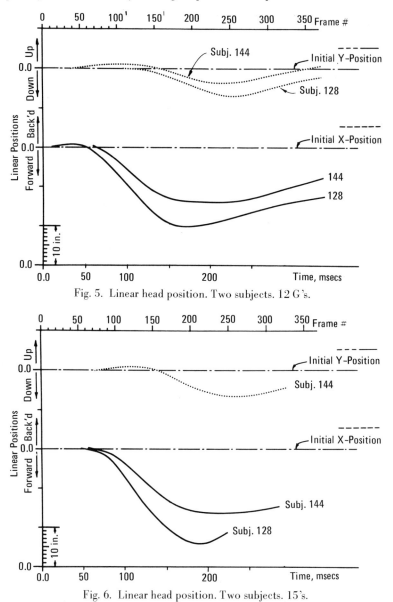

Fig. 5. Linear head position. Two subjects. 12 G's.

Fig. 6. Linear head position. Two subjects. 15's.

COMPARATIVE TEST RESULTS – ALL SUBJECTS

Data from all 12 and 15 G tests involving the six lap-belted subjects have been superimposed to show the range of motions, velocities, and accelerations which can be generated by a population of human subjects (See Fig. 7-22). Although there is a great deal of scatter, the phasing of the various physical events is similar. For example, Fig. 7 shows the forward and downward motions of the head. The maximum and minimum forward excursions occur at 210 and 220 msec respectively and all curves have the same shape. These properties are reflected throughout the rest of the graphs.

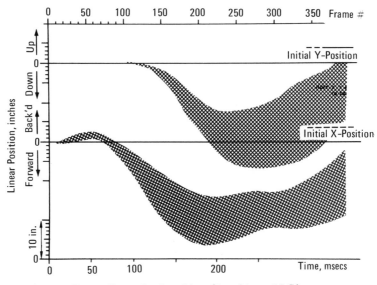

Fig. 7. Linear head position. Six subjects. 12 G's

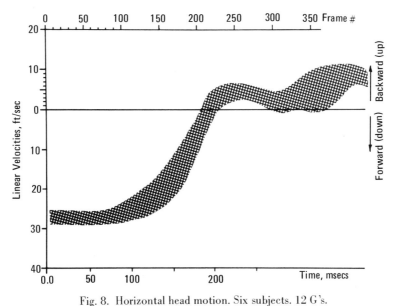

Fig. 8. Horizontal head motion. Six subjects. 12 G's.

Reference p. 393

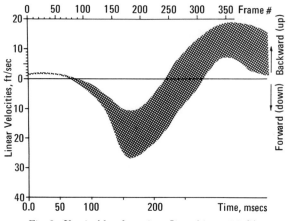

Fig. 9. Vertical head motion. Six subjects. 12 G's

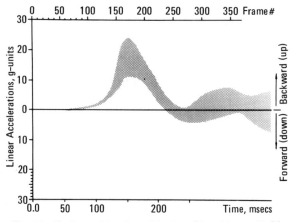

Fig. 10. Horizontal head acceleration. Six subjects. 12 G's.

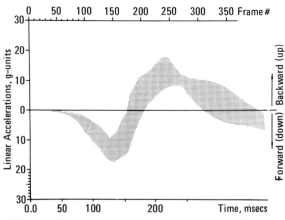

Fig. 11. Vertical head acceleration. Six subjects. 12 G's.

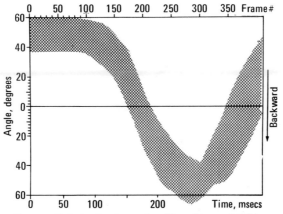

Fig. 12. Relative head rotation. Six subjects. 12 G's.

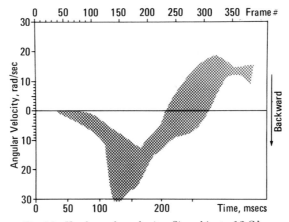

Fig. 13. Head angular velocity. Six subjects. 12 G's.

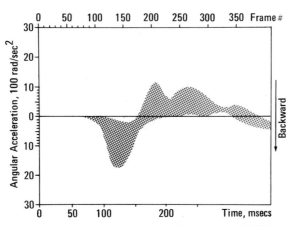

Fig. 14. Head angular acceleration. Six subjects. 12 G's.

References p. 393

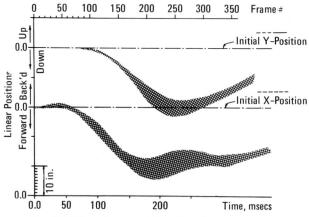

Fig. 15. Linear head position. Six subjects. 15 G's

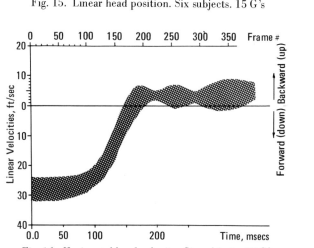

Fig. 16. Horizontal head velocity. Six subjects. 15 G's

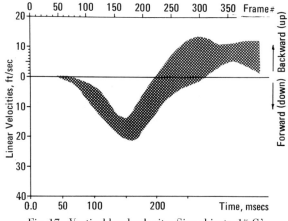

Fig. 17. Vertical head velocity. Six subjects. 15 G's.

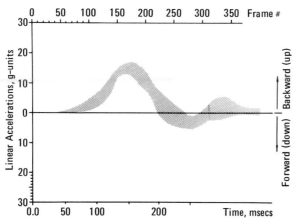

Fig. 18. Horizontal head acceleration. Six subjects. 15 G's.

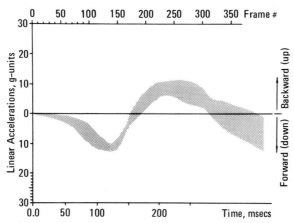

Fig. 19. Vertical head acceleration. Six subjects. 15 G's.

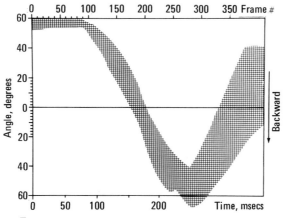

Fig. 20. Relative head rotation. Six subjects. 15 G's.

References p. 393

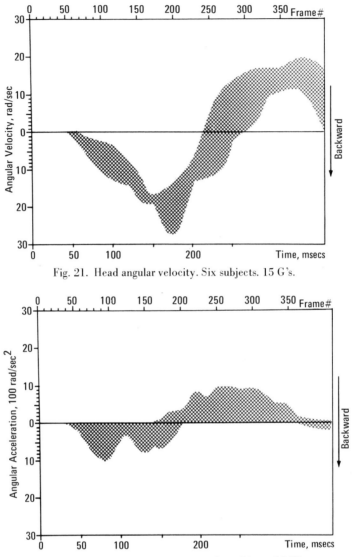

Fig. 21. Head angular velocity. Six subjects. 15 G's.

Fig. 22. Head angular acceleration. Six subjects. 15 G's.

STATISTICAL ANALYSIS OF RESULTS

As a concluding summary of the test results, with particular reference to the data scatter, the peak values for various parameters have been determined and listed in Table 3 along with standard deviation from the mean value of the peak. In several cases the standard deviation is half as great as the mean peak value of the quantity being considered. This again emphasizes the complexity and inherent difficulty of controlling the vast number of physical parameters which any human possesses. That it is difficult to obtain test reproducibility with test devices is not surprising.

TABLE 3

Statistical Comparison of Test Results

Parameter (peak value)	Units	Mean	Standard Deviation
Deceleration Pulse	G	11.6	0.81
Lap belt load	lbs.	761	173
Foot pan load	lbs.	345	147
Seat back load	lbs.	584	252
Chest deceleration	G	14.9	4.7
Head forward excursion	in.	20.8	4.5
Head downward excursion	in.	18.6	6.0
Head downward rotation	deg.	100.3	17.8
Head forward velocity	ft/sec.	34.7	3.1
Head downward velocity	ft/sec.	29.0	11.0
Head rotational velocity	rad/sec.	31.1	11.0
Head forward deceleration	G	15.3	4.9
Head downward deceleration	G	13.0	4.4
Head rotational acceleration	rad/sec.^{2}	850	567
Shoulder forward excursion	in.	15.8	3.3
Shoulder downward excursion	in.	6.8	3.1
Knee forward motion	in.	2.2	1.7
Knee downward motion	in.	3.0	1.2
Thigh forward motion	in.	2.0	1.4
Thigh downward motion	in.	1.5	0.8
Thigh rotation	deg.	12.0	4.7

REFERENCES

1. J. H. McElhaney, V. L. Roberts and D. H. Robbins, "Analysis of Daisy Track Human Tolerance Tests," Final Report on U.S. DOT Contract No. FH-11-6962, NTIS Access No. PB201265, p. 158, February 1971.
2. D. H. Robbins, R. G. Snyder and V. L. Roberts, "Mathematical Simulation of Daisy Track Human Volunteer Tests," Final Report on U.S. DOT Contract No. FH-11-6962, NTIS Access No. PB203717, p. 57, June 1971.

DISCUSSION

R. R. McHenry *(Cornell Aeronautical Laboratories)*

I realize that this wasn't used in the paper, but on the first slide there is a factor of two difference in one radius of gyration and a factor of four in another. I was wondering what the basis was for those moments of inertia of the torso segments?

V. L. Roberts

The approach we used for the torso segment moment of inertia determination appeared in a paper that Hurley Robbins authored at the last Stapp Conference, Ray. I think in there he goes into some detail about this variation. Hurley Robbins would you care to comment?

D. H. Robbins

I think that the moment of inertia is a linear function of the mass and that it has a variety of other geometrical components in it which tend to modify the mass, you might say. The calculation of the moment of inertia was based on both the weight quantity as well as a variety of anthropometric measurements. These measurements consisted of length quantities of the torso as well as circumferential quantities for the torso and all of the other elements. I would have to look at the specific sets of anthropometry for these two volunteers to be able to answer your question completely.

C. A. Moffatt *(West Virginia University)*

Regarding that same moment of inertia, I just wanted to make an observation. Looking at the numbers for the arms, the moments vary by a factor of approximately four and the weights are about equal. That would indicate that the arms of one volunteer were twice as long as the arms of the other, which confuses me too.

V. L. Roberts

We didn't have any explanations for that one.

Y. K. Liu *(Tulane University)*

On the acceleration data you acquired, how did you get them? I wasn't clear on that.

V. L. Roberts

Photometric analysis. Our process was very similar to the procedure that Mark Haffner and Gerry Cohen explained this morning. We used double differentiation.

Y. K. Liu

I guess you noted the comment which Dr. Goldsmith made on the efficacy of that technique.

R. R. McHenry

I'd like to make one comment on the variability that appears to be due to muscle tone effects. It seems that from a practical point of view the loose condition would be the most realistic condition to use for a compliance test of a vehicle. In other words, occupant contacts with the interior would be made maximum by the loose adjustment of the dummy as opposed to the tensed dummy. There is some directly related material, I believe, in the next paper by J. A. Bartz on the Validation of a Three-Dimensional Mathematical Model of the Crash Victim.

PARTICIPANTS

M. B. Achorn
Ford Motor Company
Dearborn, Michigan

S. H. Advani
West Virginia University
Morgantown, West Virginia

P. D. Agarwal
Research Laboratories, GMC

W. G. Agnew
Research Laboratories, GMC

T. Aibe
Nissan Motor Company
Englewood Cliffs, New Jersey

S. W. Alderson
Alderson Biotechnology Corporation
Los Angeles, California

C. Arthur
Sierra Engineering
Sierra Madre, California

S. H. Backaitis
Department of Transportation
Washington, D.C.

D. I. Bailey
Research Laboratories, GMC

J. A. Bartz
Cornell Aeronautical Laboratory
Buffalo, New York

S. E. Beacom
Research Laboratories, GMC

E. Becker
Naval Aerospace Medical Research Laboratories
New Orleans, Louisiana

J. V. Benedict
Technology Incorporated
San Antonio, Texas

R. J. Berton
Ford Motor Company
Dearborn, Michigan

F. J. Biluk
Medical Director
GM Technical Center

T. L. Black
Design Staff, GMC

W. Boehly
Department of Transportation
Washington, D.C.

R. R. Bonnington
Vauxhall Motors Limited
England

R. Bouche
Endevco Corporation
Pasadena, California

P. C. Bowser
Buick Motor Division, GMC

D. B. Breedon
Department of Transportation
Washington, D.C.

H. G. Brilmyer
Ford Motor Company
Dearborn, Michigan

J. Brinn
Chrysler Corporation
Highland Park, Michigan

R. Brocklehurst
British Leland Motors
England

P. J. Brown
Department of Transportation
Washington, D.C.

A. L. Browne
Research Laboratories, GMC

W. E. Brunsdon
Chevrolet Motor Division, GMC

A. V. Butterworth
Research Laboratories, GMC

L. R. Buzan
Research Laboratories, GMC

G. A. Campbell
Research Laboratories, GMC

K. L. Campbell
Research Laboratories, GMC

J. D. Caplan
Research Laboratories, GMC

M. Carpenter
Engineering Staff, GMC

R. Cernowski
Hamill Manufacturing Company
Washington, Michigan

D. B. Chaffin
University of Michigan
Ann Arbor, Michigan

R. F. Chandler
Civil Aeromedical Institute
Oklahoma City, Oklahoma

P. F. Chenea
Research Laboratories, GMC

S. J. Chris
Newhouse & Chris, Incorporated
Gaithersburg, Maryland

G. B. Cohen
Department of Transportation
Washington, D.C.

J. M. Colucci
Research Laboratories, GMC

C. C. Culver
Research Laboratories, GMC

R. Davies
Research Laboratories, GMC

P. W. Davis
Department of Transportation
Cambridge, Massachusetts

N. Deleys
Cornell Aeronautical Laboratory
Buffalo, New York

D. Dresden
Fisher Body Division, GMC

E. T. Driver
Department of Transportation
Washington, D.C.

S. Elmore
Orthopaedic Surgeon
Richmond, Virginia

C. L. Ewing
Naval Aerospace Medical Detachment
New Orleans, Louisiana

N. Feles
Fisher Body Division, GMC

J. J. Fodermaier
Chrysler Corporation
Highland Park, Michigan

J. Follat
Engineering Staff, GMC

K. Foster
Consulting Engineer
Woodland Hills, California

P. G. Fouts
Chrysler Corporation Proving Ground
Chelsea, Michigan

R. H. Fredericks
Ford Motor Company
Dearborn, Michigan

C. W. Gadd
Research Laboratories, GMC

K. D. Gardels
Research Laboratories, GMC

R. W. Gibson
Research Laboratories, GMC

T. Glenn
Department of Transportation
Washington, D.C.

W. Goldsmith
University of California
Berkeley, California

C. Gulash
Environmental Activities Staff, GMC

R. C. Haeusler
Chrysler Corporation
Highland Park, Michigan

M. P. Haffner
Department of Transportation
Washington, D.C.

E. M. Hagerman
Research Laboratories, GMC

D. J. Hajduk
Cadillac Motor Car Division, GMC

E. Harris
Tulane University
New Orleans, Louisiana

J. L. Hartman
Research Laboratories, GMC

R. C. Haut
Research Laboratories, GMC

D. F. Hays
Research Laboratories, GMC

C. E. Heeden
Fisher Body Division, GMC

D. J. Henry
Research Laboratories, GMC

W. E. Hering
Research Laboratories, GMC

R. Herman
Research Laboratories, GMC

R. Hickling
Research Laboratories, GMC

V. R. Hodgson
Wayne State University
Detroit, Michigan

J. E. Hofferberth
Department of Transportation
Washington, D. C.

J. C. Holzwarth
Research Laboratories, GMC

J. D. Horsch
Research Laboratories, GMC

B. Howard
Fisher Body Division, GMC

R. P. Hubbard
Research Laboratories, GMC

S. A. Iobst
Research Laboratories, GMC

K. Itoh
Itoh Seiki Company
Japan

F. E. Jamerson
Research Laboratories, GMC

R. H. Jensen
Engineering Staff, GMC

T. O. Jones
Engineering Staff, GMC

J. W. Justusson
Research Laboratories, GMC

M. M. Kamal
Research Laboratories, GMC

T. Kato
Nissan Motor Company
Englewood Cliffs, New Jersey

A. I. King
Wayne State University
Detroit, Michigan

J. C. King
Ford Motor Company
Dearborn, Michigan

W. F. King
Research Laboratories, GMC

M. S. Koga
Ford Motor Company
Dearborn, Michigan

D. P. Koistinen
Research Laboratories, GMC

C. K. Kroell
Research Laboratories, GMC

A. S. Kuchta
Ford Motor Company
Dearborn, Michigan

W. Lange
Wayne State University
Detroit, Michigan

M. B. Leising
Chrysler Corporation
Highland Park, Michigan

T. E. Lobdell
Research Laboratories, GMC

L. C. Lundstrom
Environmental Activities Staff, GMC

T. MacLaughlin
Department of Transportation
Washington, D.C.

R. B. MacLean
Ford Motor Company
Dearborn, Michigan

T. Maeda
Nissan Motor Company
Englewood Cliffs, New Jersey

D. C. Mallett
American Motors Corporation
Detroit, Michigan

T. J. Mao
Research Laboratories, GMC

C. Marks
Engineering Staff, GMC

J. E. Martens
Allstate Insurance Company
Northbrook, Illinois

D. E. Martin
Environmental Activities Staff, GMC

H. T. McAdams
Cornell Aeronautical Laboratory
Buffalo, New York

W. E. McCarthy
Buick Motor Division, GMC

D. B. McCormick
Environmental Activities Staff, GMC

R. J. McDonald
Research Laboratories, GMC

J. H. McElhaney
University of Michigan
Ann Arbor, Michigan

R. R. McHenry
Cornell Aeronautical Laboratory
Buffalo, New York

I. K. McIvor
University of Michigan
Ann Arbor, Michigan

D. G. McLeod
Research Laboratories, GMC

D. I. McNaughton
GM of Canada, Limited

W. C. Meluch
Research Laboratories, GMC

J. W. Melvin
University of Michigan
Ann Arbor, Michigan

H. J. Mertz
Research Laboratories, GMC

S. Miller
Cornell Aeronautical Laboratory
Buffalo, New York

W. K. Miller
Research Laboratories, GMC

C. A. Moffatt
West Virginia University
Morgantown, West Virginia

E. A. Moffat
Engineering Staff, GMC

K. L. Morgan
American Motors Corporation
Detroit, Michigan

M. Mouilleron
Union of Technical Auto Companies
France

N. L. Muench
Research Laboratories, GMC

R. D. Mulroy
Orthopaedic Surgeon
Waltham, Massachusetts

D. Murray
Fisher Body Division, GMC

R. V. Myers
Department of Transportation
Canada

D. A. Nagel
Stanford University
Stanford, California

A. M. Nahum
University of California
San Diego, California

K. Nakajima
Toyota Motor Company
Lyndhurst, New Jersey

R. F. Neathery
Research Laboratories, GMC

H. L. Newhouse
Newhouse & Chris, Incorporated
Gaithersburg, Maryland

G. W. Nyquist
Research Laboratories, GMC

L. Ovenshire
Department of Transportation
Washington, D.C.

B. Parr
Engineering Staff, GMC

L. W. Parr
Chrysler Corporation
Highland Park, Michigan

L. M. Patrick
Wayne State University
Detroit, Michigan

G. W. Pearsoll
Duke University
Durham, North Carolina

V. Phillips
Vauxhall Motors Limited
England

L. E. Popp
Sierra Engineering
Sierra Madre, California

R. A. Potter
Engineering Staff, GMC

B. C. Prisk
Research Laboratories, GMC

T. Rasmussen
Oldsmobile Division, GMC

E. A. Ripperger
University of Texas
Austin, Texas

D. L. Rivard
Ford Motor Company
Dearborn, Michigan

D. H. Robbins
University of Michigan
Ann Arbor, Michigan

S. B. Roberts
University of California
Los Angeles, California

V. L. Roberts
University of Michigan
Ann Arbor, Michigan

R. P. Rodriguez
Orthopaedic Surgeon
New Orleans, Louisiana

J. W. Rosenkrands
Engineering Staff, GMC

J. L. Roshalla
Sierra Engineering
Sierra Madre, California

R. W. Rothery
Research Laboratories, GMC

R. Rubenstein
Alderson Research Laboratories
Stamford, Connecticut

G. R. Ryder
GM Overseas Operations, GMC

C. B. Salan
Endevco Corporation
Pasadena, California

R. J. Salloum
Research Laboratories, GMC

H. Schmidt
Department of Transportation
Washington, D.C.

D. Schwartz
Research Laboratories, GMC

J. C. Scowcroft
Motor Vehicle Manufacturers Association
Detroit, Michigan

J. Silver
Proving Ground, GMC

G. R. Smith
Environmental Activities Staff, GMC

J. Smrcka
Alderson Research Laboratories
Stamford, Connecticut

R. G. Snyder
University of Michigan
Ann Arbor, Michigan

R. L. Stalnaker
University of Michigan
Ann Arbor, Michigan

J. P. Stapp
Rome, New York

E. S. Starkman
Environmental Activities Staff, GMC

J. D. States
University of Rochester
Rochester, New York

G. F. Stofflet
Proving Ground, GMC

J. Takada
Takata Kojyo, Limited
Japan

Y. Tanaka
Toyota Motors Company, Limited
Lyndhurst, New Jersey

K. D. Taylor
Buick Motor Division, GMC

L. Thibault
National Institutes of Health
Bethesda, Maryland

A. M. Thomas
Department of Transportation
Washington, D.C.

R. F. Thomson
Research Laboratories, GMC

J. Tomassoni
Department of Transportation
Washington, D.C.

R. V. Trax
Pontiac Motor Division, GMC

K. Trosien
Wayne State University
Detroit, Michigan

F. J. Unterharnscheidt
University of Texas
Galveston, Texas

N. K. Van Allen
Research Laboratories, GMC

R. J. Vargovick
Ford Motor Company
Dearborn, Michigan

H. E. VonGierke
Air Force, Wright Patterson
Dayton, Ohio

N. M. Wang
Research Laboratories, GMC

P. Warnor
Ogle Design Limited
England

H. P. Waters
Department of Transportation
Washington, D.C.

E. F. Weller
Research Laboratories, GMC

M. L. Wenner
Research Laboratories, GMC

C. Wentzel
GM Overseas Operations, GMC

W. E. Whitmer
GMC Truck & Coach Division, GMC

R. A. Wilson
Proving Ground, GMC

F. J. Winchell
Engineering Staff, GMC

E. L. Windeler
Pontiac Motor Division, GMC

J. A. Wolf
Research Laboratories, GMC

F. A. Wyczalek
Engineering Staff, GMC

AUTHORS AND DISCUSSORS INDEX

Bold Face Type refers to papers

SUBJECT INDEX